About Island Press

Island Press, a nonprofit organization, publishes, markets, and distributes the most advanced thinking on the conservation of our natural resources—books about soil, land, water, forests, wildlife, and hazardous and toxic wastes. These books are practical tools used by public officials, business and industry leaders, natural resource managers, and concerned citizens working to solve both local and global resource problems.

Founded in 1978, Island Press reorganized in 1984 to meet the increasing demand for substantive books on all resource-related issues. Island Press publishes and distributes under its own imprint and offers these services to other nonprofit organizations.

Support for Island Press is provided by The Geraldine R. Dodge Foundation, The Energy Foundation, The Charles Engelhard Foundation, The Ford Foundation, Glen Eagles Foundation, The George Gund Foundation, William and Flora Hewlett Foundation, The James Irvine Foundation, The John D. and Catherine T. MacArthur Foundation, The Andrew W. Mellon Foundation, The Joyce Mertz-Gilmore Foundation, The New-Land Foundation, The Pew Charitable Trusts, The Rockefeller Brothers Fund, The Tides Foundation, and individual donors.

FORCING THE SPRING

FORCING THE SPRING

The Transformation of the American Environmental Movement

ROBERT GOTTLIEB

ISLAND PRESS

Washington, D.C.  *Covelo, California*

Library of Congress Cataloging-in-Publication Data

Gottlieb, Robert.
Forcing the spring : the transformation of the American environmental movement / Robert Gottlieb.
p. cm.
Includes bibliographical references and index.
ISBN 1-55963-123-6 (cloth : alk. paper)
ISBN 1-55963-122-8 (pbk : alk. paper)
1. Environmental policy—United States—History. 2. Green movement—United States—History. I. Title.
GE180.G68 1993
363.7′0525′0973—dc20 93-3513
CIP

Printed on recycled, acid-free paper.

Manufactured in the United States of America

10 9 8 7 6 5

To my kids:
Casey, with his love of the earth,
and Andy, with her passion for life

Contents

FORCING THE SPRING

Introduction

Where We Live, Work, and Play

There was tension and excitement in the conference room at the Washington Court Hotel on Capitol Hill when Dana Alston stepped to the podium. Alston was to address an audience of more than 650 people, including 300 delegates to the first national People of Color Environmental Leadership Summit. Delegates included grassroots environmental activists from across the country: African-Americans from "cancer alley" in Louisiana; Latinos from the cities and rural areas of the Southwest; Native American activists such as the Western Shoshone, who were protesting underground nuclear testing on their lands; organizers of multiracial coalitions in places such as San Francisco and Albany, New York. The purpose of the summit, held on October 24–27, 1991, was to begin to define a new environmental politics from a multiracial and social justice perspective. The delegates sought to address questions of agenda, organizational structure, movement composition, and social vision: issues central to the definition of environmentalism in the 1990s.

The delegates had just heard speeches by leaders of two important national environmental groups. The first, Michael Fischer, executive director of the Sierra Club, admitted to the summit delegates that his organization had too often been "conspicuously missing from the battles for environmental justice," but argued that the time had come for groups to "work and look into the future, rather than to beat our breasts about the past." "We national organizations are not the enemy," Fischer claimed, warning summit participants that conflict between grassroots activists and national groups would only reinforce the divide-and-conquer approach of the Reagan and Bush administrations. "We're here to reach across the table and to build the bridge of partnership with all of you," Fischer insisted.[1]

Fischer's remarks paralleled the comments of John Adams, executive director of the Natural Resources Defense Council, a prominent, staff-based group of lawyers and other environmental professionals. Adams recited how the NRDC, during its twenty-year history, had "relentlessly confronted the massive problems associated with air, water, food and toxics" and had challenged the "disproportionate impacts on communities of color" of a wide range of environmental problems. "I believe the efforts we've engaged in are significant," Adams declared, and he offered, like Fischer, to facilitate a "partnership" between the national and grassroots groups. "You can't win this battle alone," Adams concluded, underlining Fischer's warning about the consequences of disunity.[2]

Many of the delegates felt that the speeches by these environmental chief executive officers, or CEOs, were not responsive to the criticisms they and their communities had directed at these groups. Activists had complained about the absence of people of color in leadership and staff positions of the national groups, the failure of these groups to incorporate equity or social justice considerations in selecting the issues they fought, and the disregard for local cultures and grassroots concerns in the positions these national groups took with regard to environmental conflicts. But beyond these specific complaints, delegates to the People of Color Environmental Leadership Summit were seeking a redefinition of environmentalism to place the concerns, methods of organizing, and constituencies of the grassroots groups at the center of the environmental discourse. They wanted to redefine the central issues of environmental politics, not just to join a coalition of special interest groups.

When Alston, a key organizer of the summit, took the stage to respond to the speeches of the environmental CEOs, there was a great deal of anticipation about what she might say. Alston symbolized the new kind of environmentalist the summit had sought to attract. Born in Harlem, she first became active in the mid-1960s in the black student movement, addressing issues of apartheid and the Vietnam War. Pursuing an interest in the relationship between social and economic justice issues and public health concerns, Alston completed a master's degree in occupational and environmental health at Columbia University. She subsequently took a series of jobs that extended those interests, from the Red Cross, where she dealt with the new emergency issues associated with toxics and nuclear power problems, to Rural America, where she organized conferences on pesticide issues. As part of her work, she frequently encountered staff members from the national environmental groups. It was at these meetings and strategy sessions that Alston, an African-American woman, was struck by how she was consistently the only person of color in attendance and the only participant to press such issues as farmworker health or the discriminatory effect on communities of color of the location of hazardous waste sites.

In February 1990, Alston joined the staff of the Panos Institute, an organization that deals with the intersection of environment and development issues from the perspective of Third World needs and concerns. Alston was hired to develop a program related to the rise of domestic people-of-color organizations concerned with environmental justice. In that capacity, she was invited to be part of the planning committee organizing the People of Color Environmental Leadership Summit. A thoughtful speaker, Alston had been asked to respond to the presentations of the environmental CEOs, given her background and familiarity with both the national and grassroots groups.

As she began to talk, Alston told the delegates and participants that she had decided not to respond to the speeches by Fischer and Adams. Instead, she would try to "define for ourselves the issues of the ecology and the environment, to speak these truths that we know from our lives to those participants and observers who we have invited here to join us." Alston engaged her audience, responding to their appeal for self-definition. "For us," she declared, "the issues of the environment do not stand alone by themselves. They are not narrowly defined. Our vision of the environment is woven into an overall framework of social, racial, and economic justice." As Alston spoke, many in the audience talked back to her, shouting their agreement. "The environment, for us, is where we live, where we work, and where we play. The environment affords us the platform to address the critical issues of our time: questions of militarism and defense policy; religious freedom; cultural survival; energy-sustainable development; the future of our cities; transportation; housing; land and sovereignty rights; self-determination; employment—and we can go on and on." Turning to the environmental CEOs, Alston declared that what she and the delegates wanted was not a paternalistic relationship but a "relationship based on equity, mutual respect, mutual interest, and justice." This required a vision of the future. In pursuing these goals, Alston concluded (restating a dominant theme of the summit), "we refuse narrow definitions."[3]

The question of definition lies at the heart of understanding the past, present, and future of the environmental movement. Today, the environmental movement, broadly defined, contains a diverse set of organizations, ideas, and approaches: professional groups, whose claims to power rest on scientific and legal expertise; environmental justice advocates concerned about equity and discrimination; traditional conservationists or protectionists, whose long-established organizations have become a powerful institutional presence; local grassroots protest groups organized around a single issue; direct-action groups bearing moral witness in their defense of Nature.

Environmental organizations range from multimillion-dollar operations led by chief executive officers and staffed by experts to ad hoc neighborhood associations formed to do battle concerning a local environmental issue. Some environmental groups speak the language of science; others criticize the way science is used to direct policy. There are groups concerned with improving efficiency in existing economic arrangements and those that seek to remake society; groups that promote market solutions and those that want to regulate market failures; conservative environmentalists hoping to strengthen the system and radical environmentalists interested in an agenda for social change.

Given the diverse nature of contemporary environmentalism, it is striking how narrowly the movement has been retrospectively described by historians. In all the standard environmental histories, the roots of environmentalism are presented as differing perspectives on how best to manage or preserve "Nature," meaning Nature outside the cities and the experiences of people's everyday lives. The primary figures in numerous historical texts—the romantic, unyielding, Scottish mountaineer John Muir and the German-trained, management-oriented forester Gifford Pinchot are the best-known examples—represent those perspectives to the exclusion of other figures not seen as engaged in environmental struggles because their concerns were urban and industrial. There has been no place in this history for Alice Hamilton, who helped identify the new industrial poisons and spoke of reforming the "dangerous trades"; for empowerment advocates such as Florence Kelley, who sought to reform the conditions of the urban and industrial environment in order to improve the quality of life of workers, children, women, and the poor; or for urban critics such as Lewis Mumford, who spoke of the excesses of the industrial city and envisioned environmental harmony linking city and countryside at the regional scale.

In part because of these historical omissions, scholars offer sharply divergent views about the origins, evolution, and nature of contemporary environmentalism. Most common explanations place the beginning of the current environmental movement on or around Earth Day 1970. The new movement, they emphasize, came to anchor new forms of environmental policy and management based on the cleanup and control of pollution. These histories review how this movement influenced and was shaped by legislative and regulatory initiatives focused on environmental contamination rather than on the management or protection of Nature apart from daily life. This explanation thus provides a convenient way to distinguish between an earlier conservationist epoch, when battles took place concerning national parks, forest lands, resource development, and recreational resources, and today's environmental era, when pollution and environmental hazards dominate contemporary policy agendas.

The problem with the story historians have told us is whom it leaves out and what it fails to explain. Pollution issues are not just a recent concern; people have recognized, thought about, and struggled with these problems for more than a century in significant and varied ways. A history that separates resource development and its regulation from the urban and industrial environment disguises a crucial link that connects both pollution and the loss of wilderness. If environmentalism is seen as rooted primarily or exclusively in the struggle to reserve or manage extra-urban Nature, it becomes difficult to link the changes in material life after World War II—the rise of petrochemicals, the dawning of the nuclear age, the tendencies toward overproduction and mass consumption—with the rise of new social movements focused on quality-of-life issues. And by defining contemporary environmentalism primarily in reference to its mainstream, institutional forms, such a history cannot account for the spontaneity and diversity of an environmentalism rooted in communities and constituencies seeking to address issues of where and how people live, work, and play.

Forcing the Spring offers a broader, more inclusive way to interpret the environmentalism of the past as well as the nature of the contemporary movement. This interpretation situates environmentalism as a core concept of a complex of social movements that first appeared in response to the urban and industrial changes accelerating with the rapid urbanization, industrialization, and closing of the frontier that launched the Progressive Era in the 1890s. The pressures on human and natural environments can then be seen as connected and as integral to the urban and industrial order. The social and technological changes brought about by the Depression and World War II further stimulated environmental points of view. And if Earth Day 1970 is seen not simply as the beginning of a new movement, but as the culmination of an era of protest and as prefiguring the different approaches within contemporary environmentalism, it is possible to more fully explain the commonalities and differences of today's complex environmental claims.

This book offers that broader interpretation by reconsidering and reconstructing the analysis of historical and contemporary environmentalism. The book is organized into three parts. Part I explores the historical roots of this movement from the 1890s through the first half of this century. Chapters 1 and 2 trace the rise of the social movements that sought to address how a new urban and industrial order influenced different environments, whether in the industrial cities of the East or the resource-rich and rapidly developing West. These chapters reevaluate the traditional debates concerning resources and recreation and identify the new movements addressing the hazards of urban and industrial life. At some points these movements briefly intersected, both through the particular insights and efforts of figures such as Robert Marshall and Benton MacKaye, who were concerned with the relationships between natural and human environments, and in various efforts to define a new progressive politics consisting of rational decision making, social justice, and resource management in both city and countryside. For much of the time, however, the histories of the movements for the preservation of nature and for the remaking of everyday life were distinctive and separate, based on different constituencies and differing ways of responding to the new urban and industrial order. Chapter 3, the final chapter in this section, reviews the post–World War II era up through the late 1960s, situating these differences in the context of further economic and social change. It discusses the new ideas and movements that arose to challenge what writer Paul Goodman called the "organized society" and the search of these 1960s movements for environmental alternatives in the midst of social rebellion.

Part II situates the rise and consolidation of the contemporary environmental groups in the period between Earth Day 1970 and Earth Day 1990. It distinguishes between mainstream environmentalism—those groups and individuals involved in the framing of, and conflicts concerning, contemporary environmental policy—and alternative environmentalism, which has directly challenged many of the assumptions of that policy. The contrast between groups such as the Sierra Club, the Natural Resources Defense Council, and other national environmental organizations, and grassroots or direct-action groups such as the Citizen's Clearinghouse for Hazardous Wastes and Greenpeace highlights the very different conceptions of politics, process, and participation that distinguish these two manifestations of the environmental movement. Chapter 4 explores the professionalization and institutionalization of the mainstream groups, including their reliance on lobbying, litigation, and the use of expertise as defining organizational features. Chapter 5 analyzes the community-based and direct-action groups, several of them populist in orientation, including the ways they have formed a countermovement of transformation. These alternative groups advocate environmental justice and environmental democracy as opposed to environmental management through greater system efficiency.

Part III takes up the importance of gender, ethnicity, and class, which have significance both in how movements have historically formed and defined themselves and in how current movements identify their issues and constituencies. Gender, ethnicity, and class questions are thus pivotal to both an analysis of the contemporary environmental movement and anticipation of its future directions.

Gender is central to how environmental issues are identified and is discussed in Chapter 6. Penny Newman, who organized her own community of Glen Avon to fight against the contamination seeping from the nearby Stringfellow Acid Pits east of metropolitan Los Angeles, is one of many women drawn into environmental groups by the discovery that her family and her community are at risk. These women claim authority based on everyday experience, on common sense as well as science, and on the moral position that those who may suffer the consequences of environmental hazards most directly in their lives, their communities, and the futures of their children must have a loud voice in the environmental decisions that affect their lives.

Ethnicity is also a clear factor in environmental problems and in the debates about their resolution, as explored in Chapter 7. People of color, in their jobs and in their communities, are subjected to the most intense environmental hazards. New toxic disposal sites are located in poor African-American rural communities in the South; farmworkers face pesticide poisoning in the fields; Native Americans, offered no other resources for economic development, provide a labor force for uranium mines, while their reservations become sites for the disposal of toxic and nuclear wastes. People of color face environmental problems in circumstances where they are also challenged by other manifestations of discrimination and disadvantage: unemployment, economic vulnerability, and exclusion from political power. The experience of ethnicity, especially through the experience of racial discrimination, connects environmental justice to social justice immediately and intimately.

Class, the subject of Chapter 8, is also crucial to the framing of the environmental movement. Every toxic environmental hazard is encountered first by the workers who work in toxic industries and who use, transport, and dispose of hazardous chemicals. It is known that DBCP, a soil fumigant, is hazardous because it sterilized the men who were employed to formulate it. Asbestos is known to be hazardous because those who worked with it developed cancer. The connections among the workplace health and safety movement, the labor movement, and the environmental movement are made in the industrial choices and practices out of which contemporary environmental problems arise. This is true as well of the issue of employment and environmental quality, although the debate on this has focused on the effects of environmental regulation on employment, whether in the forests of the Pacific Northwest or the furniture factories of Los Angeles. But activists in the environmental and labor movements continue to discover that the environment-versus-jobs issue is a false opposition, used to divide the victims of economic decisions who have yet to find a common voice.

Chapter 9, the concluding chapter of the book, offers a perspective on the future of this broadly defined environmental movement. It analyzes additional factors crucial to the future of environmentalism: specifically, the end of the Cold War and the uneven but renascent development of environmentalism among students and young people, and how such factors may shape the direction of the movement. It discusses the future directions that environmentalism may take, centering on the role of this complex movement in the contemporary urban and industrial order and its potential influence on social and environmental change. Finally, it identifies the possibilities of a renewed activism in the context of the current emphasis on the need for change at every level of American society.

When I began research for this book more than five years ago, I immediately confronted what was to emerge as my central research question: What was and is environmentalism? Which individuals and groups make up the environmental movement? Is environmentalism a new kind of social movement? In seeking to answer these questions, I encountered, among other important figures, the extraordinary Alice Hamilton, the mother of American occupational and community health, who so clearly anticipated many contemporary environmental themes. This pivotal figure is not found in any of the environmental history texts. Yet Hamilton is clearly as much an environmentalist as John Muir, the much celebrated defender of Yosemite and passionate advocate of wilderness. I mentioned my interest in Hamilton to a staff member of one of the leading mainstream environmental groups who was curious about my project. "But who's Alice Hamilton?" he asked in a puzzled manner. When I recounted the story to another friend involved in public health issues, I explained that the response by my environmentalist friend was equivalent to ignorance about John Muir. "Who's John Muir?" my public health friend replied.

Through its effort to broaden the definition of environmentalism, this book shifts environmental analysis from an argument about protection or management of the natural environment to a discussion of social movements in response to the urban and industrial forces of the past hundred years. Defining environmentalism in this broad way draws attention to the commonalities and connections among segments of complex and varied movements for change. It includes groups focused not just on wilderness or resource management but on issues affecting daily life. And while the agendas, organizational forms, and political biases of environmental groups can differ significantly, they still share a common search for a response to the dominant urban and industrial order. Whether this search leads to a new direction and a new vision for environmentalism relates back to the question Dana Alston posed about definitions at the People of Color Environmental Leadership Summit. *Forcing the Spring* seeks to answer that question by providing a more comprehensive view of where environmentalism comes from within American experience and whether environmentalism is capable of transcending its narrow definitions to change the very fabric of social life.

Part I

COMPLEX MOVEMENTS, DIVERSE ROOTS

Chapter 1

Resources and Recreation: The Limits of the Traditional Debate

A "Green Utopia"?

To understand a complex movement with diverse roots, it might be best to begin with a paradoxical figure within environmentalism. Passionate about his hiking and climbing, a champion of the poor and powerless, deeply committed to wilderness, and equally forceful about the need to make nature a direct part of people's lives, Robert Marshall is an enigmatic figure for those who have sought to define environmentalism in narrow and limiting terms. Yet this intense, engaging, always smiling, always curious radical forester proposed a common thread for a movement split between those focused on the management and/or protection of Nature and those who defined environment as the experience of daily life in its urban and industrial setting. The liberation of society, Marshall proclaimed, was a condition for the liberation of Nature, and the liberation of the natural environment from its would-be exploiters was an essential condition for social liberation. The absence of such a common thread serves as the environmental movement's actual point of departure.

The son of a well-known lawyer who was a senior partner in the prestigious Washington, D.C., firm of Guggenheimer, Untermeyer, and

Marshall, Robert Marshall grew up steeped in liberal values, including defense of civil liberties, respect for minority rights, and the fight against discrimination. Encouraged by his father, whose strong interest in forest conservation led him to make a large endowment to the Forestry School at Syracuse University, Marshall decided to attend the program at Syracuse to launch a forestry-related career. After graduation, he worked in various capacities for the U.S. Forest Service, where he began to develop strong feelings about forests as a necessary retreat "from the encompassing clutch of a mechanistic civilization," a place where people would be able to "enjoy the most worthwhile and perhaps the only worthwhile part of life." Marshall quickly became a strong critic of development pressures on forest lands and the activities of private logging companies, which had led to a decline in productivity, increase in soil erosion, and "ruination of the forest beauty."[1]

Marshall most loved to hike and explore. He was constantly on the move, a pack on his back, entering and discovering new lands, new environments, new wilderness. In *Arctic Village*, a 1930 bestseller describing his activities in the Arctic wilderness area, Marshall spoke of a "vast lonely expanse where men are so rare and exceptional that the most ordinary person feels that all the other people are likewise significant." His compassion for people and powerful desire to be in touch with wilderness eventually led Marshall to adopt two distinct, yet, for him, compatible positions about wilderness protection. On the one hand, Marshall feared a loss of the wild, undeveloped forest lands in both their spectacular western settings and in the less monumental forest areas of the East, such as the Adirondacks. In a February 1930 article for the *Scientific Monthly*, Marshall laid out this concept of wilderness as a "region which contains no permanent inhabitants, possesses no possibility of conveyance by any mechanical means, and is sufficiently spacious that a person crossing it must have the experience of sleeping out." To achieve that goal, Marshall urged a new organization be formed "of spirited people who will fight for the freedom of the wilderness" and be militant and uncompromising in their stance.[2]

At the same time, Marshall argued that wilderness belonged to all the people, not simply to an elite who wanted such areas available for their own use. Already by 1925, Marshall was writing that "people can not live generation after generation in the city without serious retrogression, physical, moral and mental, and the time will come when the most destitute of the city population will be able to get a vacation in the forest."[3] Marshall was particularly critical of the policies of the National Park Service with their expensive facilities and concessions. Though he argued against more roads and increased development in either park or forest lands, Marshall nevertheless wanted wilderness accessible to "the ordinary guy." During the New Deal era of the 1930s, this was a particularly appealing position to the Forest Service, which convinced Marshall to head a new outdoors and recreation office. Through this office, the Forest Service hoped to contrast itself as a kind of blue-collar alternative to the Park Service.[4]

Despite his agency role, Marshall remained a critic of both the Forest Service and the Park Service, blunt in his attack on the prodevelopment posture of the Forest Service as well as the Park Service's recreation-oriented policies, which ended up destroying wilderness. His criticism of the Forest Service, laid out in his best-known work, *The People's Forests*, was tied to Marshall's overall critique of private forestry and its role both in destroying wilderness and in injuring the work force, the community, and the land itself.

In *The People's Forests*, Marshall distinguished between private ownership of forest land (where the vast bulk of private lands had already been overcut), private ownership with public regulation, and full public ownership, which Marshall strongly endorsed. With public ownership, Marshall argued, "social welfare is substituted for private gain as the major objective for management." To Marshall, that meant a new labor and rural economic development strategy and careful land use planning, more research and science, and safeguarding recreational values from "commercial exploitation." His concept of linking protectionist objectives within a social policy framework was, according to one reviewer from *The Nation*, the best assurance for future generations that the forests could provide "a green retreat from whatever happens to be the insoluble problems of their age."[5]

This search for a green retreat, or a "green utopia," became a continuing passion for Marshall, both in his governmental activities and advocacy work. After his return to the Forest Service in 1937 following a stint with the Bureau of Indian Affairs, Marshall laid out this combined social and environmentalist vision. It included subsidizing transportation to public forests for low-income people, operating camps where groups of underprivileged people could enjoy the outdoors for a nominal cost, changing Forest Service practices that discriminated against blacks, Jews, and other minorities, and acquiring more recreational forest land near urban centers. At the same time, he sought to designate wilderness as places "in which there shall be no roads or other provision for motorized transportation, no commercial timber cutting, and no occupancy under special use permit for hotels, stores, resorts, summer homes, organization camps, hunting and fishing lodges or similar uses. . . ."[6]

Marshall also sought to integrate some of these ideas into the approach of The Wilderness Society, an organization he helped found and finance in its first years of operation. In 1937, Marshall enlisted his close friend Catherine Bauer, a leader in the regional planning movement, to explore the issues of wilderness, public access, and social policy. In a long letter to Marshall, Bauer noted that wilderness appreciation was seen as "snobbish," but that a great many people, even the majority, could enjoy the wilderness, given a chance to experience it. Bauer suggested that "factory workers, who experience our machine civilization in its rawest and most extreme form" were the ones who could most benefit from wilderness and that by doing so they could broaden wilderness' political base.[7]

Though Bauer's suggestions reflected Marshall's own approach, they caused some concern and consternation among other key figures in The Wilderness Society, especially its executive director, Robert Sterling Yard. Yard, whose salary was largely paid out of Marshall's funds, worried that the New Deal forester might interest too many "radicals" like Bauer to attempt to influence wilderness policy. Yard and others were also sensitive to the redbaiting that Marshall himself became subject to during the late 1930s, with its possible taint for The Wilderness Society as well. Most of these attacks, led by members of Congress associated with the House Un-American Activities Committee, sought to tar Marshall through his non-wilderness activities and financial contributions. Wilderness Society leaders such as Yard feared that Marshall's activities might reflect on the organization and worried that his advocacy of a "democratic wilderness" policy could undermine their preservationist concept of "protection." In response to the democratic wilderness concept, key Wilderness Society figure Olaus Murie would later write in an essay that "wilderness is for those who appreciate" and that if "the multitudes" were brought into the backcountry without really understanding its "subtle values," "there would be an insistent and effective demand for more and more facilities, and we would find ourselves losing our wilderness and having these areas reduced to the commonplace."[8]

Murie's pivotal essay was written a few months after Robert Marshall unexpectedly and tragically died in his sleep during an overnight train ride from Washington to New York. His death (Marshall was only thirty-nine when he died on November 11, 1939; some attributed his mysterious death from unknown causes to his hard-driving, passionate wilderness hiking) brought an end to the idea of combining a social and protectionist vision. In his will, Marshall divided his $1.5 million estate into three trusts: one for social advocacy, including support for trade unions and for promoting "an economic system in the United States based upon the theory of production for use and not for profit"; a second to promote civil liberties; and a third for "preservation of wilderness conditions in outdoor America, including, but not limited to, the preservation of areas embracing primitive conditions of transportation, vegetation, and fauna." It was this last trust that came to be controlled by key officials of The Wilderness Society, including Yard, whose approach was more narrowly conceived (in terms of membership and constituency) and politically limiting (in terms of resource policy) than Marshall's own inclinations. Ignoring his social vision, Wilderness Society leaders focused instead on Marshall's protectionist ideas, successfully lobbying to have a wilderness area in Montana named after him. Over time, Robert Marshall's life and ideas began to undergo reinterpretation, with the suggestion that his love for wilderness had really been an exclusive and separate concern. With his death, Robert Marshall, the "people's forester" whose life's mission had sought to link social justice and wilderness protection, would become an ambiguous historical figure representing environmentalism's divide between movements, constituencies, and ideas.[9]

From Resource Exploitation to Resource Management

The standard histories of environmentalism in the United States almost invariably begin in the West. In the vast, spectacular landscapes, in the breathtaking vistas, powerful mountain ranges, and sharp-cutting rivers, in the West's abundance and scarcity of resources, in its aridity and fertility, the forces of urbanization and industrialization created some of the most dramatic changes in environment. It was in the West that the best-known mainstream environmental group, the Sierra Club, was formed, and where some of the most bitter urban and industrial conflicts took place involving hard-rock miners or resource and development disputes over access to water sources. It is also in the West that much of the traditional interpretation of environmentalism is grounded.

It was in the West that the idea of a land ethic was first put into practice by the Mormons, followers of a quasi-utopian, theocratic movement who settled throughout the Colorado River Basin. Influenced by the land use approaches of the Utes and other southwestern Indian tribes, the Mormons sought to organize on a cooperative basis to benefit the group and the community as opposed to individuals acting separately from one another. From these principles a concept of "stewardship," applied to land ownership and resource use, was derived. One of the earliest questions the Mormons confronted along these lines was control of water resources in an arid and unpredictable environment. Mormon leaders rejected the prevailing riparian doctrine, which defined water rights as property rights based on ownership of the land adjacent to the water, as inappropriate to the irrigation requirements of the Colorado River Basin. Instead, a community value to the water was established, based on community ownership of dams and ditches designed to direct the flow of water.

The stewardship approach to water was applied to other resources as well. Timber harvesting required access roads to the canyons where the forests were located and were constructed under the direction of the Mormon Church, with provisions for use placed under the jurisdiction of county courts controlled by church officials. The act establishing these courts explicitly identified the stewardship requirement to "best preserve the timber" and to "subserve the interests of the settlements" in timber cutting and in the distribution of water for irrigation and other purposes.[10]

By the 1880s, the Mormon stewardship approach had come to be considered both controversial and exceptional. In his journals and later writings, John Wesley Powell, who first explored the Colorado River, headed the U.S. Geological Survey, and became the West's first great resource analyst, spoke admiringly of the Mormon experiment in cooperation and stewardship. Powell feared that settlement in the West would be inappropriately organized through private control of land and related water rights with a tendency toward monopolies or large-scale government intervention. Mormon stewardship, including cooperative management of land and water and irrigation on a more limited scale to serve the needs of planned communities, appealed to Powell's instinctive environmentalism. Yet it was quickly becoming apparent to the railroad companies, cattle owners, wheat farmers, mining and timber companies, and their financial backers that the West and its resources provided an extraordinary source of new wealth, ripe for exploitation, not cooperation.[11]

Already by the 1870s and 1880s, resource exploitation of the West dominated development patterns. Massive overgrazing, timber cutting, land monopolization, boom-and-bust mining practices as well as the industrialization of mines, speculation in land and water rights, and monocrop plantings overwhelmed limited efforts at cooperation and more orderly resource development. Gifford Pinchot, a young, wealthy forester recently returned from Germany, where he had begun to learn the principles of forest management, and soon to become a pivotal figure in the emergence of a conservationist movement, would write of this period that "the Nation was obsessed, when I got home, by a flurry of development. The American Colossus was fiercely intent on appropriating and exploiting the riches of the richest of all continents—grasping with both hands, reaping where he had not sown, wasting what he thought would last forever. New railroads were opening new territory. The exploiters were pushing farther and farther into the wilderness."[12]

The problems of resource exploitation seemed most pronounced with respect to land, water, and forests: an exploitation of the natural environment that paralleled the exploitation of labor in early industrialization, as historian William Cronon has argued. The 1890 U.S. Census Report called attention to dwindling supplies of timber and arable land, which in turn was seen as a function of increased concentration of ownership and intensified development. The concentration of land ownership had escalated rapidly during the 1870s and 1880s, creating major landholdings in California and other parts of the West. This was typified by the creation of the huge Miller and Lux holdings, two competing interests involved in a major water rights ruling in 1886, who between them controlled more than a million acres in the San Joaquin Valley of California. Government entities such as the General Land Office housed in the Department of Interior and local land offices responsible for the management and allocation of public lands were easily subject to political manipulation and abuse, often fronting for powerful private interests adept at using existing legislation to obtain additional landholdings.

These tendencies toward land monopolization also established political tensions in the region. Semipopulist countercurrents began to advocate government-backed irrigation projects, hoping that the availability of public capital would offset the accumulated power of the land monopolists. By the 1890s, groups such as the National Irrigation Congress had formed to push for federal legislation as a way to establish a kind of irrigation revolution in the West. The Irrigation Congress included not only populists who wished to impose restrictions on land ownership, but railroad interests who saw irrigation as a further opening of the West, politicians concerned that an immigration outlet for the East remain open, and larger western landowners who welcomed the availability of public capital for the construction of water development projects.

Passage of the Reclamation Act in 1902 failed to resolve the political tensions inherent in the irrigation coalitions. While the legislation provided land ownership restrictions for receiving federally reclaimed water, the Reclamation Service, which was mandated to carry out the provisions of the law, itself became subject to pressures that undermined the social vision associated with public support for irrigation. These included multiple requests for exemptions from acreage limitations, differing interpretations of what constituted land ownership, repayment provisions that clouded the question of socially designed subsidies, and proposed projects that stretched the original definition of beneficiaries.

The emergence of the Reclamation Service (which became the Bureau of Reclamation in 1923) as a key institution in western resource development was perhaps most significant as part of the push for organized, "scientifically based" resource management as opposed to the chaotic resource exploitation associated with land monopolization and private resource development. Under the leadership of Arthur Powell Davis (John Wesley Powell's nephew), the Reclamation Service became a leading advocate of applying the principles of science and engineering to the orderly management of resources. Resource development projects, through the application of science, would be designed to maximize the efficient use of a resource while preventing its overexploitation. Combining irrigation, storage, and possibly flood control and power generation suggested the potential for regional economic development and provided the Reclamation Service with its social engineering aura.

This engineering-based utility principle was also associated with the rapid emergence of the Forest Service and its ambitious plans for restructuring timber production along scientific principles. Even more than the private land and water speculation that characterized the West prior to the Reclamation Act, unregulated timber cutting had become a major scandal in the West and elsewhere.[13] The rapid depletion of forests within just a few decades created fears that a "timber famine" was imminent and that only through some dramatic government intervention could the timber lands be saved. These fears led to passage of the Forest Reserve Act of 1891, which temporarily protected certain forests against further development. The intent of the legislation was to halt destructive and unregulated timber cutting rather than to "lock up" forest reserves as a form of wilderness protection. During the next decade, pressures to reopen forest lands shifted the debate from the effort to stop overuse to the development of appropriate logging or forest management approaches as defined by Gifford Pinchot and others. Even John Muir, foremost champion of wilderness values, would write as late as 1901 that American logging should draw on the Prussian approach to forest management, whereby "the state woodlands are not allowed to lie idle" but are made to "produce as much timber as is possible without spoiling them."[14]

By the turn of the century a broad consensus had emerged, extending from Muir's own California-based Sierra Club, founded in 1892 as an outdoor recreation and advocacy group, to the Boone and Crockett Club, an elite hunting and wildlife protection association whose members included Theodore Roosevelt and Pinchot.[15] These groups put forth a complementary vision of what Muir called "preservation" and "right use" of resources and wilderness, arguing against the waste and spoilation associated with unregulated private development while suggesting that "right use," or the application of science and technique, would enhance the values of preservation, or "the necessity" of wilderness.

The assassination of William McKinley in 1901 and elevation of Theodore Roosevelt to the presidency pushed forward these new resource strategies. Less than two months after his inauguration, Roosevelt delivered his first message to Congress directly on the question of resource development, a speech that would become a benchmark in the rise of conservationist politics. "The fundamental idea of forestry," Roosevelt proclaimed on December 2, 1901, "is the perpetuation of forests by use. Forest protection is not an end in itself; it is a means to increase and sustain the resources of our country and the industries which depend upon them. The preservation of our forests is an imperative business necessity."[16]

Central to this concept of economic utility lay the role of the government in capturing the tools of science and establishing the principles of regulation and management beyond the single-purpose focus of the existing bureaucracies. Pinchot would later say that the "heart and soul" of Roosevelt's first message to Congress was the establishment of the new U.S. Forest Service, pieced together from several different existing agencies and provided with a mandate to coordinate private development through government regulation and management. Similarly, Roosevelt strongly promoted the creation of the Reclamation Service to accomplish "the reclamation and settlement of the arid lands."

These two bureaucracies became the cornerstone of conservationist politics over the next half-century. Pinchot, for one, was continually seeking to consolidate existing organizations or establish new ones, such as the American Conservation League and the National Conservation Association, to promote this utilitarian and expertise-driven vision. Most of these efforts at movement-building were unsuccessful, as were the attempts to maintain the irrigation-related coalitions once the Reclamation Service was organized. Instead, specific regional or industry-related interest groups emerged to serve as lobbying organizations and agency support groups, further situating conservationism less as a social movement than as a development strategy linked to government action based on the principles of efficiency, scientific management, centralized control, and organized economic development.[17]

During the seven years of the Roosevelt administration, when conservationism emerged as the country's dominant resource strategy and when the conservationist agencies became central to the formation of these resource strategies, the first sharp divisions between those primarily focused on "preservation" and those linked to "right use" also emerged. These debates over preservation versus development also exposed a lack of consensus within organizations such as the American Civic Association and the Sierra Club and the absence of a clear vision concerning how to contend with the forces of urbanization and industrialization. At the same time, the leadership within the government agencies set out to establish their own mandates, which became the heart of the conservationist world view. That process culminated in the May 1908 Governors' Conference on Conservation, aimed at consolidating the gains of conservationism and establishing it as a permanent fixture within the policy process regarding resources. While the focus of the conference was the marriage of science and development, conference organizers specifically sought to downplay the principle of preservation. Pinchot, in fact, specifically excluded John Muir from this gathering. By then, the author and preservationist champion had become the most visible and effective advocate of the notion that wilderness maintained a separate value as a "fountain of life," independent of its utility as a resource.[18]

This division between Muir and Pinchot, much celebrated in the history of environmentalism, has tended to obscure the crucial role of the government agencies and their resource strategies in the framing of conservationist politics. It has also served to obscure the contending pressures among the conservationist-oriented agencies themselves. Pinchot's departure from the Forest Service in 1910, for example, can be seen as less significant in terms of the specific circumstances of his departure—a leasing dispute over Alaskan resources within the Department of the Interior—than in terms of the uncertainties surrounding overall resource policy and the role of the agencies.

By 1912, with the founding of the Progressive Party and the attempt to restate conservationist politics, conservationism as an ideology also seemed harder to define. This group of former and present government officials, industry critics, resource development advocates, and professionals sought to incorporate not only the principles of efficiency and science, but the remaking of civil society as well. The passing of resources under monopolistic control, former Pinchot aide W. J. McGee wrote in 1910, was creating a generation of "industrial dependents."[19] To break that linkage required not just better management, some conservationists argued, but social transformation.

This conservationist coalition, which had become an amalgam of very different movements, ultimately failed to last much beyond the 1912 election. The process of creating a unifying vision for remaking society—from strategies with respect to resource management to issues of urban and industrial reorganization—splintered before the new Progressive Party had any chance of cohering. The long-standing historical argument over whether conservationism represented a more exclusive effort that relied on science and efficiency in the management of resources or of industry itself or whether, in fact, it provided a forum for those concerned more with "economic justice and democracy in the handling of resources than with mere prevention of waste," as one historian put it, was never fully resolved. By World War I, those distinctions had faded, as the country entered a new age in which growth and expansion again became linked to resource availability rather than resource management.[20]

This emphasis on new development also tended to divide the government agencies at the center of the conservationist approach. The Forest Service split between advocates of "cooperation," who supported allowing market forces to fuel the drive for forest management, and the defenders of public regulation. By 1920, the postwar shortage of lumber, the upward spiral of lumber prices, and the extension of logging into new areas had led Pinchot and many of his followers to fear that without renewed regulation and government intervention, another cycle of overexploitation could undermine conservationist gains.[21] The new leadership of the Forest Service, including one-time Pinchot ally William Greeley (chief forester during the Harding and Coolidge administrations), argued forcefully against intervention, declaring that the Forest Service's "real objective" was, as Greeley put it, "the actual production of timber." This shifting agency leadership role was best symbolized when Greeley himself, on resigning from his position as chief forester, became the timber industry's chief political advocate and spokesman as secretary and manager of the West Coast Lumberman's Association.[22]

The prominent role of the timber industry in the affairs of the Forest Service paralleled changes at other agencies. This included the Bureau of Reclamation, which continued to drift from its original social vision toward a supportive role in the development of a western water industry. By the 1920s, BuRec activities were being framed less in relation to a conservationist "science" and more as a tool for development on behalf of particular private interests, whether landowners or urban development interests.[23]

Through the 1920s and into the 1930s, the language of conservationism was increasingly appropriated by the resource-based industries and other industrial interests attracted to the concepts of efficiency, management, and the application of science in industrial organization. Groups organized to monitor and influence the agencies, such as the industry-dominated American Forestry Association and the National Water Resources Association, increased the pressure on the agencies to redefine their mission as incorporating the techniques of science and management to support private development. Industry interests were also able to adopt the principles of multiple use as justification for their own environmentally destructive activities, such as the discharge of untreated wastes into streams or other water sources defined as "nature's sewers."[24]

With industry occupying a central role in defining and interpreting conservationism by the end of the 1920s, the historic tension between conservationism as an anticorporate social movement and as an effort to rationalize a resource-based capitalism had disappeared. By the close of the Progressive Era in the 1920s, conservationism as expertise and rational management of resources for business uses had emerged as the movement's dominant ideology, an ideology eagerly embraced by the very industries an earlier generation of conservationists had so forcefully challenged.

NATURE SET APART: THE SEARCH FOR PROTECTION

If utility became the byword of the early conservationists, the setting apart of Nature became the focus of those who emphasized that wilderness needed to be protected from urban and industrial influences. This preservationist or protectionist movement provided an even less coherent vision and organizing principle than conservationism. It included such diverse approaches as nationalism (Nature as a national treasure); commercialism (wilderness available for tourism and recreation); spiritualism (wilderness as regeneration in an urban and industrial age); ecology (Nature as biological richness and diversity); and a kind of elite aestheticism (Nature as beauty and experience, especially for those presumed to be most capable of appreciating it).

Wilderness advocacy has a long tradition in the United States, dating back to the early and middle years of the nineteenth century when urban expansion and resource activities in the East and middle border regions transformed much of the natural environment. It was the westward movement, however, that elevated wilderness as a key issue, both in terms of the scenic impact of the West's natural wonders and the ambiguities associated with expansion and the closing of the frontier. By 1872, with the setting aside of Yellowstone as a "preservation," there emerged the notion that western wilderness was distinctive, even as the policies designed to address wilderness remained bound by other considerations.

In the case of Yellowstone, the area set aside was considered worthless in economic terms, except insofar as it offered an opportunity for tourism with its related economic benefits. The wonders of Yellowstone, such as its geysers, were deemed important in part because they highlighted what was absent in other parts of the country: spectacular and monumental Nature. As a result, Yellowstone provided America with an opportunity to compete culturally with the Old World, with an image of Nature frozen in time, made possible by the absence of competing interests.

During the period that Yellowstone began to be celebrated as a national monument, the country also experienced a revival of interest in and sentimental attachment to the cause of the Indians, whose tribal organizations and lands, including the Yellowstone area, had largely eroded in the face of military action by the government and settlers. Yellowstone National Park, in fact, had been established from lands belonging to the Shoshone, Bannock, Blackfoot, and Crow tribes. The conception of the park as a cultural monument further reinforced the notion that preservation was specifically *not* about protecting living environments subject to the land uses and activities of organized societies, but rather about safeguarding cultural artifacts.[25]

The most significant issue facing early wilderness advocates was the overriding influence of resource development. The 1890 Yosemite Act, for example, referred to the newly established Yosemite National Park as "reserved forest lands." Wilderness protection became feasible only after resource development was defined as remote from any given area or where the area needed protection as a hedge against overexploitation of resources. Pinchot and Muir worked closely together during the 1890s, in part because Muir, who valued wilderness primarily as a spiritual resource, had decided that wasteful industry practices were a major cause of the decline of wilderness and that there existed what he called the "legitimate demands on the forests" tied to their economic utility.[26]

However, as the outlines of a new resource-based conservationist approach took hold during the Roosevelt administration, Muir and some of his allies became increasingly dismayed by the logic of the conservationist argument. The linkage of efficiency and science with maximum use could also mean the sacrifice of wilderness, particularly where urban or industry interests were involved. Thus, the decade-long battle over the construction of a dam in the Hetch-Hetchy Valley twenty miles to the northeast of Yosemite to meet the city of San Francisco's water needs has rightfully been identified as a critical event demonstrating the basis for a conservationist/preservationist dispute where the claims for development and protection competed in the same arena.

For the preservationists, grouped in part around Muir and his supporters in the Sierra Club, the defense of Hetch-Hetchy represented the first clear delineation between the need for protection versus the logic of resource development. After more than thirty years of often brilliant writing and advocacy about the "beauty, grandeur, and all-embracing usefulness of our wild mountain forest reservations and parks" in such publications as *Century* and the *Atlantic Monthly*, Muir became his most eloquent and inspiring in his defense of Hetch-Hetchy as wilderness. His famous "Dam Hetch-Hetchy! As well dam for water-tanks the people's cathedrals and churches" was complemented by numerous other tracts and writings less well known but nevertheless magnificently capturing his vision of wilderness as its own life force.[27]

When resource development was not seen as conflicting with this kind of scenic wilderness value, the preservationists stayed away from the issue. This occurred in the Owens Valley, where a proposed water resource development pitted conservationists allied with Los Angeles business and development interests against local Inyo County forces as well as urban Angelenos contesting the spiral of expansion proposed for their community. On the sidelines stood the preservationists, fixed on their definition of wilderness as scenic resource, and for whom neither rural development nor urban growth issues were seen as relevant.[28]

The Owens Valley area, this "land of little rain," was for those who knew it a special place, where "to understand the fashion of any life, one must know the land it is lived in and the procession of the year," as that other great essayist of the era, Mary Austin, wrote of her home.[29] But to the preservationists, interested primarily in the monumentalism of nature, this semi-arid valley held little interest as a natural or scenic environment. Most important, the issue of the urban environment, so critical to the events that unfolded around the securing of Los Angeles' water supply, was even further removed from the preservationist frame of reference.

The preservationist position, in retrospect, appears particularly poignant, given how the issues of urban growth and sustainability emerged between 1908 and 1913 during the construction of the Los Angeles aqueduct. Though the problem was perceived in part as a labor issue (immigration represented a source of cheap labor), key aqueduct opponents, including the Los Angeles Socialist Party and its leader, Job Harriman, questioned whether the region could support an indefinite cycle of expansion based on the concept of an unlimited water supply. The Socialists offered an alternative vision of a democratic community organized to live and grow within its existing resource base, with real estate subdivisions organized according to plan rather than through speculation.

After narrowly losing the mayoralty election in 1911, the Socialists ultimately lost the battle over disposition of the water supply by 1913. Los Angeles would grow in a rapid and crazy-quilt fashion via access to this new, imported water supply, and an annexation policy would make surplus water available to areas willing to annex to the city as a precondition for expansion. Between 1913 and 1928, when the next imported water supply was secured, Los Angeles grew fourfold in land area, establishing a pattern of development for the southern California region and ultimately for the country as a whole. And while John Muir became best remembered for his defense of Hetch-Hetchy and scorn for San Francisco, Job Harriman became a forgotten symbol of the effort to define Los Angeles as a place of limits, equating environment with the conditions of urban life.

Recreational Politics

The preservationists' disinterest in the urban environment was reinforced by the anti-urban biases that prevailed among the most radical and forceful wilderness advocates, including Muir and wildlife defender William Hornaday. Although he wrote for urban, cosmopolitan publications that allowed him to establish urban support for wilderness protection, Muir was nevertheless especially hostile to urban living. As his biographer, Stephen Fox, pointed out, Muir distinguished between the urban "lowland" and the wilderness high ground, which provided a kind of spiritual replenishment for daily life. Returning to Yosemite after a visit to San Francisco, Muir wrote how he experienced his own physical regeneration in the wilderness, "sufficient to shake out and clear away every trace of lowland confusion, degeneration and dust."[30]

The anti-urban attitudes of the preservationists were also linked to their attitudes about class. The issue of hunting, for example, pointed to preservationist biases about class. By the mid- to late nineteenth century, hunting for food was being strongly criticized in the name of wilderness protection. The terms *pot hunting* and *pot shot*, or hunting for food, entered the language as depicting acts of lower-class cowardice and ill-breeding, as distinguished from the upper-class "sportsman [who] pursues his game for pleasure . . . [and] shoots invariably upon the wing and never takes mean advantage of bird or man," as Theodore Roosevelt's uncle, Robert Roosevelt, put it.[31] William Hornaday, the strong-willed director of the New York Zoological Society and impassioned defender of wildlife, made similar distinctions in his writings and specifically called for legislation that would discourage those who "sordidly shoot for the frying pan." Much of Hornaday's argument was structured as an appeal to his upper-class supporters, whom he urged "to take up their share of the white man's burden and bear it to the goal."[32]

Where Hornaday and other wildlife defenders differed with this elite constituency was over its preoccupation with sport hunting as upper-class recreation. Similar to Muir in the kind of absolute protectionist position he adopted, Hornaday became especially suspicious of those groups and individuals willing to ally with antiprotection groups such as the gun and ammunition businesses and elite hunting and commercial duck clubs. Hornaday, in fact, engaged in a bitter conflict that lasted more than two decades with T. Gilbert Pearson, the head of the National Association of Audubon Societies. During his lengthy tenure as leader of one of the oldest preservationist groups (until his forced resignation in October 1934), Pearson centralized the administration and sought to shift the group toward a middle ground between protection and accommodation to hunting and commercial interests.[33]

The hunting disputes also reinforced the strong perception that preservationist debates were primarily disputes among elites—between those who wished to leave the natural environment in a pristine state and those who viewed it as a place for recreation and pleasure. These disputes were most directly associated with the question of the national parks. When the National Park Service was established in 1916, it had all the trappings of an institution run for and by the elite. Its first head, Stephen Mather, a wealthy businessman with strong ties to the railroads, ran the Park Service as a kind of fiefdom, a playground for the wealthy, whose support for the park system he hoped to secure.

Mather's strategy to develop support for the Park Service was directly tied to the development of the parks as recreational resources. In pursuing this new approach, Mather initially relied on the railroads, which had already been instrumental in linking the concept of tourism to park management. During the late nineteenth and early twentieth centuries, the railroads had become the primary source of capital for new concession businesses in such parks as Yosemite, Yellowstone, Zion, Bryce, the Grand Canyon, and Mount Rainier. Since the railroad companies saw tourism primarily as an upper-class activity, they sought to establish "proper" tourist facilities, including grand hotels catering to the wealthy. Hotel management, in turn, served as an important adjunct to passenger train service to park areas, primarily designed to attract wealthy patrons.[34]

Mather extended this tradition of service for the wealthy, particularly through promotional activities such as the publication of glossy, expensive portfolios and picture books (some financed by Mather as well as by western railroad interests). These promotions were designed, as Mather's biographer, Robert Shankland, wrote, "to reach a hand-picked elite, capable, they hoped, of passing the habit of park travel down from above." Central to Mather's conception, already implicit in the Hetch-Hetchy conflict, was that the economic rationale for preservation lay in the growth of the tourism industry. "Our national parks are practically lying fallow," Mather wrote shortly before his appointment, "and only await proper development to bring them into their own."[35]

This economic argument—"making a business of scenery" as one article put it—became the dominant park policy: greater park access was needed in order to stimulate the tourism trade. To promote the parks, Mather also understood that the long-standing nationalist appeal about cultural monuments, such as the railroads' promotional slogan of "See America First," had to be integrated into a broader appeal of experiencing nature. This "back to nature" appeal was part nostalgia and part therapy, an arcadian myth in an increasingly urbanized and industrial society. It also permeated the life and thinking of a good portion of the urban middle and upper classes during the first two decades of the twentieth century. The Park Service refined those sentiments, both in attempting to protect Nature by setting areas apart and then by opening protected areas for larger and larger numbers of people to enable them to imagine life as it was "lived before the call of the city was insistent," as one 1912 writer put it.[36]

The revolution in access that Mather envisaged became most feasible with the rise of the automobile as the primary mode of transportation into the parks. Mather saw the automobile as a crucial stimulant to park travel and framed policies such as road construction, decisions about access, and joint promotional efforts to encourage car visitors. Long active in the American Automobile Association, Mather was a park-and-auto booster. He promoted and in large part financed, for example, the reopening of the Tioga Road in Yosemite for auto traffic. "The automotive and corollary industries counted him safely among their friends at court and reciprocated by advertising the parks almost gratuitously," Mather's biographer wrote, describing one instance in which a tire manufacturer widely distributed a three-sheet billboard poster exhibiting Yosemite Valley from Glacier Point while superimposing, with Mather's sanction, the company's tire in the center of the poster.[37]

Automobile access to the parks, far more than the earlier railroad traffic with its elite constituency, fundamentally transformed the nature of the park system. Tourism emerged as the dominant concept driving park policy. In Yellowstone, for example, the Park Service sought to increase the elk population as a showpiece to attract tourists and helped accomplish this by creating an open war on elk predators such as the mountain lion. Auto traffic into the parks, meanwhile, increased dramatically through the 1920s, with the number of cars entering Yosemite alone jumping more than tenfold in less than a decade.[38]

These figures also indicate that tourism had expanded the park constituency in terms of class as well. Mather's successors, among them Horace Albright, continued these pro-automobile, protourism policies, even as the Park Service's strongly elite character and ties remained in effect. The divisions within the preservationist movement that emerged in the late 1920s and 1930s over park policy focused less on those elite ties or even the political necessity for encouraging tourism than on the presumably unavoidable consequences of tourism on wilderness areas. Correspondence between *Saturday Evening Post* publisher George Horace Lorimer and J. Horace McFarland of the American Civic Association, two key wilderness advocates and park supporters, highlighted this dilemma for preservationists. In November 1934, Lorimer wrote that the growing role of the automobile had caused him to lose enthusiasm for the national park system. "Motor roads and other improvements are coming in them [the parks] so fast that they are gradually beginning to lose some of their attraction for the out-of-door man and the wilderness lover," Lorimer lamented. In response, McFarland acknowledged that while automobile access helped generate political support for the parks, he still felt, like other key preservationists, a condescending attitude toward those he called the park-going "dear public."[39]

These park tourism critics, dismissed as purists by Park Service defenders, became increasingly vocal during the 1920s and 1930s at a time when tourism was increasing and wilderness protection was becoming a contested policy arena. Former Park Service officials such as Robert Sterling Yard joined with a new generation of wilderness advocates, many of them tied to the Forest Service bureaucracy, in promoting a conception of wilderness as separate from its tourism-derived economic utility. A key figure in this evolving definition of wilderness was Aldo Leopold, a Forest Service employee who wrote in 1921 of wilderness as "a continuous stretch of country preserved in its natural state." Although suggesting that such areas could be open to lawful hunting and fishing, Leopold still asserted that wilderness necessarily had to be "kept devoid of roads, artificial trails, cottages, or other works of man."[40]

This concept of wilderness as distinct from the urban and industrial environment, as having value in its own right, and as insulated from the pressures of both resource development and the tourism trade became directly linked with the emergence of what Leopold called a "land-based ecology." Leopold's own evolution, in his activities and writings, charts this shift toward ecology. Early in his career, Leopold accepted the dominant conception of forestry as a science that allowed for a certain level of timber cutting and logging. He also sought to distinguish between scientific forestry and scientific game management, which required a level of protection for "the perpetuation of every indigenous species," with the significant exception of predators and other nongame wildlife. This approach drew heavily on Leopold's allegiance to hunting as a form of adventure and cultural replenishment to counter the "tragedy of prescribed lives" embedded in the urban and industrial culture. "The hunting instinct is a fixed character, and will continue to appear in a certain proportion of all normally developed individuals," Leopold wrote in 1919. This core interest in hunting shaped his search for a minimal impact use concept, allowing for such activities as hunting, fishing, canoeing, and camping while providing for what Leopold called "some logical reconciliation between getting back to nature and preserving a little nature to get back to."[41]

Through the 1920s and early 1930s, Leopold tried to steer a middle road. He was sympathetic to wildlife advocates such as Hornaday and the newer, more contentious conservationist organizations such as the Izaak Walton League (whose Wisconsin chapter Leopold joined in 1925), while also seeking common ground with his former colleagues in the Forest Service over the question of scientific management of forestry and wildlife. But Leopold's studies in the science of game management increasingly led him to the conclusion that most game advocates, such as sportsmen, naturalists, and outdoor writers, had less interest in conservation as a science of living environments than as a method of keeping sufficient numbers of wildlife game alive for their economic or scenic utility. Though Leopold still sought to reconcile his continued love of hunting with his increasing concerns about the precarious state of various wildlife environments, he began to despair that growing population and economic expansion pressures were making the two positions irreconcilable. It was crucial, Leopold argued in an unpublished manuscript from the early 1930s, to find a workable synthesis between expansion and protection.[42]

During the New Deal years, Leopold grew increasingly pessimistic, believing population and expansion pressures had become too powerful. The essays that constitute his most famous and posthumous collection, *Sand County Almanac*, written during the last decade of his life, reveal his growing hostility toward "mass recreation," which he contrasted with "rudimentary grades of outdoor recreation" linked to "recreational ethical restraint." At the same time, Leopold's insistent call for a "revision of the national attitude toward land, its life, and its products" was being increasingly perceived as too radical, too removed from the mainstream conservationist and even preservationist ideas concerning economic and scenic utility. As early as 1935, National Wildlife Federation founder Jay "Ding" Darling wrote Leopold that "you are getting us out into the water over our depth by your new philosophy of wildlife environment" based on Leopold's insistence that ownership and use of the land created "obligations and opportunities of trans-economic value and importance." "The end of that road," Darling wrote Leopold, "leads to socialization of property which I could only tolerate willingly if I could be shown that it would work."[43]

Leopold, however, was far from a socialist or even a supporter of the New Deal. Leopold's biographer, Curt Meine, characterized this advocate of a new land ethic as an "anti-ideologue." "His experience of urban problems," Meine wrote of Leopold, "was vicarious at best, naive at worst. He appreciated the problems of urbanized man, but he was not a social activist." Leopold focused more on what he called the "individual responsibility" of the private landowner and was hostile to the notion of government responsibility for "land health."[44]

Within this individualist credo, nevertheless, was contained a crucial, radical idea: the desire to infuse the industrial culture with what Leopold called "ecological conservation." "To change ideas about what land is for," Leopold wrote in 1940, "is to change ideas about what anything is for." Within the year, he would extend that argument to proclaim that the essential value of wilderness was not recreation and its corollary economic interests but its value as a "science of land-health," presenting a "base-datum of normality." This "picture of how healthy land-maintenance [exists] itself as an organism" ultimately became a crucial objective in its own right. The concept of protecting wild areas, Leopold wrote in a Wilderness Society publication, had to extend beyond the "spectacular scenic resources" of some park areas to include such places as "low altitude desert tracts heretofore regarded as without value for 'recreation' because they offer no pines, lakes, or other conventional scenery." A conservationist deeply entrenched in the tradition of rugged individualism, Aldo Leopold, in his last years, could also be seen as this country's first deep ecologist.[45]

The Technological Imperative

For conservationists and preservationists, the main order of business during the 1930s and 1940s continued to be the development of policies related to the management of resources or protection of the natural environment. The Roosevelt administration maintained an active interest in resource management and brought many key conservationists to the center of its resource policy making. These included most prominently Chicago lawyer and Progressive Party leader Harold Ickes, whose term as secretary of the interior (thirteen years) lasted longer than that of any previous or subsequent head of Interior. Though Ickes failed in his quest to create a unified Department of Conservation by combining such agencies as the Bureau of Reclamation and the Forest Service, he was successful in maintaining a political balance between protectionist approaches, such as the creation of Olympic and Kings Canyon national parks, and development-oriented policies, such as construction of the Colorado–Big Thompson and Central Valley water projects. Both approaches, however, were largely subsumed under the economic development policies established to deal with the overriding question of unemployment and the depression.

With the end of World War II and its military-induced economic recovery, balancing preservation and development objectives seemed more problematic. That became especially true after Harold Ickes' resignation in 1946 as new resource development plans among such bureaucracies as the Bureau of Reclamation came to the fore. These initiatives had the potential to reopen the divide within the conservationist movement (the term *preservationist* had largely disappeared from use during the Ickes era). Tensions were emerging over questions of population growth and potential resource scarcity, the offsetting role of technology, and differing protest tactics and strategies regarding whether and how to protect the natural environment.

In the late 1940s two books written by prominent conservationists launched a debate about the state of the postwar order and the problem of population. These books—*The Road to Survival*, written by ornithologist and one-time editor of the Audubon Society magazine, William Vogt, and *Our Plundered Planet*, authored by the prominent head of the New York Zoological Society, Fairfield Osborn—became key conservationist tracts, helping reelevate a neo-Malthusian perpective within the movement.

The two authors were different in background and temperament. Vogt, the more impatient and radical in his approach, framed *The Road to Survival* as a continent-by-continent survey of land and population issues. He argued that the relationship between human populations and the supply of natural resources necessary for daily life had become highly unstable. This was due in part to "free competition and the application of the profit motive." "Free enterprise—divorced from biophysical understanding and social responsibility," Vogt wrote, "must bear a large share of the responsibility for devastated forests, vanishing wildlife, crippled ranges, a gullied continent, and roaring flood crests." Vogt saw the problem as global in nature, stimulated by a "sanitary revolution" that had spread beyond the industrialized countries into less developed continents, such as South America, thus causing a population explosion. The resulting breakdown in the ratio between population and resources had the potential to create social disorder and possible starvation, a "meeting at the ecological judgement seat." With the postwar population explosion, Vogt gloomily concluded, "the handwriting on the wall of five continents now tells us that the Day of Judgement is at hand."[46]

Vogt's unabashed Malthusianism and especially his anticapitalist tone contrasted with Osborn's *Our Plundered Planet.* Osborn, the son of New York Zoological Society founder Henry Fairfield Osborn, also sought to present population and resource management concerns but argued that a free enterprise system could be mobilized to correct potential system abuses. At the same time, Osborn criticized "technologists [who] may outdo themselves in the creation of artificial substitutes for natural subsistence" as a way to avoid the possibility of loss of resources. The only solution, Osborn warned, was to recognize "the necessity of cooperating with nature."[47]

Even more than *Road to Survival*, the publication of *Our Plundered Planet* set off an intense debate in the late 1940s and early 1950s about population, resource, and technology issues. In a 1949 MIT forum, "The Social Implications of Scientific Progress—An Appraisal at Mid Century," Osborn's position, characterized by *Time* magazine as the "familiar Malthusian bogy of ever-shrinking resources, ever-increasing population," was attacked by several conference participants, including Vannevar Bush, the wartime director of the Office of Scientific Research and Development. In a panel session entitled "The Problem of World Production," Bush argued that both population and scientific discoveries might be "bursting upwards" simultaneously, but that "science gets there first." Both Bush and fellow panelist Nelson Rockefeller, then head of the International Basic Economy Corporation, spoke glowingly of the power of American technology to stretch resources and allow for a worldwide increase in the standard of living through such new technologies as insecticide sprays.[48]

By the early 1950s, the debates over resource, population, and technology issues had been joined. On the one hand, technology advocates, such as Thomas Nolan, director of the U.S. Geological Survey, spoke of the "inexhaustible resource of technology" and how potential resource shortages would simply "inspire the research and technical advances that will make it possible to resolve such problems well in advance of the doom we are often prone to foresee."[49] In contrast, the "new conservationists," as Osborn and his assistant, Samuel H. Ordway, Jr., called themselves, argued that the demand for resources combined with population increases would outstrip technological advances. "The future promises of technology are potential promises," Ordway argued, and, as such, remained " 'pie in the sky' in the laboratory and the unplumbed seas."[50]

With few exceptions, both the technology-oriented optimists and the conservationist pessimists shared the belief that the private sector rather than government intervention would best correct abuses or provide answers. Both sides also sought to address growth issues, a central facet of postwar urban and industrial ideology. One key document, the 1952 President's Materials Policy Commission report, which primarily focused on the problems of possible materials shortages, nevertheless also asserted a fundamental belief in "the principle of Growth." "Granting that we cannot find any absolute reason for this belief," the commission declared, "we admit that to our Western minds it seems preferable to any opposite, which to us implies stagnation and decay." The postwar growth debate thus centered on whether there existed "any unbreakable upper limits to the continuing growth of our economy," as the commission put it. Even Samuel Ordway, who fretted about the role of advertising and complained about "faster automobiles, radio and television sets blaring imprecations to buy more machine products, 90-page newspapers, pulp magazines, and Mickey Spillane by the millions," still argued that growth could be maintained as long as it was no longer based on consuming "more than the earth produces." Such a shift in approach would require, according to the conservationists, less a faith in technology than a willingness to incorporate the values of conservation—efficiency, wise (or temperate) use, better management—into how the urban and industrial order operated. Warning that a failure to adopt such changes could lead to widespread government intervention in everything from land use planning to price controls, the conservationists maintained a belief that the "free enterprise system" was capable of making such adjustments.[51]

How those adjustments could be made became the central mission of the two most important organizations of the new conservationists: the Conservation Foundation and Resources for the Future. Founded in 1948, the Conservation Foundation (CF) defined its goals in terms of research, education, and reports that addressed resource and population issues. Led by Osborn and Ordway, the organization's leaders during its first two decades, the CF relied heavily on a tight circle of officers, conservationist allies, financial contributors (including, most prominently, Laurance Rockefeller), and favored experts commissioned to write about key CF issues. The CF placed population and resources at the center of the conservationist discourse, while seeking to make linkages between industry, government, and universities in promoting better resource management. Unlike the Sierra Club and The Wilderness Society, which became embroiled in specific resource conflicts, the Conservation Foundation deliberately removed itself from a direct advocacy role, preferring instead to emphasize through conferences and publications how government and industry practices could be made more rational. The organization primarily focused on promoting expertise for its educational and policy roles and thought policy changes were best achieved through the proper application of expert knowledge.[52]

The establishment of Resources for the Future (RFF) four years after the founding of the Conservation Foundation furthered this expertise-oriented view of conservationism. The original impulse behind the creation of RFF was the effort by several leading conservationists, including Osborn and former National Park Service director Horace Albright, to undertake a major "midcentury" conference on resource and conservation issues. After extensive maneuvering that caused the new group to disassociate itself from New Deal–style conservationism and its emphasis on national planning, the Mid-Century Conference was able to secure sponsorship from several conservative businessmen and resource-oriented trade associations. Lewis Douglas, chairman of the board of the Mutual Life Insurance Company, served as conference chairman, and the conference steering committee consisted of executives from cattle companies, the Farm Bureau, the American Petroleum Institute, Standard Oil, Newmont Mining, and Monangahela Power, with only Ira Gabrielson of the Wildlife Management Institute representing any of the conservationist advocacy groups.[53]

On December 2, 1953, President Dwight Eisenhower welcomed the 1600 Mid-Century Conference participants who had filled Washington, D.C.'s Shoreham Hotel in anticipation of this signal event in the evolution of conservationism. Eisenhower set the tone for the conference when he declared that conservation was not about "locking up and putting resources beyond the possibility of wastage or usage," but involved "the intelligent use of all the resources we have, for the welfare and benefit of all the American people."[54] Conference participants emphasized the need for population control, technological innovation, and appropriate resource development strategies to best address the key problems of population growth, an expanding economy, and a need for materials for military preparedness, which were the primary topics of the conference.

This attempt to combine the need for resource management with support for the development of new technologies characterized much of the early work of RFF. "Progress toward greater control [of the environment]," the group's first annual report declared, "reflects basic discoveries in pure science and the ability to adapt new principles for the benefit of society through improved patterns of organization and cooperation."[55] Along these lines, one of the first major grants by the organization involved exploring the "productive uses of nuclear energy. . . . Though nuclear power is not yet economic in most places," RFF president Reuben Gustavson reported in the group's 1955 annual report, "man has in the nucleus of the atom an almost inexhaustible source of energy, which he will be able to control and make available at prices that will decrease with advances in technology."[56]

Throughout the 1950s, RFF emphasized its problem-solving approach. "The strategy of our efforts to enlarge understanding of the role of natural resources in the growth of the American economy and the welfare of the American people," Gustavson's successor, Joseph Fisher, commented, "is to start with a basic and critical problem, stated usually in economic, political or social terms, and then to pursue it by means of research, fundamental or applied as may be indicated."[57] Much of the research focused primarily on resources, especially water resources, whose management had become central to the conservationist agenda in the 1950s. During this period, RFF researchers also worried about a breakdown in the conservation ethic caused by a powerful "Iron Triangle" of congressional leaders, government agencies, and local development interests who set the framework for policy on the basis of (sometimes marginal) economic interests rather than efficiency criteria.

By the 1960s, the focus on resource development for groups such as Resources for the Future and the Conservation Foundation had shifted from materials shortages to the externalities of resource development: inefficient projects, water pollution, waste discharges, air emissions. Led by a team of welfare economists, RFF raised the possibility that a failure of the market was occurring because "decisions concerning the use of natural resources do not always take into account the effects of that use."[58] By correcting such abuses, the RFF economists emphasized they were interested not in "abolishing adverse unfavorable effects" as such, but "reducing them in some cases where investigation shows that on balance such a reduction is worthwhile."[59]

To carry out such corrections, RFF researchers became strongly wedded to two related approaches: cost-benefit analysis to see where corrections would be most cost-effective; and a reliance on cleanup or pollution control technologies when their benefits outweighed the costs involved. In cases where externalities impacted the natural environment, such as polluted streams, RFF analysts argued that the benefits side of the ledger had to incorporate the recreation value of the resource, underlining the increased importance of recreation for the conservationist argument. And while the shift in emphasis from materials shortages to inefficiency and externalities preoccupied the conservationist experts, the concern for safeguarding and setting aside pristine areas continued to be the dominant focus for conservationist advocacy groups, at a time when resource development was once again becoming identified as synonymous with an attack upon Nature.

The Fight for Wilderness

The consolidation of the Iron Triangle during the 1950s was best exemplified by the Bureau of Reclamation's ambitious projects to ring the Colorado River with storage dams, hydroelectric plants, and various other facilities. As early as 1946, Department of Interior officials laid plans linking water reclamation with regional economic development strategies. They hoped to stimulate agricultural and resource extraction activities as well as overall urban and industrial growth in the West, especially the Mountain States and the Southwest. By the late 1940s, an initial package of Iron Triangle–related projects serving the Upper Basin states of Colorado, Utah, New Mexico, and Wyoming was introduced as legislation. One of the sites proposed in the legislation, a dam and hydroelectric facility at Echo Park within Dinosaur National Monument at the intersection of the Green and Yampa rivers on the Colorado-Utah border, immediately became a focus of concern for certain conservationist groups. The subsequent battle over Echo Dam, which would last nearly a decade and would be seen by some as a "turning point of historic significance" for environmentalism,[60] revitalized several of the advocacy groups and set one wing of the conservationist movement off on its search to find a place for wilderness in the postwar order.

The two leading organizations in the Echo Park fight, the Sierra Club and The Wilderness Society, seemed at first ill-equipped to undertake a major public battle against the entrenched interests of the Iron Triangle. The Sierra Club had become strongly associated with such activities as mountain climbing, skiing, and backpacking, and its leadership tended to be drawn heavily from the business and professional worlds. Even during the New Deal period, the Sierra Club, many of whose top officers were conservative Republicans, played a relatively passive role in the development of resource policy and preservationist goals.[61]

If the Sierra Club did not appear to be a strong candidate for confrontation, the same could be said for The Wilderness Society, particularly during the reign of its first executive director, Robert Sterling Yard. But when Yard died at the age of eighty-four in 1945, his replacement, Howard Zahniser, a former employee of the U.S. Biological Survey, defined his mission as safeguarding and extending wilderness designations. Zahniser also saw the need for The Wilderness Society to develop working alliances with other conservationist groups, including the Sierra Club, which became more interested in direct advocacy after the new position of executive director was filled by university press editor and long-time Sierra Club member David Brower.

As Bureau of Reclamation plans to develop projects within the Colorado River Basin began to unfold, both organizations realized they had reached a point where a high-profile, public campaign to save Dinosaur National Monument was necessary. With the decision to pursue such a campaign, including lobbying and major promotional efforts designed to highlight this scenic resource, the Dinosaur fight served to distinguish between those conservationists still wedded to reclamation and other multi-use projects and those groups such as the Sierra Club and Wilderness Society that sought to recapture their preservationist or protectionist roots by emphasizing the need for protection. Yet these same protectionist groups were also afraid of being characterized as "just aginners" and thus sought to support an alternative facility in the Upper Colorado River Basin as long as it did not "adversely affect Parks, Monuments, or Dedicated Wilderness," as Brower and other leaders repeatedly declared. This position pushed the protectionist groups to accept a proposal to eliminate the Dinosaur facility but significantly expand a dam to generate additional hydroelectricity at Glen Canyon at the Arizona-Utah border to the east of the Grand Canyon. Unlike Dinosaur, which was quickly becoming an attractive recreation site popularized by the protectionist groups as a mobilizing tactic, the rugged and isolated but spectacular Glen Canyon was largely unknown to the groups' leaders and members. Most important, it had never achieved park status, so the protectionist leaders, including Brower, reluctantly agreed to the compromise as the best way to save Dinosaur, even though they privately acknowledged that the Glen Canyon facility was an inefficient and questionable project.[62]

This compromise also had an unintended but substantial impact on the protectionist groups, especially the Sierra Club. On the one hand, the Dinosaur campaign had placed the club at the center of resource politics through its lobbying and mobilization tactics. This, in turn, had transformed the club's public identity, attracting in the process a growing number of middle-class members, which helped to shift the organization beyond its more exclusive, upper-class frame of reference. At the same time, the loss of Glen Canyon forced the group to further reconsider its approach concerning wilderness, particularly where development plans threatened potential scenic resources.

One of these plans, the Park Service's proposed Mission 66 project, immediately became controversial for the protectionist groups. Mission 66 plans called for improved roads and services in the national parks and a major expansion of parking facilities to ultimately accommodate more than 155,000 cars in park areas. Although parts of the plan, including those for Yosemite, the Sierra Club's most valued scenic resource, had involved input and approval of top Sierra Club leaders in earlier years, the new, more aggressively protectionist-oriented club leaders, especially Brower, became sharp critics of Mission 66 and led the lobbying and public protest campaigns to curtail it.[63]

For the Sierra Club and, especially, The Wilderness Society, the best way to achieve renewed protectionist goals seemed to be the legislative route. In the wake of the Dinosaur fight, the groups felt the need to be more proactive in safeguarding wilderness by making protection responsibilities subject to legislative mandate. Toward that end, in 1955, Howard Zahniser drafted the first of eventually dozens of versions of a Wilderness Act calling for the permanent creation of 50 million acres of wilderness with no commercial activities, such as mining or hydroelectric generation, to be permitted.

During the next nine years, the protectionist groups found themselves immersed in lobbying and deal making. Facing significant opposition from resource development interests, government agencies, and certain congressional figures, and obliged to trek through eighteen hearings and sixty-six redraftings before a final compromise was reached in 1964, the groups were able to salvage only a sharply reduced National Wilderness Preservation System that consisted of 9.1 million acres and included provisions to review another 5.4 million acres over a ten-year period. Though the legislation identified the mechanisms to set apart future wilderness areas, it also allowed significant exemptions for mining, water development, livestock grazing, and recreational uses.

While passage of the Wilderness Act was hailed at the time by protectionists as a "benchmark in our civilization," it quickly became apparent that the legislation would not, as Brower put it, "be the end of a series of problems, but the beginning."[64] The review process was slow and cumbersome, tying up enormous conservationist resources. Both the Park Service and Forest Service resented the dilution of their decision-making powers to designate and administer wilderness areas and dragged their heels in getting the process under way. The first parcels to be placed in the system were not completed until 1968, and countless hours were spent by Sierra Club and especially Wilderness Society staff monitoring the process. Scientific and lobbying expertise had to be developed to comment on the unit-by-unit procedures established. "Whereas the long campaign for the Wilderness Act was a propagandistic tour de force," the Sierra Club's Michael McCloskey wrote in 1972, "the subsequent efforts to implement it have been dissipated by bureaucratic technicalities. Problems of administration, statutory interpretation, field studies and reviews, record building, schedules, and congressional tactics have become the grist of the effort."[65]

This immersion in bureaucratic rule making contrasted with the highly charged and publicly visible battles that swept up and ultimately polarized the Sierra Club during the mid- to late 1960s. One fight centered on a proposed dam and power-generating facility at Marble Bridge at the edge of the Grand Canyon, part of a major water-development legislative package introduced in the early 1960s. The intense conflict that ensued was an even more contentious affair than the earlier Dinosaur fight. Brower especially saw the battle as a crusade, complete with impassioned rhetoric (the proposed dam was compared to flooding the Sistine Chapel in one memorable ad, intentionally recalling John Muir's rhetorical defense of Hetch-Hetchy), mobilization of the club's membership, and an apparent no-compromise posture. Whereas park status had been the cutting-edge issue with Dinosaur, and the protectionists had declared themselves "neither pro-reclamation, nor anti-reclamation," Brower and his supporters now defined their position as absolute opposition to all water development and hydro facilities in scenic resource areas. And while some Sierra Club directors fretted over the tone and direction of the Grand Canyon campaign and worried about the retaliatory actions of the Johnson administration (the club lost its tax-exempt status in a much-publicized move by the IRS, though it also immediately gained several thousand new members), the powerful symbol of the Grand Canyon allowed Brower to adopt a more aggressive approach.[66]

Once again, as with Dinosaur, the telling moment for the protectionists was the decision to agree to a compromise that removed the threat to the

Grand Canyon but allowed the expansion of a coal-fired plant in the Four Corners region, where the states of Utah, New Mexico, Colorado, and Arizona meet. On September 30, 1968, the new water development package, without a Grand Canyon facility, was signed into law. It turned out to be the last of the Bureau of Reclamation's big water projects, with the protectionists applauding their victory while privately feeling wary of the outcome.[67]

The Grand Canyon fight, similar to the Dinosaur battle, represented an ambiguous conclusion. A few years after the bill was signed, the Sierra Club would initiate litigation against the Four Corners power plants for their enormous air pollution—emissions also affecting the Grand Canyon. The issue of what constituted a "scenic resource," a concern raised twenty-five years earlier by Aldo Leopold, also remained unresolved. Nevertheless, the Grand Canyon campaign had become a critical battle for the Sierra Club. The organization had become a new and potent political force in the water policy arena, a public interest (as opposed to special economic interest) group able to mobilize an unrepresented public that valued the area for its own sake. Even more striking was the nature of the campaign, a protectionist version of 1960s-style, direct-action tactics. Letter writing, expressive ad campaigns, and demonstrations at congressional hearings became protectionist equivalents of civil rights and antiwar sit-ins and protests. Brower was in his element during this campaign, willing to take on all comers. But Brower's critics within the Sierra Club saw him as a charismatic and unsettling force acting in an arbitrary and volatile manner. In the wake of the Grand Canyon fight, the Sierra Club began to move toward a decisive showdown over tactics and leadership that culminated in the fight over a siting proposal for a nuclear power plant.

During the 1950s and 1960s, the siting of nuclear power plants and their potential impact on scenic resources emerged as a major source of controversy for protectionist groups such as the Sierra Club. In this period, the proposal by the Pacific Gas & Electric utility for a Diablo Canyon power plant began to preoccupy both the club's national leadership as well as several of its local chapters in terms of where the plant was to be located. Some of the club's leaders worked closely with PG&E to find an acceptable site, eventually deciding to support PG&E's choice of the Nippomo Dunes, a relatively undeveloped area along the California coast south of San Luis Obispo. Debates within the club erupted over the scenic value of Nippomo, the group's decision-making process, and whether to seek tradeoffs or otherwise participate in allowing certain projects to proceed. When

Brower protested the club's endorsement of Nippomo, a bitter organizational conflict ensued, which eventually led to the firing of Brower and a shake-out of the club's leadership.[68]

For Brower and some of his supporters, the lessons of Diablo Canyon became linked to their mistrust of the deal-making process itself involving large government or corporate-sponsored development projects. Brower saw the outcome of the Diablo conflict and the earlier battles around the Dinosaur and Grand Canyon facilities as reflecting two different directions for conservationist and protectionist groups: one implacable in the defense of wilderness and the other seeking to balance the demands of development and protection. But Brower's supporters never came to articulate a different vision nor organizational approach other than the deal making, lobbying, and use of expertise that characterized all of the conservationist and protectionist groups. The Diablo fight ultimately had to be seen as a personality dispute, not an ideological divide. And although the firing of David Brower was written about as a sign of the times, the influence of the 1960s in redefining environmentalism, it turned out, would emerge more directly from other sources and issues.

By the 1960s, the search for a common frame of reference, embedded in either management or protectionist language, had reached a certain impasse regarding the natural environment. With a complex relationship to the resource bureaucracies, an unresolved debate over technology and population, a growing expertise and lobbying focus, and an increasingly visible profile, conservationist and protectionist groups had managed to carve out for themselves a major role in the policy arena regarding resources and the natural environment. Yet these groups had failed to identify a common agenda even in the resources area, nor had they been able to respond to the urban and industrial realities that marked the Progressive Era and subsequently transformed the post–World War II order. Other groups—of reformers, professionals, and radicals—had emerged to confront the urban and industrial environmental realities of the Progressive Era, the New Deal, and the postwar period. These were separate and distinctive groups, defining issues and constituencies from a different starting point than their conservationist and protectionist counterparts. These groups also provide a different though essential lens through which a complex social movement with diverse roots and contending perspectives can best be understood.

Chapter 2

Urban and Industrial Roots: Seeking to Reform the System

Exploring the Dangerous Trades

A tenacious reformer, a compassionate advocate, a cautious and careful researcher, Alice Hamilton was this country's first great urban/industrial environmentalist. Born in 1869 in New York City and raised in Fort Wayne, Indiana, Hamilton decided to study medicine, one of the few disciplines available to this first generation of women able to enter the universities and embark on a professional career. "I chose medicine," Hamilton would later say in her autobiography, *Exploring the Dangerous Trades*, ". . . because as a doctor I could go anywhere I pleased—to far-off lands or to city slums—and be quite sure that I could be of use anywhere."[1]

Even prior to entering medical school at the University of Michigan, Hamilton thought of combining her interest in medicine and science with humanitarian service and social reform. She found the ideal outlet when she moved into the Hull House settlement in Chicago while accepting a position as professor of pathology at the Woman's Medical School of Northwestern University. During the 1890s and through the first decade of the new century, Hull House became an extraordinary meeting ground for reformers, humanitarians, and urban activists of all kinds. For Hamilton, as her biographer Barbara Sicherman noted, Hull House "was an ideal place from which to observe the connections between environment and disease." She began a whirlwind of activity, organizing a well-baby clinic, looking into the cocaine traffic endemic in the neighborhood and the city, taking part in efforts to improve the quality of health care for the poor, and investigating a serious typhoid epidemic. The typhoid epidemic was particularly instructive, since Hamilton's investigation eventually helped reveal that a sewage outflow (an episode covered up by the Board of Health) bore a direct relationship to the outbreak of the disease in certain neighborhood wards. It became Hamilton's first experience with how the issues of health, the environment, and politics intersected.[2]

Hull House also became a staging ground for Hamilton's growing interest in the little understood and poorly treated area of industrial disease. At the settlement house, Hamilton heard countless stories about "industrial poisoning": carbon monoxide in the steel mills, pneumonia and rheumatism in the stockyards, "phossy jaw" from white phosphorus used in match factories. Though industrial medicine had become an accepted discipline in Europe, its detractors in the United States suggested, as Hamilton wryly noted, that "here was a subject tainted with Socialism or with feminine sentimentality for the poor."[3] Through her exploration of industrial poisons, Hamilton was able to combine her passion for reform and her desire to pursue a real world–based science.

During the first decade of the new century, there had been few investigations of occupational health and even fewer reforms of industrial practices in the United States, though occupational hazards were present in a wide range of industries. In 1908, Hamilton's interest in the subject was stimulated by an encounter at Hull House with John Andrews, the executive secretary of the American Association for Labor Legislation (AALL). Andrews had investigated more than 150 cases of phossy jaw, a debilitating and disfiguring disease prevalent in American match factories. Jane Addams of Hull House had been familiar with phossy jaw, having attended during the 1880s a mass meeting in London where several people had shown their scars and deformities. Until 1908, little had been done even to explore the problem in this country, since the American medical establishment argued that American factories were cleaner and less susceptible to occupational hazards. Andrews' report, however, not only documented the problem but pointed to a reasonably inexpensive substitute for white phosphorus and helped set in motion a high-profile campaign around the issue. It eventually led to the passage in 1912 of legislation that effectively eliminated all white phosphorus use through taxes and regulatory requirements.[4]

The success regarding phossy jaw, Hamilton later noted, proved to be an exception rather than the rule in investigating and addressing occupational hazards. Already initiating her own investigations of specific hazardous industries, Hamilton was appointed to the Illinois Commission on Occupational Diseases in December 1908 by Illinois reform governor Charles S. Deneen. Hamilton (the only female participant) and the commission's eight other members issued a report that identified a number of industries potentially exposing their workers to serious hazards and concluded that a far broader and lengthier study was needed. After some delay, the Illinois Legislature provided funding for a nine-month survey, with Hamilton as its medical investigator. Hamilton was discouraged by the difficulties in gathering information at the factory level, where "the foremen deny everything and the men will not talk and they live in all parts of the city and employ any number of physicians," but nevertheless agreed to participate in what would be the first investigation of its kind in the country.[5]

One of the key aspects of the survey was the investigation of the lead industries. Lead was a widely recognized industrial poison, known to be responsible for convulsions, abdominal pain, paralysis, temporary blindness, extreme pallor, loss of weight and appetite, indigestion, constipation, and numerous other problems. Hamilton sought to identify which industries used lead and the kinds of health problems associated with them. In pursuing her research among the lead companies, Hamilton frequently encountered the belief (a kind of ideological rationale for inaction) that worker unwillingness to do things such as "wash hands or scrub nails" was the primary cause for occupational lead poisoning and its occurrence was therefore "inevitable." Hamilton also quickly came to realize that lead hazards and health impacts were underreported by workers, who concealed their illnesses out of fear of losing their jobs.[6]

Though Hamilton was faced with a lack of documentation and information, few resources, company resistance, and workers' fears, her investigations, first with the Illinois survey and subsequently with the Bureau of Labor within the U.S. Department of Commerce, demonstrated an extraordinary resourcefulness and persistence as she pursued her "shoe-leather epidemiology." Her search for data led her to undertake numerous interviews, home visits, and discussions with physicians and apothecaries, undertakers, charity workers, visiting nurses, and countless others. It required long hours and uncertain information, but it was a duty, she felt, "to the producer, not to the product." Hamilton recognized that her compassion as a woman for the victims of the dangerous trades gave her certain advantages in soliciting information through more informal settings. "It seemed natural and right that a woman should put the care of the producing workman ahead of the value of the thing he was producing," Hamilton remarked. "In a man it would have been [seen as] sentimentality or radicalism."[7]

During the next several decades, Hamilton became the premier investigator of occupational hazards in the United States. Her research and advocacy ranged over a number of industries and toxic substances. Her insights and investigative techniques broke new ground in the areas of worker and community health and anticipated later interest in the occupational and environmental problems associated with such substances as heavy metals, solvents, and petroleum-based products. Forty years prior to the major environmental debates about the uses of science and technology and the nature of risk, Hamilton was already warning that workers were being used as "laboratory material" by industrial chemists who were introducing new products such as petrochemicals and petroleum distillates "about whose effect on human beings we know very little." This rush to introduce new industrial products, such as solvents, she argued, represented new hazards in the workplace and the general environment. "The quicker the solvent evaporates the greater the contamination of the air," Hamilton said of these new products, "which means that a coating or a degreaser which is advertised as a powerful, quick-drying fluid is one that must be regarded by a physician with suspicion."[8]

Hamilton was also convinced that control techniques, such as respirators or other protective devices, were far from adequate, anticipating similar debates within OSHA (the Occupational Safety and Health Administration) during the 1970s and 1980s. She focused on the impacts from even low exposures or emissions of substances such as lead, anticipating the recognition that for certain toxic substances there is no acceptable threshold. During debates over the decision by the automotive industry to introduce tetraethyl lead in gasoline during the 1920s, Hamilton became a key critic of the claim that the small amounts of lead involved were not significant. In a 1925 article, Hamilton wrote, "I am not one of those who believe that the use of this leaded gasoline can ever be made safe. No lead industry has ever, under the strictest control, lost all its dangers. Where there is lead, some case of lead poisoning sooner or later develops. . . ." The question of when environmental or public health factors needed to be considered was critical for Hamilton. "It makes me hope," Hamilton said of the tetraethyl lead controversy, "that the day is not far off when we shall take the next step and investigate a new danger in industry before it is put into use, before any fatal harm has been done to workmen . . . and the question will be treated as one belonging to the public health from the very outset, not after its importance has been demonstrated on the bodies of workmen."[9]

As her research began to receive attention, Hamilton's standing grew in an area that had largely failed to elicit interest in academic, industry, or government circles. In 1919, she was appointed assistant professor of industrial medicine at Harvard University following Harvard's decision to initiate a degree program in industrial hygiene. The appointment attracted attention, since she was the first woman professor in any field at Harvard and the university did not admit women to its medical school. But Hamilton was chosen partly because there weren't any men interested in the position and the medical field and academic world in general still viewed occupational and environmental issues with little interest.[10]

By the 1920s, with the publication of her classic text *Industrial Poisons in the United States*, her increasing prominence in issues of occupational and environmental health, and her participation in organizations such as the Workers' Health Bureau, Alice Hamilton had become the country's most powerful and effective voice for exploring the environmental consequences of industrial activity. Her interest touched on issues of class, race, and gender in the workplace and the long-term hazards of the production system. She was concerned not only about the visible, acute problems of occupational hazards but also generational issues associated with "race poisons," reproductive toxins such as lead whose "effects are not confined to the men and women who are exposed to it in the course of their work, but are passed on to their offspring."[11] She was able to communicate effectively with industry and government figures because of her sincerity in the goals and substance of her research, while developing a sympathetic relationship with workers due to her compassion and her commitment to change. A reformer in the mold of the settlement house worker (even while at Harvard, Hamilton maintained a "home" at Hull House), and a powerful environmental advocate in an era when the term had yet to be invented, Alice Hamilton situated the question of the environment directly in its urban and industrial context.

The Environment of Daily Life in the Industrial City

The rise of what Lewis Mumford has called the "industrial city" of the nineteenth and early twentieth centuries represents a chronicle of new and pervasive forms of environmental degradation. Though celebrated as a period of great economic expansion, industrial growth, and urban reconfiguration, the period from the 1860s through World War I can be seen as laying in place the contemporary environmental hazards of daily life in urbanized and industrialized America. The triumph of fossil fuels as dominant energy sources, the diminishing of wilderness areas and open urban spaces through the railroads and then with gasoline-powered vehicles, the growth of new industries such as steel, rubber, and chemicals, and the discharge and disposal of vast urban and industrial by-products contributed to the massive human and environmental toll accompanying the more celebrated changes in this period. H. L. Mencken, writing about the industrializing Monongahela Valley in the late nineteenth century, expressed this ironic juxtaposition directly: "Here was the very heart of industrial America, the center of its most lucrative and characteristic activity, the boast and pride of the richest and grandest nation ever seen on earth—and here was a scene so dreadfully hideous, so intolerably bleak and forlorn it reduced the whole aspiration of man to a macabre and depressing joke."[12]

The hazards of the workplace paralleled the hazards of the city in communities adjacent to the factories subject to their stench, noise, and foul air, as well as in neighborhoods overcrowded with immigrants seeking to work and survive in the great urban centers of the East and Midwest. Jane Addams, in a famous passage describing Halsted Street on Chicago's west side, captured this quality of the urban environment near the turn of the century: "The streets are inexpressibly dirty," she wrote in *Twenty Years at Hull House*, "the number of schools inadequate, sanitary legislation unenforced, the street lighting bad, the paving miserable and altogether lacking in the alleys and smaller streets, and the stables foul beyond description. Hundreds of houses are unconnected with the street sewer."[13]

The environmental problems endemic to the period—water quality, sewage and sanitation, solid and hazardous waste generation and disposal, ventilation and air emissions, occupational and public health issues—parallel the items of a contemporary environmental agenda. Water was a particularly critical issue. It was needed for urban expansion, used for municipal and industrial discharges, and was a deadly source of infection and disease once contaminated. With the expansion of the industrial city, urban areas became dependent on fresh supplies of water, once existing sources such as local wells or stream beds became contaminated. Water quality and water supply issues became intertwined. Better-quality sources allowed for more growth, while a sufficient supply helped improve the quality of urban life through such tasks as hosing down the streets and settling the dust, thereby reducing the threat of pestilence and disease in main thoroughfares and back alleys.

Most cities decided to confront the problems of degraded water supplies by opting to secure new, more pristine supplies, often imported from distant, nonurbanized areas. As a consequence, original water sources, such as local rivers, were allowed to become a kind of nineteenth-century Superfund site. Municipal leaders assumed that industries should be largely exempt from discharge controls and that property owners, as Pennsylvania governor Daniel Hastings complained in 1897, "had the right to develop anything they wished in water bodies fronted by their property."[14] Lewis Mumford, in his epic study *The City in History*, wrote about how the early industrial city subordinated all details of daily life to the requirements of the factory, leading to urban and industrial discharges that fouled the waters of rivers and streams. "There are myriads of dirty things given it to wash, and whole wagonloads of poisons from dye houses and bleach yards thrown into it to carry away," Mumford quoted one nineteenth-century source about an urban river. "Steam boilers discharge into it their seething contents, and drains and sewers their fetid impurities; till at length it rolls on—here between tall dingy walls, there under precipices of red sandstone—considerably less a river than a flood of liquid manure."[15]

As urban rivers, streams, and wells came to resemble open sewers, community leaders were forced to confront the enormity of the public health hazards and environmental degradation such practices represented. Infectious disease epidemics—yellow fever, then cholera and typhoid—plagued the industrial city up through the 1890s, striking seemingly at will. The problems were exacerbated by an unwillingness to control industrial and municipal discharges, because to do so might, as Philadelphia's chief engineer put it, "interfere with the large manufacturing interests which add so greatly to our permanent prosperity."[16] Instead, municipalities searched for ways to accommodate industry and locate better-quality water sources. During the 1870s, the City of Newark, for example, sought to prevent discharges into the Passaic River, its primary source of drinking water. This included restrictions on upstream sources, such as a large paper mill, and such downstream practices as dumping dead animals and refuse directly into the river. But by the 1880s, the increase in industrial activities, particularly factories along the lower stretch of the river that poured out their own "peculiar filth," overwhelmed the local water departments. By the end of the decade, the Passaic River had become subject to uncontrolled discharges, its waters unfit to drink or even to be used by those industries, such as Newark's large breweries, that relied on fresh water sources. In 1889, city officials, under pressure from local industrial interests, contracted with a private company to deliver water from an upland tributary of the Passaic, bypassing and ultimately abandoning the main stem of the river to continuing contamination.[17]

Water discharges represented only one type of environmental hazard: releases into the air and disposal onto the land were also poorly regulated and widely prevalent. In the industrial city, the minimal, often disorganized street collection of solid wastes and the lack of an effective and comprehensive sewer system for liquid wastes intensified the environmental problems of daily life already magnified by contaminated water. The city's streets were strewn with horse manure, mounds of uncollected garbage in alleyways, and slag and rubbish heaps from factories. Solid-waste disposal practices at the turn of the century involved locating the most convenient area for dumping wastes—often open pits directly within the urban area, or available bodies of water, such as the ocean for coastal communities. Garbage collection remained uncertain, with rotting mounds of garbage and ash a common sight in immigrant neighborhoods and industrial zones. The sewer system, where available, represented yet another uncontrolled source of wastes released by industries and cities.

The absence of any effective policy mechanism to distinguish between wastes also meant that a large volume of what we would today call hazardous wastes were entering the environment without restrictions. By World War I, heavy metals such as lead, organic chemicals such as benzene and naphtha, and a variety of other toxic substances used in steel mills, dye and munitions manufacture, paper and pulp mills, metallurgical processes, and mining were being discharged in significant amounts into the air, water, and land.[18] These, in turn, were causing a wide variety of occupational and community-based environmental and public health problems largely ignored by industry and government officials alike. Environmental degradation was seen as a necessary by-product of urban and industrial growth, with industrial and civic leaders seeking to shift the burden of response to the worker and community resident rather than the industry or municipality. Sanitation, occupational hazards, and environmental and industrial poisoning all became defined as results of poor individual habits, creating a kind of environmental victimization. The experience of environmental degradation thus took on a class dimension. On the one hand, the industrial city became a source of great wealth and a symbol of progress for those directly benefiting from the industrial and urban expansion of the period. On the other hand, this very same expansion, with its belching factories, polluted waterways, and untreated and sometimes uncollected wastes, became, for many poor and industrial workers, an environmental nightmare that seemed impossible to escape.

Environmental Order: The Rise of the Professionals

Despite the great power of industrial interests in shaping the industrial city, the urban environment was still a protracted political battleground during the Progressive Era (1880s–1920s). Efforts to address environmental degradation were slow to develop at first, but grew in scale after the turn of the century, becoming especially significant by World War I. By then, two urban environmental approaches had emerged: professionalization and reform. Paralleling the conservationist and preservationist movements of the same era, professional and reform groups defined themselves in response to the environmental conditions brought by changes in the urban and industrial order.

First and foremost, urban and industrial conditions raised new public health problems. The environmental problems of the industrial city—limited and contaminated water supplies, inadequate waste and sewage collection and disposal, poor ventilation and polluted and smoke-filled air, overcrowded neighborhoods and tenements—provided the means to spread disease, the social and physical environment for it to flourish, and the arena in which problems were contested and solutions sought. These environmental and public health issues generated new social movements, most situated outside the medical profession and formed in response to specific disease epidemics. Seeking to link such issues as sanitation and public health, the new groups pushed for the establishment of municipal and statewide boards of health and other municipal and national public health and environment agencies. They hoped to initiate sanitary campaigns, check food and water supplies, and institute special measures, such as community quarantine programs.

These public health advocates, including those who came to call themselves "sanitarians," were a disparate group. They included a few physicians; businessmen who feared that an unhealthy environment hindered community growth; various professionals such as engineers and chemists interested in improving water and sewer systems as well as more systematically testing, analyzing, and gathering information about food and water sources; middle-class reformers, including members of various women's groups; and radical activists within the labor movement and the "sewer socialist" faction of the Socialist Party. These public health advocates often faced a hostile medical establishment, felt resistance from both industrial interests and municipal leaders, and had to contend with an overall lack of emphasis at the national level on social programs designed to address environmental and public health questions.

The research breakthroughs beginning with London physician John Snow's famous epidemiological findings of 1854 about the contaminated Broad Street pump and its link to the geographical incidence of cholera and culminating in the bacteriological discoveries of the 1880s ultimately undercut the social basis of advocacy while encouraging professionalization within the new public health movements. Simultaneously, the developing concept of "conservationism" as a science-based, expertise-driven approach was also extended to what the National Conservation Commission characterized as "the grandest of our resources, human health."[19]

A new public health agenda, based on a new science and a developing profession, seemed ready to emerge. Ironically, early resistance to such an agenda came from the medical profession itself. While bacteriology was revolutionizing the concepts of medical treatment and prevention, the medical profession was still responding cautiously. Some elite physicians continued to argue that contagious diseases and other public health disorders were a result of the "filthy habits of the poor" and resisted a bacteriology-oriented public health agenda. At the same time, physicians were scornful of the social nature of a public health approach that focused on environment, preferring instead to focus on individual health care as the cornerstone of their profession.[20] In contrast to the physician groups, advocacy and community-based groups, such as the certified milk movement, anti-tuberculosis associations, organizations seeking to eradicate hookworm in southern textile mills, settlement workers engaged in health reform, child health groups, and various labor and farmer organizations, were in the forefront of the effort to reform the environment as the most effective way to accomplish public health goals.

By 1912, at the height of the Progressive Era, a new public health agenda seemed more imminent. With the acceptance and initial implementation of certain disease treatment and prevention programs, the position of the growing professional class of physicians and other public health "experts" effectively shifted. More powerful coalitions incorporating the public health and sanitary movements as well as the insurance industry began to promote a broad package of public health reforms. With the passage of legislation establishing the U.S. Public Health Service in 1912 and the development of its bacteria-based water quality standards two years later, the institutionalization of public health seemed secure.

With the triumph of bacteriology and the professionalization of public health, the policy focus shifted from occupational and environmental issues and social reform efforts toward individual, bacterial causes of disease based on the new science of diagnosis and treatment. This occurred at a time when urban and industrial environments were themselves subject to new and unprecedented hazards. And whereas the sanitarian movements had included both professionals and constituent groups seeking to address social conditions in the industrial city, the newly restructured public health profession of the Progressive Era was explicitly controlled by physicians and other trained specialists seeking to apply science exclusively to deal with disease and other public health problems. The application of bacteriology was therefore seen as lessening the need for environmental reform, since diseases could be treated on a case-by-case basis rather than by "cleaning up" the industrial city. This concept of a public health science offered health departments a way to "promote health without the cost of major social changes," disassociating the public health agenda from the question of whether a city's streets "are clean or not, whether its garbage is removed promptly or allowed to accumulate, or whether it has a plumbing law," as one early twentieth-century public health official put it.[21]

The professionalization of public health also coincided with the rise of professional associations and science- and engineering-based approaches in the new environmental arenas of air pollution control, sanitation, and water quality. The first professional group concerned with air pollution issues was organized in Chicago in 1891 to help develop programs for reducing the level of "smoke" and other air emissions. The organization was established in advance of Chicago's 1893 Columbian Exposition, a signal event in defining that city and its celebration of urban and industrial change. In 1907, the first nationwide professional association—the International Association for the Prevention of Smoke—was formed in the area of air quality. It defined its tasks explicitly as "pollution control," including efforts to "attach mechanisms to boilers and fireboxes which will cure the evil [i.e., smoke]."[22]

Unlike the early volunteer public health movements, the air pollution control associations were defined from the outset as professional bodies. They consisted primarily of industrial engineers seeking engineering and technical solutions to the problems of emissions. The associations also attracted wealthy civic reformers for whom "smoke control" was a form of civic improvement. Through the early years of the Progressive Era, smoke control was also a highly charged political issue, since the system of "smoke" inspections was frequently subject to graft and manipulation by industry interests. In this context, the professional agenda also became part of a larger reform agenda aimed at cleaning up an outmoded and corrupt system dealing with infrastructure issues.

The rise of professional groups in the areas of sanitation and solid waste collection and disposal also reflected the development of a reform agenda focused on the inadequacies and corruption associated with sanitation efforts. The first major sanitation reformer, George Waring, the commissioner of sanitation for the City of New York during the 1890s, was an engineer who sought to rationalize and make more efficient (and equitable) the city's nearly nonexistent garbage collection system. At the same time, Waring explored new disposal technologies first developed in Europe and subsequently introduced in this country. Similar to Waring's interests, engineering and public health journals as well as popular publications such as *Cosmopolitan* printed articles during the 1890s and early 1900s about fusing engineering concerns with reform objectives in the municipal solid waste collection and disposal area. "Cleaning up the city" became at once an engineering strategy and a political metaphor.

Like the resource-oriented conservationists, solid waste reformers in the Progressive Era, partly through newly organized professional associations of sanitary engineers, emphasized concepts of efficiency and "multiple use," including the idea that garbage in its reusable forms was "gold." New technologies, such as reduction (the heating of garbage to extract raw materials or resources), reinforced the conservationist theme that production could be more "productive," as opposed to the wastes and excesses associated with early industrialization. Solid waste management, in this context, became not only a professional objective but constituted one element in the overall attempt to rationalize the production system.[23]

The professionalization theme was perhaps most pronounced in the area of water quality, where the introduction of new forms of treatment and engineering became linked to both the science-based bacteriological and sanitary revolutions. By 1914, filtration and chlorine disinfection were seen as the best means of reducing exposure to bacterial agents of infectious disease. These new, successful methods of water treatment, advocated by and increasingly enforced by sanitary or civil engineers, not only reshaped municipal water department responsibilities for securing "clean" water supplies but effectively decoupled the issue of contaminated streams and other bodies of water from the quality or purity of drinking water after treatment.

This separation of environmental and public health impacts created uneasiness among some sanitarians. A few conservation-oriented professionals, such as Marshall O. Leighton of the U.S. Geological Survey, raised the concern that industrial wastes still remained the "great pollution problem" of the day, due to their overall environmental impacts on rivers and streams and potential economic losses, such as declining fisheries resulting from contamination. But the vast majority of water quality engineers and related professionals were willing to tolerate a significant level of industrial wastes because of the presumption, in an era of bacteriological controls, that such wastes no longer had any "public health significance." Ultimately, the water quality changes associated with filtration and disinfection became most symbolic of the sanitary revolution. This was a revolution defined by the introduction of pollution control technologies, the professionalization of groups that had helped to craft and carry out this revolution, and an accommodation to industrial interests parallel to the reshaping of the conservationist agenda in the areas of forestry and water reclamation.[24]

The rise of treatment and control technologies and the dominant role of the professional groups in addressing public health issues had also played a part in diminishing the significance of Progressive Era social reform. By narrowing the issue of water quality to one of treatment, the sanitary/civil engineers successfully removed themselves from the more controversial arena of industrial and urban regulation. What industries produced and how they produced it—the direct concern of Alice Hamilton—had become, in the era of professionalization, a question of managing and controlling the by-products of industry rather than changing its processes and outcomes. In contrast, the social reform agenda in the urban and industrial environmental arena came to be associated more with passion than technique, the "producer" rather than the "product." These positions ultimately led to a different frame of reference for two quite distinct, yet often linked movements.

Socializing Democracy: The Settlement Idea

When Jane Addams established Hull House on the west side of Chicago in 1888, new movements for social and environmental reform appeared ready to emerge as a major social force in the industrial city. Chicago, a major manufacturing, resource, transportation, and ethnic crossroads, seemed a particularly appropriate choice for a social and environmental reform movement. For many reformers, it had become symbolic of the new urban and industrial order, a city "first in violence, deepest in dirt, loud, lawless, unlovely, ill-smelling, new; an overgrown gawk of a village, the teeming tough among cities," as muckraker Lincoln Steffens characterized it.[25]

Addams wanted Hull House to become a center for Chicago's reform movements, a neighborhood home at once part of and an alternative to the urban and industrial order. Her notion of the "social settlement," though partly derived from the upper-class orientation of the English settlement house movement of the 1880s, was quickly adapted to the conditions of the industrial city and the neighborhoods in which settlements located. Centers for ideas and reform, settlements would provide a "higher civic and social life," as the Hull House charter put it, an arena for advocates "to investigate and improve the conditions in the industrial districts of Chicago."[26] Hull House itself quickly became a "colony of efficient and intelligent women living in a workingmen's quarter with the house used for all sorts of purposes by about a thousand persons a week," as one key Hull House figure wrote in a letter at the time.[27] Though the settlement house was by no means the only force for social and environmental reform in the Progressive Era, it became both the meeting ground and a key symbol of the movements for change contesting the urban and industrial order of the period.

Focused on the conditions of daily life in their neighborhoods, the settlements immediately confronted questions of housing, sanitation, and public health. Concerns about overcrowded tenements, inadequate paving, and lack of crucial services such as garbage collection dominated the discussions and framed the actions of settlement workers. The immigrant wards and industrial neighborhoods adjacent to factories seemed most especially at risk, given the frequent breakdown of the urban infrastructure in those areas.

The initial response to this breakdown in services at the neighborhood level was to try to correct abuses by exposing the problems and bringing them to public attention. The dissemination of solid research and information along with greater accountability were central to the settlement concept of reform. Hull House resident Alice Hamilton's typhoid investigation, which pointed to the corruption and failure to act on the part of board of health personnel, was considered a model of success, as it led to the dismissal of eleven of the board's twenty-four inspectors and pressure for improved sanitary measures.

Similarly, Hull House figures focused on inadequate garbage collection as one key symptom of environmental breakdown. The garbage problem seemed so severe—"the greatest menace in a ward such as ours" according to Jane Addams—that the settlement workers sought a major investigation into the city's overall garbage system as a first step toward reform. As a result of that investigation, undertaken by the Hull House Woman's Club, Addams decided to submit her own bid to collect the garbage in the nineteenth ward. Though the bid was thrown out on a technicality, the ensuing publicity forced the mayor to appoint the Hull House founder to the position of garbage inspector for the nineteenth ward.[28]

The appointment proved to be an extraordinary affair. The forceful, well-known Addams, along with her assistant, Amanda Johnson, a young University of Wisconsin graduate student, formed their own garbage patrol, driving around in what the children in the ward called "Miss Addams' garbage phaeton." Up at 6:00 A.M., the two would follow the huge trucks to the dump, keep charts and maps, make citizen arrests of landlords, complain to contractors, and raise enough of a stink that restructuring the garbage collection system quickly rose to the top of the agenda for both city hall and reform movement alike.[29]

The garbage issue was also a central concern to the settlement workers in Chicago's "back of the yards" industrial and residential section, where the city's largest garbage dump was located. The settlement workers from the area were led by Mary McDowell, daughter of a prosperous Chicago steel man and a one-time kindergarten teacher at Hull House who had also helped organize and lead the Hull House Woman's Club. Known as the "garbage lady" because of her continuous advocacy of better disposal practices, McDowell sought to mobilize neighborhood and civic organizations, women's clubs (who formed their own "garbage committees"), and other public sanitarians to explore more effective scientific and technical solutions, such as reduction. McDowell also insisted that industry take responsibility for its trash and sewage disposal practices by funding some of the costs of developing new disposal and treatment techniques.[30]

The garbage question was just one of several environmental problems of daily life preoccupying settlement workers. The settlement movement tended to be spatially oriented, concerned about such issues as housing and other living environments for the crowded industrial districts. The improvement of housing, a key reform objective in the era of the tenements, was seen as a social/spatial reconstruction of neighborhoods and an effort to improve the total environment of the industrial city. Similarly, the campaign to establish community playgrounds, another key settlement movement objective, was designed to reconstitute the physical space for children in an otherwise bleak urban environment. A key Hull House participant, Florence Kelley, helped lay the groundwork for Chicago's first children's playground. Kelley had initiated a campaign about the unsanitary and unsafe conditions of several tenements near the Hull House area and had complained to the press about one particular piece of property. The property had been recently inherited by William Kent, later a leader in the Progressive movement. In response, Kent negotiated an arrangement with Kelley and Addams to demolish the tenements in order to design a public playground. Kent subsequently paid taxes on the property while the Hull House workers administered the facility.

The playground concept soon became a central component of the reform agenda, particularly with respect to children. Settlement workers in New York, for example, helped establish the Outdoor Recreation League, the major advocate for such facilities. This group, along with other reform organizations, challenged the dominant tendency among park supporters, many drawn from elite and professional circles, who wished to recreate "rural, sylvan, and natural scenic" park areas in the midst of "the evils which attend the growth of modern cities and the factory system," as Harvard University president Charles Eliot put it. The settlement reformers, on the other hand, saw this "keep off the grass" approach as meeting the needs of the middle and upper classes but not of factory workers and the children of the poor. A better "environment," the reformers argued, could be cast in cement as well as green space.[31]

The reform movement was particularly focused on the needs of children in the urban and industrial order. Degraded environments appeared particularly harmful to children for their physical or health-related impacts and their social space and individual needs. Along these lines, settlement workers and other reformers became active in the establishment of "milk stations" or "baby health stations" to provide free or low-cost milk to tenement families, an idea that eventually evolved into the concept of child and maternal health centers. Some reformers saw the milk station idea, with its potential for health education and related social services, as a kind of settlement in miniature.[32]

The most active child reform advocate was Florence Kelley, who linked the problems of degraded environments with workplace issues. The daughter of a well-known, socially progressive congressman, the divorced mother of three children, a translator and long-time friend of Friedrich Engels, and an early socialist activist, Kelley was an intelligent, strong-willed, and compassionate defender of her neighborhood residents, particularly its women and children. Kelley was instrumental in mobilizing Hull House interest in the industrial conditions for these residents. By confronting such issues directly, she became a major figure in the settlement movement. "She brought magnificent weapons to bear on her enemy," Hull House resident Edith Abbott said of Kelley. "Sleepless, tireless, indefatigable, she was always on the alert. Life was never dull and the world was never indifferent where she lived and moved."[33]

While at Hull House, Kelley became involved in a study of social conditions in the industrial city sponsored by the U.S. commissioner of commerce. Canvassing a square mile radiating from Hull House, Kelley detailed environmental conditions in the neighborhoods and highlighted significant problems associated with neighborhood sweat shops, or the "sweating system," as it was then called. Located in the tenement houses within the area, the sweats were new kinds of garment factories or home-finishing workplaces and maintained some of the worst imaginable working and living conditions. "Every tenement house," Kelley declared, "is ruinous to the health of the employees," almost all of whom were women and children. Kelley's research, much of which was later published under the title *Hull House Maps and Papers*, helped stimulate an 1893 state legislative commission of inquiry into the employment of women and children in manufacturing, including the "sweaters." This, in turn, led to eventual passage of legislation establishing a state factory inspection system, sanitary standards for home workshops for the production of such items as clothing, artificial flowers, and cigars, and provisions for the eight-hour day for women, teenage girls, and children. Following its passage, Kelley was appointed to the position of chief inspector for the law by Illinois reform governor John Peter Altgeld. Through this position, she emerged as one of the country's foremost champions of workplace reform for women and children.[34]

Establishing a connection between community and environmental issues and the problems of the workplace (especially, though not exclusively, for children and women) became a prominent part of the settlement approach. The first women's labor groups in the country, such as the National Women's Trade Union League and trade unions for the shirtmakers and cloakmakers, were initiated by settlement workers or at meetings held in settlement-house living rooms. Figures such as Florence Kelley, Ellen Starr, Mary McDowell, and Crystal Eastman played critical roles in the organizing and advocacy around labor issues for women, including environmentally hazardous work conditions. The most radical of the settlement workers, Crystal Eastman, a prominent figure within the Eugene Debs–led Socialist Party, became best known for her participation in the "Pittsburgh Survey," a comprehensive review of working conditions in the steel industry in one industrial city. Published in six volumes between 1909 and 1914 (including Eastman's work on "Work Accidents and the Law"), this massive study revealed, as settlement movement chronicler Allen Davis put it, "the cost and consequences of low wages and poor pay, of preventable diseases and industrial accidents, of ramshackle tenements and little planning."[35]

The settlement movement also became adept at helping bring about broad reform agendas by linking workplace hazards with community and consumer concerns. This was best exemplified by the scandals surrounding the meat-packing industry that erupted a few years prior to the Pittsburgh Survey. Settlement reformers first became involved with the conditions in "Packingtown," as the stockyards community was known, through the establishment of a settlement house at the University of Chicago near the stockyards. This settlement house was headed by the ersatz "garbage lady," Mary McDowell. In her new location, McDowell was immediately appalled at the back-of-the-yards environment and sought to develop what one reform publication characterized as a "neighborhood consciousness" about the surrounding area.[36]

In this small, dense, urban neighborhood was an extraordinary mix of environmental hazards. It was bounded on the north by the stagnant backwater of the South Fork of the Chicago River, where putrefying refuse filled in the river bed and formed "long, hideous shoals along the bank." The decaying organic matter released quantities of carbonic acid gas, which continually broke through the "thick scum of the water's surface," causing this section of the river to be named "Bubbly Creek." To the west lay the city's garbage dumps, four huge holes from which the clay had been dug for neighboring brickyards. These vast, open pits were fed daily by horse-drawn wagons carrying trash from throughout the city. This included waste from the packers that was then burned, creating a permanent stretch of fire surrounded by a moat to keep it from spreading. To the east was vacant land used as "hair fields," containing animal hair and other incidental slaughterhouse wastes that putrefied while drying. And, finally, to the south lay an open prairie. Without paved streets, without trees, grass, or shrubbery, with no sewer connections or regular trash pickup, and with its densely polluted air and powerful odors, Packingtown had become an urban environmental catastrophe by the turn of the century. "No other neighborhood in this, or perhaps in any other city, is dominated by a single industry of so offensive a character," wrote two settlement figures in 1911.[37]

Once established in the stockyards neighborhood, McDowell and her settlement workers rapidly became active concerning local environmental and social conditions. Beginning with her efforts to improve garbage disposal practices, McDowell pushed much of the settlement agenda: sanitation reform, support for municipal bathhouses, playgrounds, summer schools, kindergartens, adult education, and overall neighborhood improvement. But McDowell's most significant activities stemmed from her recognition of the industrial character of her community and the link between the conditions of work and the conditions of daily life in the community. In that context, she became a pivotal supporter and figure of solidarity during the famous (and unsuccessful) stockyards strike of 1904 over union recognition.

In this same period, McDowell's Northwest Settlement also became a retreat for muckraker journalist Upton Sinclair, who ate his meals at the settlement, discussing life in the back-of-the-yards neighborhood with McDowell and others while preparing the material for his thinly fictionalized account of stockyard conditions, *The Jungle*. Published in 1906, the book immediately caused a sensation, not so much about the working and living conditions in the stockyards as about the quality of the meat provided the consuming public. For Sinclair, the inadvertent focus on impure food undercut his intention to expose environmental conditions in the workplace. But for the settlement reformers, including McDowell and Jane Addams, the impure food controversy and other consumer issues provided an effective way to bring about the reform of overall living environments, beginning with passage of the Pure Food and Drug Act in 1906. As progressive (though not necessarily radical) reformers, the settlement workers sought to define and legislate what they considered to be the "certain minimum requirements of well-being" in an industrial city. This could be brought about by making available information to help empower the poor and working classes, and by creating efficient, bureaucratic, managerial structures to help implement reform programs.[38]

By 1912, settlement workers had become involved in each of the different strands of the reform movement emerging at the national level: the Socialist Party, with its focus on class divisions, working conditions, and the empowerment of labor; the Woodrow Wilson–led Democratic Party, which incorporated a number of industrial and municipal reform positions; and the newly created Progressive Party, which nominated Theodore Roosevelt as its standard-bearer for the presidential election. The Progressive Party especially attracted members of the settlement movement to its cause, including the indefatigable Jane Addams, who, in fact, delivered one of Roosevelt's nominating speeches at the Progressive Party convention and became the figure of the moment, receiving the loudest and most sustained applause for her speech. After Woodrow Wilson was elected, the reformers continued to advocate social justice and environmental reform positions for the party and helped establish, after the 1912 election, a Progressive Party reform advocacy commission—the Progressive Service—with a Department of Conservation headed by Gifford Pinchot and a Department of Social and Industrial Justice headed by Jane Addams. This attempt to construct a coherent reform agenda was short-lived, however. After 1914, with U.S. entry into the war looming, the reform agenda began to splinter. By the 1916 presidential election, the Progressive Party would collapse when Roosevelt himself abandoned its cause to support the Republican nominee, Charles Evans Hughes.[39]

The restructuring of the economy and urban communities through World War I and the postwar period was ultimately most responsible for the eclipse of the reform agenda. Changes in the industrial city reflected both changing demographics as well as improvement of the urban infrastructure and the professionalization and legitimation of the reform agenda. The settlement house concept gave way to the idea of social work, associated with professional social work schools and the consolidation of "apolitical" charitable organizations such as the Community Chest, which abandoned the language of reform in favor of professional "service." Environmental reform was also supplanted by science and planning techniques in the reorganization of the urban infrastructure, further abandoning the social equity framework essential to the settlement movement's approach.

Still, at its core, the settlement movement was passionately concerned about social change. For Addams, the social settlement represented a form of "socializing democracy." Despite their leaders' privileged backgrounds, their support and initiation of professional solutions, as well as their reliance on expertise and contribution to what one key analyst of the period characterized as the "search for order," the settlement movement remained a unique and powerful blend of social advocacy, neighborhood living, and community empowerment. As a movement for democracy, it provided an organizing model for a later generation of activists, including the 1960s civil rights movement.[40]

Over time, the settlement reformers and their ideas would become subject to much simplification and romanticization, transforming Jane Addams especially into a kind of national myth. Still, women such as Addams, Alice Hamilton, Florence Kelley, Grace Abbott, and Julia Lathrop had been able to force their issues onto the national agenda, making them "objectively important in the life of the American people," as one Hull House participant put it.[41] Their issues cut to the heart of urban and industrial change. And while the settlements themselves faded or were absorbed into less threatening, apolitical forms, the issues of daily life and urban and industrial environments that had preoccupied the movement remained as crucial and unresolved as ever.

Utopian Dreams and Urban-Industrial Realities: The Radical Impulse

As the stockyards situation so directly revealed, environmental conditions in the industrial city involved issues of labor and capital as well as urban form. The changing structure of industries such as textiles and mining as well as heavy manufacturing industries such as iron and steel and pulp and paper affected both the environmental and social conditions of work and the organization of the workplace. Hard-rock mining in the West, for example, was transformed by industrializing forces in less than a generation from small, cooperative enterprises with limited hazards to highly dangerous, environmentally hazardous work. Some of the sharpest conflicts between labor and capital occurred in the western mines in the 1880s and 1890s, with the conditions of work influencing the development of militant unions such as the Western Federation of Miners (WFM) and the Industrial Workers of the World (IWW).

During this period, the discovery of rich deposits of silver and lead in places such as the Idaho Panhandle spurred the growth of the mining industry (lead output, for example, increased from 14,100 tons in 1861 to 198,363 tons in 1891) and brought about the introduction of new technology and concentrated (often absentee) ownership of the mines. The new technology, such as drilling equipment, significantly magnified environmental hazards, including lead and silica dust poisoning. The problems of elevated dust emissions also defined such Progressive Era jobs as glass-cutting, sandblasting, and foundry work. A dust-related disease such as tuberculosis, described by Alice Hamilton as a "disease of the working classes," was seen as related to both occupational hazards and home surroundings.

Certain trades, Hamilton argued, had an "abnormally high death rate from consumption" because of the very "conditions of the trade." A number of worker movements, led by the WFM, focused on these conditions, and certain key strikes, such as those undertaken by the Coeur d'Alene lead and silver miners in the 1890s, directly involved occupational health and environmental questions. One response by the unions—to establish their own network of hospitals—also addressed some of the environmental risks, especially those associated with workplace exposures.[42]

Similarly, in the mills and sweat shops transformed by the industrialization of textiles at the turn of the century, the work force became subject to significant environmental hazards. These ranged from hazardous work environments in the textile mills (what one analyst called their "hot, humid, noisy, and dusty setting"), unsanitary conditions in the surrounding company towns, and unsanitary, overcrowded sweat shops in the cities. Inside the sweats, one could find, as Florence Kelley discovered, dangerous fumes from gasoline stoves and charcoal heaters, as well as "dye from cheap cloth goods sometimes poisonous to the skin, and the fluff from such goods inhaled by the operators [that] is excessively irritating to the membranes and gives rise to the inflammations of the eye and various forms of catarrh."[43]

These issues were compounded by the changing character of the work force in the garment trades. Their operations no longer relied entirely on skilled cutters and dressmakers, but increasingly utilized immigrant and female labor in the northern states and poor white and black workers in the South. Resistance to these conditions often took on a social and cultural dimension, with the mobilization of mill and garment workers protesting the hazards of the workplace directly associated with gender or ethnicity.

By the 1890s, new organizations, including the Working Women's Society and the New York City Consumers League, had emerged to deal with the growing numbers of children and women employed in industries with hazardous work environments. These groups sought to identify the most oppressive and hazardous work sites as targets for mobilization through working-class organization, selective consumer boycotts, and educational campaigns. Two key national organizations formed around these efforts: the Women's Trade Union League, founded in 1903, and the National Consumers League, which was established in 1899 by consolidating existing groups in several cities. The Women's Trade Union League broadly intersected with reform and socialist movements during the Progressive Era. It focused on trade union organization for women and only indirectly addressed work conditions and related environmental hazards. In contrast, the National Consumers League (whose first executive director, the redoubtable Florence Kelley, helped shape league policy for three decades) defined itself more explicitly as an environmental and consumer-based organization addressing work conditions, especially as they affected women and children.[44]

Under Kelley's leadership, the National Consumers League served as a visible and effective advocacy counterpart to research work in the area of "industrial poisons" undertaken by Alice Hamilton. In 1917, for example, the league set up a special committee on industrial diseases after Hamilton's investigations of munitions plants revealed that hazardous substances such as benzol and carbolic acid were impacting workers in these plants. Similarly, in the 1920s, the League and Kelley, along with Hamilton and other environmental researchers, became embroiled in consumer and workplace-related issues associated with radium dial painting in the watch industry and tetraethyl lead in gasoline. The league gathered information on workplace environmental hazards while challenging the prevailing scientific "expert" analysis, much of it bought by industry, which dismissed the significance of environmental impacts. The organization became particularly effective in publicizing these issues through the press, particularly in Hamilton and Kelley's case with the *New York World* and its editorial director, Walter Lippman, with whom they maintained a close relationship.[45]

The most important environmental advocacy organization for workers during the 1920s was the Workers' Health Bureau (WHB). Organized in 1921 by women reformers with roots in consumer and labor organizations, the bureau sought to link the expertise and research of public health professionals such as Alice Hamilton and C.-E.A. Winslow (a leading figure in the profession) directly with the trade union movement. Unlike the National Consumers League, the WHB defined its activities exclusively in relation to the unions, a "research adjunct to the union movement for health and safety," as one of its leaders put it.[46]

As this research arm, the WHB sought to make its investigations of industrial hazards and working conditions part of the organizing and negotiating approach of the unions. "Health is an industrial and class problem, deserving the same place in his [the worker's] program as hours, wages and working conditions," argued Harriet Silverman, the WHB's education secretary. Degraded work environments, according to WHB organizers, involved structural issues, whether the introduction of new technologies, concentration of ownership, and expansion of production, or the lack of review of new processes and products that created potential environmental hazards. While research and expertise were seen as important in exposing work conditions, the WHB also argued that only through the actions of labor in challenging such production choices could major environmental changes be accomplished. "Organized labor, through the instrument of the trade union," Harriet Silverman wrote of this position in 1923, "possesses the power to carry out a health program to prevent disease."[47]

The WHB's program consisted of three types of activities: establishing and supervising union health departments; undertaking occupational and environmental research and analysis; and making the information obtained from such research accessible to union members and other workers. The concept of the WHB-organized and union-run health department was a unique and controversial idea for a labor movement primarily focused on wages and hours. Still, several unions were interested in the idea. Through the WHB, a local of the Painters Union established a laboratory to help deal with health problems experienced by painters and a research program designed to gather information about worker health conditions. This research, in turn, was integrated into the demand of the union for a five-day work week in order to limit exposure to the "fumes and dusts of the highly poisonous materials constantly used by painters." WHB researchers were also able to demonstrate that the introduction of certain technologies, such as the pneumatic drill in the granite stone-cutting industry (which produced substantially greater exposure to silica dust than the manual drill), created new kinds of hazards. Through this research work, the Workers' Health Bureau hoped that union contract demands could be extended into the occupational health arena and, in the process, become a significant vehicle for environmental change.[48]

By the late 1920s, however, the trade union movement, similar to other reform movements and organizations at the time, increasingly defined itself in narrower institutional terms. Production issues and even working conditions were avoided, as unions limited their demands to wage and occasional status issues. Despite an impressive and successful national conference on labor health organized by the WHB in 1927, the leadership of the American Federation of Labor (AFL) decided to pressure individual unions to withdraw support from the bureau. Since the WHB's mission (and funding sources) were dependent on union ties, the withdrawal of AFL support forced the WHB leadership to dissolve their organization in 1928. It became a bitter disappointment for those most wedded to the concept that empowerment was also a function of making information more accessible and available. Despite its short life, however, the Workers' Health Bureau, similar to the role played by Alice Hamilton, helped situate environmental concerns in the context of a rapidly changing industrial order.

The radical environmental critique of the workplace was complemented during the early part of the century by a developing environmental critique of urbanization, initiated by movements seeking to address questions of planning and urban form. This critique was underscored by the dramatic demographic transformation that had concentrated a majority of Americans in urban areas by 1920. Such a shift had also become synonymous with industrialization and its impact on daily life in the industrial city. The Progressive Era reform movements composed of settlement workers, tenement critics, sanitarians, and public health advocates thus came to be joined by urban planners challenging the urban excesses of an expansionary and speculative capitalism.

The urban planning movement, initially associated with the City Beautiful Movement in the 1890s and the first decade of the twentieth century, experienced many of the same contradictory impulses of other reform movements in the Progressive Era. On the one hand, urban planners sought to reshape the urban environment through urban parks, open and green spaces, and more healthful and aesthetically pleasing urban settings. At the same time, many planners eagerly embraced their professional status as urban experts and tied their efforts to conservationist themes of science and efficiency. Similar to the resource management field, this reliance on expertise pushed the urban planning field in a less challenging, more conservative direction. The first organized planning association meetings, concerned primarily with such issues as urban congestion and zoning matters, never fully reconciled those tendencies.[49]

During World War I, concerns over inadequate housing stimulated a renewed interest in planning reform and urban change among architects, planners, and urban critics and theorists. Less concerned with aesthetics and social engineering than equity and community life, several of these urbanists, led by architects Clarence Stein and Henry Wright, sought to link their interest in public housing to an overall approach to regional development and the urban environment. These critics of rapid urbanization, including writer and urban theorist Lewis Mumford and Benton MacKaye, a leading conservationist and forester, coalesced by 1923 into a small but vibrant and influential urban planning movement that called itself the Regional Planning Association of America (RPAA).[50]

Though the RPAA never fully developed a coherent organizational vision nor unified body of ideas, its impact within urban planning and the movements for urban reform was significant. This stemmed as much from its members' environmental and social criticism of the industrial city and their alternative visions as from the group's own activities. Radical in its approach regarding the need for urban restructuring, the RPAA represented less a specific urbanist ideology than an eclectic blend of ideas and criticism.

For many RPAA members, the idea of the region stood directly in contrast to the industrial city. The expansionary and congested industrial city, according to its RPAA critics, was simply becoming too big, too expensive to operate, too difficult to police, too inadequate in decent housing, too burdensome a place to work, and too difficult to escape from. A regional approach could redistribute the population, industrial base, and civic facilities "so as to promote and stimulate a vivid, creative life throughout a whole region—a region being any geographic area that possesses a certain unity of climate, soil, vegetation, industry and culture," as Mumford put it.[51] For RPAA members, regionalism meant "greening" or "ruralizing" cities by designing urban centers at a smaller scale (anywhere from a population of 30,000 to 300,000). To accomplish this urban environmentalist transformation, RPAA advocates relied primarily on the key concepts of the garden city and integrating the natural environment into the daily life of urban dwellers.

The garden city concept was first put forth by the British urban theorist Ebenezer Howard in his 1898 book *Garden Cities of Tomorrow*. Howard raised the idea of linking "town and country" through population relocations, creation of greenbelts and parklands, and ultimately the development of "a new civilization."[52] The garden city concept came into vogue in the United States during the 1920s and 1930s, owing in part to the RPAA and the growing interest in remaking urban areas. For the garden city to succeed, it needed to be structured, according to one RPAA commentary, as a "town planned for industry and healthy living of a size that makes possible a full measure of social life, but no larger." In addition, such an urban area needed to be "surrounded by a permanent rural belt," with "the whole of the land being in public ownership, or held in trust for the community."[53] The garden city concept also offered a critique of the oversized city's "cycle of ecological imbalance," created by its reaching out "further and further for water, fuel, food, building materials, and sewage disposal areas." Garden city advocates contrasted this "overcity" with the "cosmopolitan city of scale," tied to the maintenance of the "primeval and the rural environment at the highest levels" developed through a regional planning process.[54]

The garden city and regional planning concepts were first popularized in the 1920s when RPAA members Clarence Stein and Henry Wright designed the new towns of Sunnyside in Queens, New York, and Radburn in New Jersey. Through private financing, these two alternative communities, built to be affordable for lower-middle-class residents, sought to incorporate more green space and pedestrian areas through tight groups of narrow houses fronting inward toward common greens, thus separating vehicular and pedestrian traffic. Plans for Radburn, the larger of the two projects with its goal of 25,000 residents, originally included industry facilities and housing for workers as well as community land ownership and a proposed ecological balance for its various land uses. But the Depression dried up the project's financing, and its more innovative features were abandoned. Radburn subsequently came to resemble an attractive suburb, reinforcing rather than restructuring existing urbanization patterns.

Complementing RPAA's garden city redefinition of urban design was RPAA founding member Benton MacKaye's focus on regional planning as a way "to make the earth more habitable." Strongly influenced by naturalist and conservationist ideas, MacKaye sought to distinguish between "mining" (e.g., forest mining or agricultural mining), defined as overproduction, and a more limited, efficient, and potentially renewable production able to sustain a more permanent community or culture (such as a rural-based forest or farming culture). MacKaye became best known for his concept of the Appalachian Trail, a regional recreation trail stretching from Maine to Georgia that would serve as the backbone of a system of wild reservations and parks linked by feeder trails. The trails would be built by local volunteer groups, including weekend campers and hikers from the cities. Some trail users might then decide to become permanent residents in these areas and thus be ready to embrace community living, cooperative food-raising, and a more direct relationship to the natural environment. MacKaye's idea, popularized both through the RPAA and annual Appalachian Trail conferences, was embraced by RPAA members such as Clarence Stein, who felt the concept would provide "a much-needed regional perspective, and bring together the conservation movement and community planning." The Appalachian Trail concept was also promoted during the mid-1930s by the newly created Wilderness Society, of which MacKaye was a charter member.[55]

The connection between conservationists and community planning advocates, however, never did extend beyond these initial ties, particularly as conservationist groups such as The Wilderness Society and the Sierra Club divorced themselves from the issues and concerns of radical urban and industrial movements during the Depression years. Even within the RPAA, there were different views of how to accomplish urban reform agendas. In the housing area, for example, RPAA members Catherine Bauer and Edith Elmer Wood focused on the need to develop housing for the poor and the working class as distinct from the regional planning and garden city concepts so central to the RPAA approach. Even prior to the formation of the RPAA, Wood had been a key advocate of government financing for new housing, defining it as central to any urban restructuring process. Bauer, Clarence Stein's research assistant and the executive secretary of RPAA in 1932, became the principal advisor and eventually executive secretary of the Housing Labor Conference. This organization was established by the American Federation of Full-fashioned Hosiery Workers Union to construct the Carl Mackley housing project in Philadelphia for working-class residents, including members of the union.[56] Through her experience with the Carl Mackley project and later involvement with the trade union movement and other New Deal radicals, Bauer began to distinguish her approach from that of Stein and Wright regarding an urban restructuring agenda. According to their one-time associate, Stein and Wright too often saw their role as training experts to design the "perfect" housing development. Bauer, on the other hand, saw her role as helping workers and community residents take an active part in their own governance, even if the outcome of such activity did not represent ideal housing designs but "a real gain in understanding, power and responsibility for the people who live in them," as she put it in a letter to Lewis Mumford.[57]

These differences between a more activist and a more conceptually oriented approach within both the RPAA and the larger urban reform movement were compounded by the influences of the Depression years and New Deal politics. Both regional planning/garden city and public housing/working-class empowerment concepts were absorbed into and redefined by such key New Deal agencies as the Tennessee Valley Authority (TVA), the Public Housing Authority, the Rural Electrification Administration, and the Resettlement Administration. For example, the TVA, for whom Benton Mackaye served as a consultant, was initially seen by the regional planning reformers as best representing large-scale regional planning and new town development. But due to its more exclusive focus on economic growth rather than social restructuring, the TVA, in the years following World War II, would become little more than a quasi-governmental entity engaged in electric power and fertilizer production activities. Similarly, the Resettlement Administration, for whom both Stein and Wright had also served as consultants, initially announced plans for the construction of fifty experimental communities based on garden city designs. Instead, the Resettlement Administration, like the overall public works bureaucracy, became tied to the job and welfare functions of the New Deal and abandoned the urban restructuring approaches of the regionalists.

At the same time, radical activists organizing around public housing, social welfare, and trade union rights were increasingly divorced from their original social agenda of worker-controlled housing, less hazardous and more democratically structured jobs, and active community life. The activists of the 1930s, inheriting the Progressive Era's agenda of the quest for urban and industrial restructuring with its broad environmental implications, began during the New Deal to focus more directly on redistribution issues within a reformed capitalist urban and industrial order. Subsequently, with the changes brought about by the economic growth of World War II and the advent of the Cold War, the debates about urban form and the social and environmental impacts of industrial activities on daily life would shift once again. A new era was about to emerge, concerned with new technologies, an even greater urbanization, and new kinds of environmental problems.

An Era of Abundance?

During the Progressive Era and the Depression, poverty and deprivation marked the conditions of environmental degradation of the industrial city. In contrast, the post–World War II and Korean War years witnessed rapid economic growth fueled by military spending, an explosion of new technologies, and the rise of an automobile-based urban culture, each of which had their own significant environmental impacts. This growth suggested that abundance had created an affluent society where the quest for "freedom from want" had given way to the search to fulfill newly created needs. The rapid expansion of advertising was the most visible expression of this shift. Though product advertising and mass marketing of goods had already emerged at the turn of the century, advertising's role was dramatically transformed during the postwar years, related to its ability to generate new values associated with the drive to establish markets for new products. Along with television, to which it was intricately connected, advertising became one of the most powerful institutions in the postwar order, exercising the kind of influence over people's habits and values formerly reserved for such institutions as schools and churches. Advertising helped create a society of consumers, its mission defined "as one of stimulating [people] to consume or to desire to consume."[58]

The expansion of markets and development of new goods (and new needs) was reinforced by the introduction of certain new technologies of production, most especially petrochemicals. In a short time during and following World War II, petrochemical products, including plastics, pesticides, additives to fuels and food, detergents, solvents, and abrasives, came to dominate a number of different markets. Plastics, especially, symbolized the new era of abundance and consumption. With its infinite capacity to reshape old products into new product lines while providing the consumer, through new packaging or redesigned products, the feeling that a new need was being met, plastics (along with other petrochemicals) became an ideal complement to an advertising-dominated system of delivering new markets for goods.[59]

The quest for abundance was also linked to resource availability, particularly the energy needed for expanding production to meet an expanded consumption. The key to energy abundance was the nuclear option, promoted by the 1960s as a source of power potentially "too cheap to meter." Intricately associated with the development of nuclear weapons technology during the post–World War II years, nuclear power was conceived as an energy counterpart to the military focus of the Cold War, especially by reducing dependence on fossil fuels from such volatile areas as Iran. Such a power source, nuclear advocates argued, could also transform the basis of production and consumption, while reducing and even eliminating pollution concerns related to energy generation. By the 1960s, the idea of future energy abundance linked to nuclear power complemented the growing electrification of urban and industrial life. Electrical energy consumption began to grow at phenomenal rates, as high as 7 percent to 10 percent annually. The affluent society was not only being plasticized, it was being electrified as well.[60]

It was also becoming a society on wheels. Changes in auto transportation had enormous impacts on urban and industrial life. The reach of the automobile, made possible by government subsidies for road construction and fuel oil prices, helped spawn an exodus from the inner cities at the same time as it expanded existing urban boundaries. A builder of highways such as New York's Robert Moses was not simply a power broker in political and economic terms, but a kind of urban potentate capable of obliterating neighborhoods, restructuring community patterns, and reshaping the urban landscape to be centered around the automobile. Reversing the RPAA concept of regionalism tied to "roadless [nonurbanized] highways," which connected resource-sufficient and environmentally designed urban centers, the auto-industrial brokers instead sought to make urban centers dependent on roads. The auto's influence on urban growth patterns became an extension of the overall claims of the automobile on daily life. The Big Three auto makers were not just economic or political powers who defined their interest as the national interest, but sponsors of a cultural message that reinforced the images of abundance and consumerism as a way of life.[61]

Along with these changes in production and consumption, the claims of abundance in the postwar order were also beginning to be associated with the unanticipated claims of pollution. Postwar environmental by-products, from wastes and emissions to urban sprawl, represented a range of hazards not experienced in the turn-of-the-century industrial city. Leaded-gasoline additives, essential to the auto culture's celebration of the high-octane, quick pickup vehicle, also produced lead emissions in the ambient environment. This issue of air pollution, a term that began to replace earlier concerns about smoke or dust, became a primary environmental reference point. The Black Monday air pollution incident in Los Angeles in 1943, the first significant environmental episode to capture widespread public attention, anticipated the concern over the deadly postwar hazard known as smog. Similarly, the subsequent air pollution episode in Donora, Pennsylvania, in 1948 and the deadly 1952 London fog, in which air contaminants were thought to have contributed to several thousand deaths, also symbolized the potential reach of these new hazards, linked to the evolving postwar urban and industrial order.

Donora, a steel town in the Monongahela Valley thirty miles southeast of Pittsburgh, housed a complex of industrial facilities, including steel plants, a zinc smelter, slag processing, a sulfuric acid plant, and wire and rod mills. A "harsh, gritty town" reminiscent of the kind of degraded environment that H. L. Mencken had described half a century earlier ("On its outskirts are acres of sidings and rusting gondolas, abandoned mines, smoldering slag piles, and gulches filled with rubbish," one commentator noted at the time), Donora was representative of those communities situated along industrial corridors subject to severe and persistent environmental hazards. The Donora episode, where an inversion layer (with its "smell of poison," as one resident put it) combined with a thick fog to cause a major health trauma, highlighted the new concerns about air pollution, a catchall term lumping acute health effects experienced at Donora with the long-term hazards associated with toxic air contaminants. Attempts to legislate or regulate this pollution had become far more difficult to accomplish than the kind of reforms generated during the Progressive Era.[62]

Air pollution was just one of the visible forms of environmental degradation generating public concern in the postwar period. Water pollution, once thought to have been controlled by the introduction of disinfection technologies, reemerged as a disturbing problem. During the late 1950s and into the 1960s, for example, a number of pristine streams, rivers, and lakes were found to be contaminated by laundry detergent runoff. The contamination appeared as foam in household water taps or as suds washing up on shores sometimes hundreds of miles from the nearest source of contamination. The concern was exacerbated by the skyrocketing sales of laundry detergent, which had, in the postwar period, displaced naturally derived soap products. Increased market share was reflected in the power of advertising to differentiate products ("cleaner than clean") that had been reshaped by new technologies. But the introduction of these products had failed to account for the pollution generated by their production and disposal.

By the 1960s, water contamination problems were being magnified by such sources as untreated municipal sewage and industrial wastes. These flowed freely into places such as Lake Erie and smaller lakes and streams, polluting them to the point of eutrophication. Untreated sewage, related to rapid urbanization, was implicated in such notorious and visible symbols of pollution as the Arthur Kill blob (the mass of untreated sewage forming off Staten Island, New York, during the 1950s), and the fouling of Boston Harbor, Santa Monica Bay, and numerous other ocean bays that became unfit to swim in or otherwise inhabit.[63]

Land disposal of wastes was another symbol of postwar pollution, reflected in rat- and vermin-infested open pits, otherwise known as landfills, whose strong odors, frequent fires, and periodic methane explosions haunted nearby urban neighborhoods and rural communities. This contamination of the land provided a striking example of an environmental hazard increasingly perceived as a public health problem rather than a natural-environment problem. Risks associated with landfills also tended to be identifiable and discriminatory, based on the location of the various landfills and discharge sites.[64]

The emergence of these myriad forms of pollution and technology-based hazards at first seemed to generate active opposition only at isolated local sites. Similarly, the postwar urban transformation, symbolized by suburbanization trends, expanding grid developments in cities such as Los Angeles and Houston, and the massive intercession of the automobile in daily life, generated only limited opposition during the 1940s and 1950s at the national level. This lack of criticism extended even to urban planners forced to confront transportation gridlock or deteriorating air quality. Popular books such as Jane Jacobs' *The Life and Death of American Cities* and Vance Packard's *The Waste Makers*, with their pleas for maintaining street and neighborhood life and contesting the shift in production toward planned obsolescence, were dismissed by planners and other technicians as lacking in scientific credibility by focusing on influences such as the automobile or advertising. For some defenders of the postwar system, industrial and suburbanization changes demonstrated a democracy of the marketplace, not the environmental hazards of daily life.

The absence of any radical critique of the urban and industrial order was most striking with respect to the postwar labor movement. Subject to unprecedented attacks stimulated by Cold War politics, environmentally oriented unions such as the Mine, Mill, and Smelter Workers were decimated by the anticommunist barrage taking place both inside and outside the labor movement. A number of industrial unions also signed contracts allowing for the introduction of new technologies and work processes that exacerbated rather than alleviated the production hazards in their plants. At the same time, unions were afforded recognition and given expanded benefits, thus creating the expectation of a lasting industrial peace, another characteristic of this new era of abundance.

Indeed, the 1950s seemed to be a time of political contentment and acquiescence to the system, where problems such as pollution were seen as minor irritants to be resolved by appropriate technological interventions. Radical critics of the postwar order despaired of the changes in the labor movement and the stifling politics of the Cold War. They spoke of reinventing their opposition by other means, to be able to "go away," in the words of novelist Clancy Sigal, from a stultifying society of apparent abundance that masked real problems and real hazards, whether social, economic, or environmental.[65] Perhaps the most pronounced oppositional movements of the 1950s were cultural in their style and substance, from the beats who protested the deadening realities of work and consumption to the rapidly emerging antinuclear movement worried about the potential of a technology out of control. These movements also suggested that the consensus politics of the 1950s would not hold. At a moment in time when the postwar order was seen as able to deliver the goods and end all ideology, tensions within the system, including but not limited to the question of pollution, had slowly but surely begun to rise to the surface, like the ooze at Arthur Kill, contaminating everything in its path.

Chapter 3

The Sixties Rebellion: The Search for a New Politics

"Elixirs of Death" and the Quality of Life: Rachel Carson's Legacy

"For the first time in the history of the world, every human being is now subjected to contact with dangerous chemicals, from the moment of conception until death." With this warning, an exploration was launched thirty years ago into the world of synthetic pesticides, particularly chlorinated hydrocarbons such as DDT that had been "so thoroughly distributed throughout the animate and inanimate world that they occur virtually everywhere."[1] While the words of Rachel Carson were perceived by the chemical industry and others as directly challenging this technology, they also anticipated a new language of environmental concern. The publication of *Silent Spring* in 1962 and the ensuing controversy that made it an epochal event in the history of environmentalism can also be seen as helping launch a new decade of rebellion and protest in which the idea of Nature under stress also began to be seen as a question of the quality of life.

Born in the small western Pennsylvania town of Springdale on the Allegheny River, Rachel Carson developed two related passions: nature writing and science research. While teaching biology courses in the evening, Carson took a job with the Bureau of U.S. Fisheries (which later became the U.S. Fish and Wildlife Service) in the late 1930s and began writing about undersea life. Though her first book, *Under the Sea Wind*, failed initially to generate interest, she continued to write about the oceans and other science-based environmental topics. During World War II, as editor of the Bureau of U.S. Fisheries publications, she became familiar with new research about the ocean environment. This research became the genesis of her 1951 book *The Sea Around Us*, which first appeared under the title "Profile of the Sea" as a three-part series in *The New Yorker*.

The Sea Around Us combined Carson's knowledge of oceanography and marine biology, her passionate concern for the harm that had been done to the sea and its life, and a readable style that made her work appealing and immediately accessible. The book was an extraordinary success. It stayed on the best-seller list for eighty-six weeks, sold more than 2 million copies, and was translated into thirty-two languages. (*Under the Sea Wind* was reissued at this time and also climbed to the top of the best-seller list, as did a follow-up book entitled *The Edge of the Sea*.) Although Carson was a private person who shunned the celebrity of a best-selling author, she was not as surprised as some about the book's public reception and its indication of a popular interest in science. In accepting the National Book Award for *The Sea Around Us*, Carson defined this interest in science as reflecting daily life concerns. "Many people have commented on the fact that a work of science should have a large popular sale," she said to the National Book Award audience. "But this notion, that 'science' is something that belongs in a separate compartment of its own, apart from everyday life, is one that I should like to challenge. We live in a scientific age; yet we assume that knowledge is the prerogative of only a small number of human beings, isolated and priestlike in their laboratories. This is not true. The materials of science are the materials of life itself. Science is part of the reality of living; it is the what, the how, and the why in everything in our experience. It is impossible to understand man without understanding his environment and the forces that have molded him physically and mentally."[2]

Carson's environmental curiosity, her willingness to pursue "the what, the how, and the why" in daily experience, made her a logical candidate to investigate the most striking petrochemical technology success story of the postwar era. By the late 1950s, when she began her work on *Silent Spring*, pesticides had already become a fixture in both agricultural production and other commercial uses. The pesticide explosion, including the development and use of chlorinated hydrocarbons such as DDT, largely dated from World War II, although a range of poisonous and potentially harmful insecticides—inorganic chemicals and heavy metal products such as lead arsenate—had been widely used prior to the war. These insecticides had also been controversial: a series of insecticide-related food poisonings during the 1920s, for example, generated significant public protest, including demands for product bans and stronger regulatory actions by the Food and Drug Administration. One best-selling book of the 1930s, *100,000,000 Guinea Pigs*, focused on the hazards of consumer and industrial products and specifically singled out lead arsenate for a possible product ban.[3]

Prewar and postwar pest control technologies, however, differed significantly both in their volume of use and their environmental impacts. During the late 1940s through the 1950s, the pesticide industry grew at a phenomenal rate. In less than a decade, sales increased by more than $300 million and continued thereafter to increase geometrically. Such a rapid expansion of the industry also had political dimensions. In states such as California, the agricultural and chemical industries became strongly interconnected, forming a potent agrichemical industrial complex that heavily influenced state legislatures and regulatory and administrative agencies. By the late 1950s, pesticides had fully supplanted all other pest control methods and insect eradication campaigns. Their use was of such magnitude that significant episodes of harm to wildlife and immediate health impacts on farmworkers began to be recorded throughout the country. One such episode drew Rachel Carson into what became her final mission.

In 1958, as part of a mosquito eradication campaign in the Duxbury, Massachusetts, area, state officials decided to spray the area with DDT. This occurred near the home of Olga Owens Huckins, a good friend of Carson's. Huckins's private bird sanctuary was immediately impacted by the DDT, bringing about the "agonizing deaths," as Huckins put it, of several of her birds. Distraught, Huckins asked Carson if she could help find someone to research the issue. Carson was then developing new material for a series of articles on ecology but felt compelled to put aside her project temporarily to make some initial inquiries about this matter. It didn't take her long to realize that independent, critical research of pesticides was nearly nonexistent.[4]

By the late 1950s, pesticides were being touted as a kind of miracle product, supported by advertising campaigns ("Better Things for Better Living Through Chemistry"), government policies designed to increase agricultural productivity, and a media celebration of the wonders of the new technology. Farmers were receptive because pesticides reduced labor costs and production risks. Public officials appeared ready to unleash, at a moment's notice, massive spraying of DDT and dozens of other synthetic organic chemicals on forests, roadways, grassy areas, and anywhere pests were seen as a threat. Although scant information was available on the environmental and public health impacts of these chemicals, enough fish and birds were being killed and farmworkers poisoned to cast doubt on industry and government claims about the pesticide revolution. These problems had also led to some initial protests against spraying campaigns, including a DDT-related lawsuit filed on Long Island that served as Carson's research point of departure.

As Carson gathered her information, she quickly realized that pesticides had far greater environmental impacts than commonly assumed. In an interview prior to the publication of *Silent Spring*, Carson told the *Washington Post* that while pesticide impacts shouldn't be considered directly equivalent to nuclear fallout, that other major environmental hazard of the period, the two were still "interrelated, combining to render our environment progressively less fit to live in."[5] Carson anticipated her information might be explosive, given the petrochemical industry's power and willingness to attack any criticisms. As a result, she built her case methodically in the form of a writer's brief in which the questions and findings of science could be used to educate and ultimately empower the public.

It is striking to read *Silent Spring* three decades after its publication. The book resonates with the continuing debates about pesticides still relevant today and reflects on issues currently facing the environmental movement. In a period when the question of pollution was only just beginning to receive significant public attention, Carson argued that public health and the environment, human and natural environments, were inseparable. Her insistence that expertise had to be democratically grounded—that pesticide impacts were a public issue, not a technical issue decided in expert arenas often subject to industry influence—anticipated later debates about the absence of the public's role in determining risk and in making choices about hazardous technologies. Carson's powerful writing style wedded a dispassionate presentation of the research with an evocative description of natural and human environments under seige from a science and a technology that had "armed itself with the most modern and terrible weapons." This technology, she declared, was being turned "not just against the insects [but] against the earth" itself. Such writing aimed not simply to present but to convince. The mission of *Silent Spring* became nothing less than an attempt to create a new environmental consciousness.[6]

To a great extent, this indomitable nature writer was successful in her task. First published as a three-part series in *The New Yorker*, *Silent Spring* generated enormous interest and controversy even prior to its publication in book form. One pesticide manufacturer, the Veliscol Corporation, even attempted to prevent the book's publication by threatening a lawsuit against the publisher. Carson's attack against the chemical industry, Veliscol's corporate counsel wrote her publisher, sought "to create the false impression that all business is grasping and immoral" and was designed "to reduce the use of agricultural chemicals in this country and in the countries of western Europe, so that our supply of food will be reduced to east-curtain parity." The Veliscol lawyer concluded that "many innocent groups are financed and led into attacks on the chemical industry by these sinister parties."[7]

Though the publisher, Houghton Mifflin, decided to proceed with the book's publication after bringing in an outside toxicologist to review Carson's material, the attacks against the book only increased in number after the book was published. These attacks, voiced by book reviewers, scientists, consultants and other pesticide "experts," and most prominently the chemical industry, were often bitter and sharp. Anticommunist innuendos were accompanied by hostile references to the sex of the author, ranging from suggestions of lesbianism to assertions that a woman was incapable of mastering as scientific and technical a subject as pesticides. Carson was accused of inaccuracies and bias and of being "hysterically overemphatic," while allusions were made to her presumed "mystical attachment to the balance of nature." A reviewer for *Chemical and Engineering News* asserted that her "ignorance or bias" threw doubt on her "competence to judge policy."[8] Edwin Diamond, a *Newsweek* senior editor and its former science editor, complained that "thanks to a woman named Rachel Carson, a big fuss has been stirred up to scare the American public out of its wits." Diamond then likened Carson's book and the public concerns it generated to the "paranoid fears" of "such cultists as the anti-fluoridation leaguers, the organic-garden faddists, and other beyond-the-fringe groups." One favorable reviewer reported how Carson had been compared to "a priestess of nature, a bird, cat, or fish lover, and a devotee of some mystical cult having to do with the law of the Universe to which critics obviously consider themselves immune."[9]

The author's thesis about reducing pesticide use was deemed controversial even among members of conservationist groups. In letters to the *Sierra Club Bulletin*, several club members employed by the agrichemical industry complained that a favorable review of *Silent Spring* in the publication did not bode well "for the future of the Sierra Club as a leading influential force in furthering objectives of conservation." The *Bulletin* editor, Bruce Kilgore, defended the review while acknowledging that "some members of the Club would disagree" since the book's subject matter was controversial within the club.[10] Privately, David Brower complained that some of the club's board members, including those tied to the chemical industry, were skeptical of *Silent Spring*.[11]

The period following the release of the book was a difficult and tempestuous time for the shy and reserved nature writer, who was also going through a debilitating bout with the cancer that caused her death eighteen months after *Silent Spring*'s publication. The book received enormous attention from politicians and policy makers as well as scientists and had a wide and passionate following among the public. In the process, Carson became an imposing figure in her own right. She strongly countered her critics by continuing to elaborate the key elements of her argument: that science and specialized technical knowledge had been divorced from any larger policy framework or public input; that science could be purchased and thus corrupted; that the rise of pesticides was indicative of "an era dominated by industry, in which the right to make money, at whatever cost to others, is seldom challenged";[12] and that the pesticide problem revealed how hazardous technologies could pollute both natural and human environments. Carson saw these environments as interrelated ecological systems, an analysis she had hoped to refine in the book she was never able to complete. Recognizing that some threats to the environment could be traced to an earlier period of industrialization, Carson still emphasized how many environmental hazards had first been introduced in the post–World War II period. Thus, while an earlier critic of the chemical industry, Alice Hamilton, laid the groundwork for discussing environmental themes in an urban-industrial context, Rachel Carson, with her evocative cry in *Silent Spring* against the silencing of the "robins, catbirds, doves, jays, wrens, and scores of other bird voices," brought to the fore questions about the urban and industrial order that a new environmentalism prepared to face.

GERMINATING IDEAS: MURRAY BOOKCHIN, PAUL GOODMAN, AND HERBERT MARCUSE

While the publication of *Silent Spring* was greeted with anticipation and controversy, other lesser-known critics of the postwar order were laying the intellectual seeds for a new kind of environmental approach to be embraced by the new kinds of social movements of the 1960s. During the early 1960s, diverse social theorists, ranging from social critic and author Paul Goodman to German philosopher Herbert Marcuse to the anarchist ecologist Murray Bookchin, developed a small but intense audience of students, intellectuals, and activists who sought to challenge what Marcuse called the "advanced industrial society." While addressing a wide assortment of topics, few of which were directly tied to the themes of the management or protection of Nature, they nevertheless focused on certain core issues of production, consumption, and urban growth that would become central to contemporary environmentalism.

One of the most developed environmental critiques of the advanced industrial society, which both paralleled and contrasted with the arguments of Rachel Carson, was put forth by Murray Bookchin. Writing under the pseudonym of Lewis Herber, Bookchin, in books, essays, and pamphlets he wrote during the 1950s and 1960s, sought to examine the new environmental problems of the postwar years. These included most prominently "new methods of food production, urban expansion, and the haphazard disposal of harmful waste materials."[13] Though Bookchin failed to attract the attention that Carson's writings had received, he still became an influential figure despite his limited audience and relative public obscurity.

Bookchin first explored the environmental questions of postwar science and technology by analyzing the issue of chemical additives in food products. Stimulated by congressional hearings on the subject during the early 1950s, Bookchin's essay "The Problems of Chemicals in Food" established a long-standing research interest in petrochemical technologies, the focus of his first major environmental work, *Our Synthetic Environment*. Published in 1962, this book surveyed the new kinds of postwar environmental hazards. It sought to demonstrate that while earlier public health problems, such as infectious diseases, had largely been alleviated, new kinds of environmentally related public health concerns, including chronic diseases such as heart disease and cancer, had also emerged in the postwar era. These health hazards, Bookchin argued in his book, represented problems of "human ecology." The development of a "synthetic environment," defined in terms of the restructuring of agriculture, the annual introduction of hundreds of new chemicals onto the market, the ominous rise of nuclear energy and nuclear military weaponry, and the rapid escalation of pollution and waste disposal concerns raised the spectre of what Bookchin called "human biological warfare." "The needs of industrial plants are being placed before man's need for clean air" while the disposal of industrial wastes gains priority over the community's need for clean water, Bookchin declared, and the most "pernicious laws of the market place are given precedence over the most compelling laws of biology."[14]

Three years later, in 1965, Bookchin/Herber published *The Crisis of the Cities*, in which he continued to draw on public health imagery to develop his critique of the postwar order. In this book, issues of air and water pollution and problems of hazardous and radioactive wastes were linked to daily stress and routinized and unrewarding work and consumption. But the environmental issue for Bookchin was preeminently an urban issue. Pollution had become "the bane of modern urban life," with the city both a source and object of the pollution. "The ecological burden the [modern] bourgeois city places on the natural environment is staggering," Bookchin later wrote. The city's continuing growth "spreads over the countryside like a rampant cancer and destroys waterways and masses of land whose preservation may well provide the indispensable agricultural margin of survival for humanity in the ages that lie ahead."[15]

Drawing significantly on the analysis of regionalists such as Lewis Mumford, Bookchin argued that the carrying capacity of the contemporary city had reached its limits. Pollution and uncontrolled growth pushed the antagonism between the land and the city to its breaking point. However, changes in technology that allowed for sustainable production had also created a "remarkable opportunity for bringing land and city into a rational and ecological synthesis."[16]

These utopian breakthroughs were particularly noteworthy in the area of energy technologies. Writing several years before the energy crisis of the 1970s and the renewed interest in alternative energy sources, Bookchin imagined a future in which solar power (including photovoltaics), wind energy, and tidal power could provide the basis of a "lasting industrial civilization." These resources were "economically inexhaustible and harmless to public health." The contemporary city, on the other hand, had become dependent on fossil fuels and, increasingly, the centralized nuclear power plant, which reinforced the "existing trend toward urban gigantism." In contrast, Bookchin's alternative energy technologies could be "tailored to the characteristics and resources of a region," especially more decentralized, smaller cities dispersed throughout the nation. Thus, the ecological argument about alternative technologies also became an argument for creating new kinds of humanly scaled communities.[17]

Bookchin's critique of the modern city and its relation to pollution paralleled the social criticism of New York intellectual Paul Goodman. A prolific writer of more than two dozen books of fiction and social and literary criticism, Goodman was a classic essayist who covered a wide variety of subjects, including urbanism and technology, sociology and psychology, education, literature, aesthetics, and ethics. Most of his books were written in the 1950s and 1960s, when academic specialization and narrowly defined technical expertise dominated both the sciences and social sciences. Goodman was frequently criticized as spreading himself too thin, a superficial generalist who allowed his radical, anarchist politics and utopian inclinations to cancel out any practical influence he might have otherwise had. Yet Goodman was able to gain an audience among disaffected students and other activists that coalesced into what began to be called the New Left. Goodman became, in his own words, an "Angry Middle-Aged Man," a utopian and critical thinker who had become "disappointed but not resigned" to the realities of the postwar urban and industrial order.[18]

Goodman's influence first emerged during the late 1940s with the publication of *Communitas*, a short, discursive text written with his brother Percival, a professor of architecture at Columbia University. This curious book included the authors' own hand drawings, trenchant and at times sardonic criticism of what they called the "American Way of Life," and commentary on such issues as urban and transportation planning as well as work and consumption-related quality-of-life concerns. Although receiving only modest attention when published in 1947, the book's influence grew over time. Reissued in 1960 with new material by the authors and a cover quote from Lewis Mumford calling it "a fresh and original contribution to the art of building cities," *Communitas* would become an influential essay of the 1960s, offering a different framework for addressing the urban environment.[19]

For the Goodmans, the core of their analysis was the criticism of the "ways of work and leisure." "Leisure does not revive us; the conditions of work are unmanly," they wrote, defining the system of distribution as "huckstering" and the system of production as one that "discourages enterprise and sabotages invention." Contemporary cities undermined any "organic relation of work, living, and play." Instead, cities had become "department stores," where the mass production of goods and changing fads and styles made consumption preeminent, while its physical and psychological separation from production activities reduced any possibility of "work becoming a way of life."[20]

The design of cities reinforced such separation. The approach toward traffic congestion, the Goodmans pointed out, was simply to build farther out, away from the center and the congestion. That approach only created "a parallel system that builds up new neighborhoods and redoubles the transit congestion, [while] no effort is made to analyze the kinds and conditions of work so that people commute less." Ultimately, an auto-industrial complex of marketing, servicing, and highway construction ends up transforming "the entire environment in unforeseen ways" by "dictating a whole way of life."[21]

This transformation was particularly unsettling, given earlier hopes that automotive technology could be liberating, as the Regional Planning Association of America (RPAA) and garden city advocates such as Patrick Geddes and Benton MacKaye had once believed. The problems of automotive technology, in fact, demonstrated how technology, in its "immodest" and intrusive forms, and science, as the "chief orthodoxy of modern times," negatively impacted the environment. Writing in the late 1960s, at the height of New Left activism, Goodman warned that technologists financed by big corporations "rush into production with neat solutions that swamp the environment. This applies to packaging products and disposing of garbage, to freeways that bulldoze neighborhoods, high-rises that destroy landscape, wiping out a species for a passing fashion, strip mining, scrapping an expensive machine rather than making a minor repair, draining a watershed for irrigation because (as in Southern California) the cultivable land has been covered by asphalt. Given this disposition, it is not surprising that we defoliate a forest in order to expose a guerilla and spray tear gas from a helicopter on a crowded campus."[22]

In contrast to these urban social and environmental realities, Goodman offered his utopian, though (as he defined it) emminently practical, alternatives. He advocated a banning of automobiles from Manhattan; a design for tackling the pollution of New York's waterfront and constructing a garden city within the inner city for work, play, and recreation; ideas about decentralizing urban functions, such as neighborhood control of grade schools; reducing commuting distances between work and home; and several other apparently quixotic, but often incisive, concepts. Given the deadly routine that Americans had sunk into, the "mere possibility of an alternative," Goodman proclaimed, "is a glorious thing." A popular, inspiring figure for the new activists for his unabashed "practical utopianism," Goodman anticipated and gave substance to a politics of the imagination. His was an effort to reinvent possibility at a time when the system, despite its unaddressed social problems and environmental externalities, appeared so confident and contained.[23]

* * *

The critique of this closed system was especially popularized by the German-born theoretician Herbert Marcuse, who became a kind of underground intellectual hero to New Left activists. Born in Berlin in 1898, Marcuse became part of the renowned Frankfurt School that had begun at the Frankfurt Institute for Social Research and reestablished itself at Columbia University after the Nazis had come to power in Germany. Though participants in the Frankfurt School differed in their contributions to what was called "critical theory," the group came to represent the effort to revise Marxist theory regarding the contemporary urban and industrial order while remaining true to a kind of Marxian spirit of method and inquiry.

Until 1964 and the publication of Marcuse's best-known and most popular book, *One-Dimensional Man*, this one-time Frankfurt School theoretician and Brandeis University professor had been a relatively obscure neo-Marxist theoretician. Interest in his three previous works—*Reason and Revolution* (an attempt to root critical theory in Hegelian and Marxist traditions), *Eros and Civilization* (a synthesis of Freudian and Marxist approaches), and *Soviet Marxism* (a neo-Marxist critique of the Soviet Union)—had been limited to academic and Marxian intellectual circles. The rapid and unanticipated popularity of *One-Dimensional Man*, published when Marcuse was already sixty-six years old, quickly changed that. Within a few short years, Marcuse would become a controversial figure in academic free speech and academic employment issues at both Brandeis and the University of California at San Diego, where he subsequently taught. Transformed into an international intellectual celebrity, his detractors scornfully characterized him as a "guru of the student rebels" and a fomentor of revolutionary change in the United States and Europe.[24]

At first glance, *One-Dimensional Man* seems an unlikely book to have achieved the influence it did. The language is awkward, the conceptual framework largely alien to an American audience, and the theoretical conclusions substantially pessimistic. Yet *One-Dimensional Man* struck a chord among the new activists, who saw his indictment of the role of technology and "the manipulation of needs by vested interests" as laying the basis for their own call to action.[25]

Marcuse's analysis, similar to Murray Bookchin and Paul Goodman's essays, situated advanced industrial society as no longer primarily defined by scarcity constraints. Such a society was now shaped by what Bookchin called "conditions of post-scarcity." The dangers of such a system, Marcuse argued, reside in its capacity to achieve the "peaceful production of the means of destruction," such as nuclear weaponry, "the perfection of waste," and a "quantification of nature," while simultaneously becoming "richer, bigger, and better" as it perpetuates such dangers. This militarized, waste-oriented economy, Marcuse states in the opening passages of *One-Dimensional Man*, "makes life easier for a greater number of people and extends man's mastery over nature." Science and technology become essential to this structure since "the industrial society which makes technology and science its own is organized for the ever-more-effective domination of man and nature, for the ever-more-effective utilization of its resources." The domination of Nature is thus inexorably linked to the domination of man, a link that "tends to be fatal to this universe as a whole."[26]

The post-scarcity society is also a consumer society with manufactured needs driving both production and individual consumer choices. For Marcuse, the manipulated desire to possess things becomes a kind of biological need, a "second nature of man." These things, or products, promote "a false consciousness which is immune against its falsehood," producing a way of life based on what Marcuse calls "one-dimensional thought and behavior."[27]

This intensely pessimistic analysis prevails in much of Marcuse's writings. His argument, he suggests, poses a choice between two contradictory hypotheses: the ability of the advanced industrial society to contain or suppress qualitative change for the foreseeable future, in contrast with the presence of forces and tendencies with the capacity to "break this containment and explode the society." Fearful of the staying power of the forces of suppression embedded in a one-dimensional society, Marcuse ends his book with a poignant quote from Walter Benjamin, the great German Jewish social and cultural critic and theoretician. Benjamin, at the onset of the fascist era (and before his own death by suicide while attempting to escape from the Nazi takeover of France), wrote, "It is only for the sake of those without hope that hope is given to us."[28]

The early 1960s themes of Herbert Marcuse, Paul Goodman, and Murray Bookchin—the critique of the consumer society and its related way of life, the prominence of science and technology in establishing such a system, and the system's new forms of domination of both human society and

Nature—paralleled the new forms of protest and social action that swept college campuses during the mid- and late 1960s. Each of these writers became participants in his own right in these upheavals: Goodman, the anarchist/utopian and pacifist, through his association with such publications and groups as *Liberation* magazine and the Institute for Policy Studies; Marcuse through his celebrated role as New Left theoretician and his informal ties with U.S. and German radical student groups; and Bookchin as a champion of New Left spontaneity and decentralist action through affinity groups (small, direct-action-oriented associations of friends) and social ecology groups (anarchist-oriented radical environmental groups). The much discussed pessimism of these writers, especially Goodman and Marcuse, became a challenge for the new movements. Activists embraced these intellectual forerunners to disprove their pessimistic conclusions about the "American Way of Life" and thus underline their utopian insights. At the center of such utopian notions stood the powerful idea that in a post-scarcity society everything is possible, even if realizing such possibilities seems distant.

The New Left and Its Unfinished Revolution

"We are people of this generation," *The Port Huron Statement* began, "bred in at least modest comfort, housed now in universities, looking uncomfortably to the world we inherit." "Our work is guided by the sense that we may be the last generation in the experiment with living," the document continued, asserting that "we ourselves are imbued with urgency, yet the message of our society is that there is no viable alternative to the present . . . that our times have witnessed the exhaustion not only of Utopias, but of any new departures as well."[29]

This sense of urgency, tied to a search for new values and utopian possibilities, was the organizational motif for one of the most significant New Left groups, the Students for a Democratic Society, or SDS. *The Port Huron Statement*, its 1962 manifesto, linked that urgency to two movements of the late 1950s and early 1960s: the antinuclear/Cold War–related protests and the civil rights movement. Both the nuclear realities of the Cold War and the institutionalized racism of the postwar society provided an important focus motivating students and others to action.

During the late 1950s and early 1960s, a powerful antinuclear movement emerged in the industrial countries, particularly the United States and

Great Britain. Increasingly, these movements mobilized around a key environmental issue tied to the Cold War: radioactive nuclear emissions and fallout associated with above-ground nuclear weapons testing programs. The growing concerns about radioactive products such as strontium 90 and iodine 131, intensified by the adamant refusal of government agencies to admit that any risks were associated with radiation exposure, fueled the protest movement. This movement included many college and, more strikingly, high school students, at a time when student apathy was defined as the dominant mood on campuses. By focusing on environmental impacts and dismissing the incessant ideological messages of the Cold War, the protesters raised the possibility of a new politics that opposed both sides of the Cold War equally for their nuclear testing.[30]

The civil rights movement also shaped the protest movements of the decade. Civil rights protesters challenged conditions of discrimination and poverty, related to the system's failure to make the "American Way of Life" available to all. Their use of nonviolence and civil disobedience contrasted graphically with the violent response of southern authorities and the vacillating role of the federal government. The organizing approach of students and other young activists within the civil rights movement—many closely identified with the emerging New Left—was tied to the hopes of creating an alternative way of life free of racism, a living example of what some movement activists called the "beloved community." This concept of a new way of life, which eventually influenced the development of cooperative and communal associations, also sought to expose the hypocrisy and the vacuousness of the "advanced industrial society" and the jobs and careers for which college students were trained and socialized. The high intensity of the civil rights experience, with its dangers and new values, heightened this distinctive search for a new kind of "vision and program in campus and community across the country," as *The Port Huron Statement* put it.[31]

Eventually the nuclear testing issue faded, particularly after the signing of the 1963 Test Ban Treaty, which eliminated above-ground nuclear testing.[32] The evolving civil rights movement, however, continued to gain momentum as it began to force whites to confront its issues in their own communities. Protest movements expanded to include other issues, particularly the war in Vietnam, reinforcing the momentum toward the creation of a New Left. The university especially became the focus of protest. At the Berkeley campus of the University of California, the Free Speech Movement, one of the largest and most dynamic of the early student protest groups, highlighted the budding New Left's critique of the university's role as caretaker and training ground for students. Although the movement arose over free speech, many of the movement's leaders, veterans of civil rights campaigns in the South and eager to introduce their sense of urgency and search for community into the university setting, sought to raise larger issues about the nature of the university system itself. Drawing on UC Berkeley president Clark Kerr's analysis of the "multiversity" and the "knowledge industry" and the blurring of the distinctions among the large research universities, the federal government (particularly the Department of Defense), and major new technology industries, the Free Speech Movement lashed out at the socialization and training directed at students. In movement leader Mario Savio's memorable words, students were no longer willing to be cogs in the machine, part of "the factory that turns out a certain product needed by industry or government."[33] Student action itself became liberating, not only by forcing confrontation concerning the definition of the university, but by suggesting an alternative mode of living and working and engaging in politics on and off campus.

While primarily based on campuses, New Left groups also began to organize within client-based professions and institutions such as those in secondary education, health, welfare, and planning. Similarly, New Leftists sought to address what were being called quality-of-life issues. These were reflected in rock music and guerrilla theater, which rejected dominant lifestyle choices regarding work, consumption, family, and (the absence of) community; through films such as *The Graduate*, which also scorned this dominant lifestyle;[34] and in the language of student protesters, who chanted at their demonstrations, "Go to school, get a job, get ahead, kill," linking lifestyle choices to the end point of the escalating Vietnam War. A new focus on the urban and industrial environment, or the "Daily Smog Smear" as one New Left tract called it, also emerged. For New Leftists, "technological progress" came to mean "urban sprawl, air pollution, smog, traffic, noise, and unpredictable changes in the job market." Two New Left analysts wrote in a working paper for a "Vocations for Radicals" conference in March 1968 that "the middle class controls very little that matters—not the conditions or purposes of their work, the myths that pour into their living room every night, the environment they live in. The sense of being manipulated and harassed, the sense of failure is a quiet, daily, muffled thing."[35] Movement protest was thus largely couched in a critique of daily life addressing both values and institutional change, with environment (referring to both daily environment as well as the natural environment) an increasingly central focus.

Though the environmental issue tended to be subsumed under the broader quality-of-life concern during the mid- and late 1960s, a series of dramatic episodes, including the 1965 power blackout and garbage strikes of New York City, the 1969 burning of the Ohio River along the industrial sections of Cleveland, and most visibly the 1969 Santa Barbara oil spill, highlighted for the New Left the darker side of a technologically centered, presumably post-scarcity system. This dark side was most harshly revealed by the war in Vietnam, where the technologies that killed people and destroyed the environment became a potent symbol of advanced industrial society. The enormity of the environmental consequences and hazards of the Vietnam War already emerged as an issue by the mid-1960s, as companies such as Dow Chemical, the manufacturer of napalm, became frequent targets of campus protesters. The term *ecocide* began to be used in describing U.S. policy, anticipating later concerns about regional and global environmental catastrophes. Even the American Association for the Advancement of Science, influenced by its younger, New Left–oriented members, adopted a resolution criticizing the use of herbicides in the war. That issue became most controversial when related to the health concerns of U.S. troops in Vietnam.[36]

By the late 1960s, a New Left environmentalist position began to take shape. This position sought to identify and challenge the central responsibility of industry in generating environmental hazards ("Where There's Pollution, There's Profit" was the title of one New Left pamphlet).[37] Military and aerospace companies, resource-based industries (such as oil companies), new chemical producers (including pesticide manufacturers), and the automotive industry became targets of New Left criticism and focal points for organizing. Industry's role was tied to the structure of economic and political power and the use of destructive and hazardous technologies. The "deterioration of the natural environment all around us," one New Left environmentalist wrote in January 1970, is "clearly a product of the nature of production and consumption, of cultural values and social relationships that today hold sway over industrial technological society—American or Soviet." For New Leftists, it was crucial not to divorce ecological politics from an overall liberatory politics.[38]

New Leftists also sought to relate the hazards of production to pressures to consume more, resulting in a "society of waste," the underside of what Marcuse called "the manipulation of needs." Several New Left ecology collectives were organized in the late 1960s to focus on the waste issue and were pivotal in the formation of community-based recycling centers.

These, in turn, came to be centers for environmental action. And while most New Left politically oriented groups focused on industry as a source of pollution, a parallel, even more powerful undercurrent of New Left environmentalism linked the question of a change in values to the question of the environment and the broader restructuring of society. Thus, for many in and around the New Left, environmentalism came to be associated with the search for alternative institutions and a new way of living.

By early 1970, as plans began to be formulated for the first Earth Day, a diffuse set of demonstrations and campus teach-ins scheduled for April 22, it was clear that the New Left had already made a distinctive mark on this emerging environmental politics. At the same time, however, New Left influence on the new forms of environmentalism quickly began to recede. Several New Left groups remained wary of the environmental label, associating it with issues such as population control, wilderness protection identified as a kind of anti-urban elitism, and personal responsibility linked to a "blame the victim" approach. The comic-strip character Pogo's famous statement—"We have met the enemy, and he is us"—was popular with groups such as the Sierra Club but sharply criticized by New Leftists, who felt it inappropriately shifted the focus away from industrial polluters. The population argument was especially disturbing to New Leftists for its politically objectionable aspects, most notably sterilization programs criticized as a form of genocide. New Leftists were also suspicious that population policies promoted by certain elite foundations such as the Rockefeller Fund were designed as an antidote to Third World upheavals, caused, according to some foundations, by high birth rates among the poor. The population agenda, New Left critics declared, was an agenda for containment.[39]

What most divorced the New Left from the new environmental politics was its own precipitous decline after 1970, which undercut its ability to influence the environmental movement at an organizational level. This decline stemmed from external pressures, such as police harassment and manipulation, as well as from changing social, economic, and political realities. By the early 1970s, the New Left's utopian and post-scarcity visions had been totally eclipsed in the wake of inner-city riots, further militarization of the economy, economic dislocations that transformed once prosperous regions into declining industrial belts, and a traumatic war that divided the country in profound ways. In response to these changes, New Left urgency turned into blatant political adventurism. Fueled by the growing fascination with violence as a form of political action, this adventurism became the coup de grâce of the organized New Left. One faction of

SDS became totally absorbed by this paroxysm of urgency-become-adventurism, rejecting the possibility of locating any viable domestic agency of change within the urban-industrial order. Similar evolutions occurred among black and Latino movements, which also faced what had become insurmountable obstacles in bringing about real change—including environmental change—in the patterns of daily life in their communities. This abandonment of political and organizational possibility, occurring at a time of the New Left's greatest potential outreach (including the Earth Day demonstrations of 1970), not only served to disperse this once promising movement but significantly altered the development of the environmental movement itself. Ultimately, the post-scarcity–oriented New Left of the 1960s, with its vision of alternative communities, its critique of daily life, and its focus on industrial-related pollution and the problems of consumption, represented a direction for environmentalism never fully explored.

THE COUNTERCULTURE INTERLUDE: A SEARCH FOR ALTERNATIVES

Even more than its incipient and interrupted relationship with the New Left, the new environmental politics of the 1960s became intertwined with what was labeled the "counterculture." This disparate collection of social movements, new forms of cultural expression, and semireligious groups and ideas connected the New Left critique of the consumer society and quality-of-life concerns with a desire to go "back to the land," or at least back to a simpler, more communal, more natural form of social life. The roots of this counterculture can be found in the beat movement of the late 1950s (and its appeal to live freer by dropping out and no longer subscribing to consumption) as well as in the "beloved community" concept of the civil rights movement and its romanticization of the rural, black South. Its approaches had more to do with its "alienation from much in the standard American culture, including its political system," as one counterculture analyst put it, than as consciously designed strategies for social renewal and transformation. Still, however diffuse and reactive its various expressions, the counterculture directly influenced a wide variety of social experiments and ideas and helped reaffirm the notion that environmentalism was also about constructing alternatives.[40]

The search for alternative forms or alternative institutions related directly to the New Left and counterculture's unresolved debate about whether to change society or, instead, change oneself. The debate, essentially about the locus of social transformation, persisted through much of the decade, despite efforts to resolve or transcend it. As early as 1965, the beginnings of a division appeared within SDS between advocates of personal and political liberation. This created an organizational tension that the SDS leadership sought to resolve by arguing that "our movement must encompass both sets of orientations."[41] The feminist movement's proclamation that "the personal is political" also only partially addressed such distinctions. Over time, these divisions led to an increasing separation between the two kinds of movements. In that context, the insistence by blacks in the civil rights movement that whites organize their own communities led to a division among the white activists about how to organize in their communities: whether to advocate for social change or instead to demonstrate how one could live differently, more alternatively, no longer dependent on dominant institutions.

By 1965, the pursuit of alternative lifestyles and alternative institutions began to take shape in concrete form. That year, for example, the Free University of New York (FUNY) was founded, to establish a model for new forms of teaching and learning. FUNY attracted both political organizers and "cultural revolutionaries," as many of the Free University participants called themselves. Such free schools were complemented by alternative publishing ventures, including underground papers and printing collectives; alternative or "free" medical clinics; retail food cooperatives, or "food conspiracies" as they were then called; and a range of other alternative institutions. Many were listed in alternative Yellow Pages or catalogues and publications such as the *Whole Earth Catalog* and *Mother Earth News*, which reached audiences of as many as 1 million readers.[42]

The underground press, which included at its height in 1969 as many as 500 papers reaching more than 4.5 million readers, was especially significant in establishing sources of alternative information—including information not being reported, or reported differently, in the established press—and as a framework for new ideas and action. A number of these papers, including *The Old Mole* in Boston, *Dock of the Bay* in San Francisco, the *Rat* in New York City, and *Great Speckled Bird* in Atlanta, sought to address political issues, such as the Vietnam War, student protests, the black power movement, and inner-city issues. But by 1967 and 1968, a number of the underground papers, led by the *Berkeley Barb*, the *East Village Other*, and even more political papers such as the *Rat* and *Quicksilver Times* in Washington, D.C., came to focus on cultural and personal liberation issues as well. For these papers, personal liberation increasingly came to be defined as a "green" activity, a term beginning to be associated with the counterculture. With a growing interest in organic gardening and food growing, urban and rural communal living, and back-to-the-land activities, the underground press contributed significantly to popularizing ecology concepts within both the counterculture and the New Left.[43]

One of the columnists for the *Rat*, for example, declared in 1969 that "revolutionaries must begin to think in ecological terms," and that "an attack against environmental destruction is an attack on the structures of control and the mechanisms of power within a society." Several underground papers carried an underground syndicated ecology column written by former Yippie activist Keith Lampe, who called for a transition to a "broader, ecologically-oriented radicalism." Drawing on tribal imagery and the romanticization of Native American culture (and ecological practices) popular with both the counterculture and the New Left, the underground press became champions of a new ecological consciousness and related way of life. The Revolution, many of the papers began to proclaim, would be an Earth Happening.[44]

The interest in ecology within the counterculture was also reinforced by the rapid development in the late 1960s of communal living experiments in urban and rural areas as political and cultural events reached a crescendo. A January 1967 "be-in" in San Francisco involving tens of thousands of young people was followed by similar events in New York, Los Angeles, and dozens of other cities. These events, or "happenings," were less an articulation of an oppositional culture (let alone an oppositional politics) than an expression of an often unfocused search for new values and a different posture with respect to the dominant institutions of society. Whether that expression took the form of budding alternative institutions, self-help groups such as the Diggers (who helped establish support systems in "dropout" neighborhoods like San Francisco's Haight-Ashbury), psychedelic drug advocacy and use, or communal living experiments, it still failed to cohere into organized action or consistent ideas to guide such action.

Although the counterculture generated enormous interest, including several books and articles seeking to define and/or celebrate it, most failed to account for the transient nature of the movement. Hugely popular books such as Charles Reich's *The Greening of America* announced a new, revolutionary form of consciousness, and even more thoughtful and insightful studies, such as Theodore Roszak's *The Making of a Counter Culture*, spoke of the search by young people for a "stance of life which seeks not simply to muster power against the misdeeds of society, but to transform the very sense men have of reality." But these books failed to extend their critique of the "technocratic society" of which Roszak spoke to include an analysis of the structure, direction, and future of the counterculture in an increasingly polarized and divided society.[45]

The established media especially helped to perpetuate and fall victim to myths about the counterculture. The 1967 "Summer of Love" in the Haight-Ashbury, for example, following on the heels of be-in celebrations and other counterculture events such as free rock concerts, also became a media happening, one of the first to sensationalize the youth culture for its sex, drugs, and generational rebelliousness.[46] While the media hype served as a magnet drawing young people to the Haight (as well as to places such as New York City's Lower East Side), it also brought about a counterreaction from the police, local business and political leaders, and law-and-order–oriented working-class and middle-class constituencies. Thus, reaction to the events of the summer of 1967 eventually caused a breakdown in the existing self-help support networks. Places such as the Haight were turned into tiny war zones, forcing the counterculture, like the New Left, into heightened rhetoric and more desperate actions.

Despite this polarization between "straight" society and the alternative groups, the counterculture continued to grow in size and influence during the late 1960s, at times in conflict and at times intersecting with the New Left. On New York's Lower East Side, for example, a new group called the Motherfuckers (after a poem, "Up Against the Wall, Motherfucker," by the black poet Amira Baraka, formerly Leroi Jones) combined both counterculture and New Left approaches. One of the group's most celebrated actions occurred during a 1967–68 garbage strike in New York City, when its members attempted to transfer bagfuls of the offensive trash (which was creating an environmental nightmare and public health hazard in poor areas such as the Lower East Side) to Lincoln Center, a performing arts center catering to the upper and middle classes. It was at Lincoln Center, the Motherfuckers declared, that "the trash really belongs."[47]

Groups such as the Motherfuckers and their more visible counterpart, the Yippies (founded and led by former civil rights participant and later environmental activist Abbie Hoffman), were central to the counterculture's entry into organized political action. Though much of this politics remained focused on issues such as the Vietnam War and racism, the question of the environment loomed increasingly large by the end of the decade. Events such as the Columbia University sit-ins (organized initially to protest the conversion of park land in a neighboring black community into a university gymnasium), the massive upheaval led by students and young working-class protesters in Paris in May 1968, and especially the People's Park protests (to prevent the bulldozing of a spontaneous community garden to make way for a university parking lot) in Berkeley in 1969 could be seen as "green" events in their symbolism, imagery, and objectives. The language of the counterculture's events and activities, from be-ins and rock concerts to new alternative institutions such as food coops, also drew heavily on Nature associations. Former beat poet Gary Snyder, one of the preeminent cultural figures and ideologues of the counterculture, called this language "earth house hold" to describe "the type of new society emerging within the industrial nations."[48]

The events of People's Park are especially suggestive of how environmental themes emerged in this period. People's Park was created by the seizure of a vacant Berkeley lot owned by the University of California. Hundreds of young people planted seeds, trees, and sod and constructed a swing set, tables, and benches. The land was declared "liberated" for the environment. The creation of People's Park became noteworthy as an act of defiance against a dominant institution and for its pursuit of the idea of environmental transformation. Some participants anticipated a violent response by police at the behest of university authorities: acts of "liberation," as the events of the late 1960s revealed, were invariably becoming acts of confrontation. But for many of these premature environmental activists, the establishment of a liberated environmental zone was itself a guide to future environmental action and transformation. It produced what New Left activist Todd Gitlin suggestively called a "conspiracy of the soil." At a teach-in on the issue, the local New Left–oriented teachers' union produced a pamphlet that linked the events at People's Park to "questions about the quality of our lives, about the deterioration of our environment, and about the propriety and legitimacy of the uses to which we put our land." Gary Snyder, who also made his presence felt during the unfolding of the events, termed People's Park a guerrilla strike on behalf of the "non-negotiable demands of the Earth" and argued that trees were like other exploited minorities, such as blacks, Vietnamese, and hippies.[49]

For the authorities, however, including Governor Ronald Reagan, People's Park was simply a matter of sixties rebellion. The issue of whether to build a parking lot for the university or create a "green" zone became incidental to the rebellion itself. Reagan's mobilization of the National Guard seemed to lead inexorably to that much-feared yet anticipated confrontation, with hundreds injured and a former student killed by a police tear-gas canister. This act of environmental liberation, like so much of the protest activity then occurring, appeared to be just one more example of the denouement of the sixties.

The year of the People's Park events, 1969, seemed to herald that explosion of society that Marcuse had foreseen. This was happening not as political transformation and renewal but as confrontation, violence, and, for the counterculture, turning inward toward alternative forms. Thus, one consequence of the collapse of the New Left was the emergence of thousands of urban and rural communes organized during 1969 and the months that followed. In December 1970, the *New York Times* suggested that as many as 2000 communes had been established in thirty-four states, while other estimates, including those by the underground press, sympathetic writers, and actual participants, indicated far greater numbers. *The New Yorker*, for example, put the number of urban communes in New York City alone at more than 1000. These were likely to have been conservative numbers, given the difficulty of determining what constituted a commune. The problem of identification was further compounded by the lack of organized networking and the desire of many communal participants to drop out quietly in order to escape the tensions of a society that seemed at the breaking point.[50]

Communes became both the symbol for and direct representation of the counterculture's search for an alternative form within an environmental or Nature-related framework. Though some communes were organized by activity or task, such as the group that compiled and published the *Vocations for Change* directory of alternative institutions and services, most communes served as cultural and environmental way stations. They were places where new values would hopefully emerge, in settings removed from dominant institutions. Yet these values and alternative living arrangements were not tested, for the most part, in any sustained way; they were ad hoc experiments rather than strategically defined alternatives. The communes, especially those in rural areas, were also not organized to challenge the existing urban and industrial order, even though they frequently clashed with local interests and became subject to economic pressures in terms of their survival. Ultimately, most participants in the communes were just passing through. They left their communes (and participation in the counterculture) not to make a transition back to a middle-class lifestyle, as many in the media suggested would inevitably happen, but to escape the force of events and the pressures and reaction of a society that had left no space for alternatives, no room for a new movement to emerge.

* * *

In September 1967, Paul Goodman wrote a science fiction essay, "The Diggers in 1984," for the New Left–oriented journal *Ramparts*. Drawing on anarchist and utopian imagery, Goodman presented a scenario of convulsion and change centered around a prototypical counterculture group he called the Diggers. Looking back from 1984, the narrator recalls the "Summer of Seven Plagues," when a transit strike led to permanent gridlock, air pollution episodes left thousands dead, power failures blackened the city, a breakdown in the water system left millions incapacitated to deal with a drought, the maiden voyage of an SST plunged into skyscrapers, and, finally, a great riot took place, stemming from the conditions of the city.

Out of those events a "rural reconstruction movement" had emerged. There were migrations to the countryside in the form of "be-ins out in the sticks with transportation provided by the CIA" to get young radicals out of the hair of the establishment. This led to the construction of rural communes. These had the blessing of liberal social engineers, who felt the trek to these "rural reservations" could solve "the young radical problem" much the way the Indian problem had been solved several generations earlier. But the Digger-led communards persisted in their ecological activism, seeking to clean the rivers, eliminate the use of insecticides, and stop the despoiling of the land. Now, in 1984, this migration, which had continued to grow during the 1970s, was threatening to become a flood of refugees. The crises that had precipitated the migrations had included several accidental nuclear explosions, such as the one that leveled the city of Akron, Ohio, blasting and poisoning 2 million people. Those refugees who had settled in the countryside in places such as Vermont had at least been able to develop "a rudimentary structure of community" compared to the urban society whose entire fabric was in shreds. Bracing for the influx of those preparing to flee the cities, the narrator wonders whether now, at last, peace will be possible and a new environmental consciousness established, or whether everyone will end up mad, casualties of the final breakdown of the urban and industrial order.[51]

Goodman's tale—a sardonic, utopian (and dystopian) vision that revealed his complex feelings about where the New Left and especially the counterculture might be heading—seems oddly prescient today. It is also a cautionary tale in what it says and fails to say about the future of those movements. A few of the communes of the late 1960s and early 1970s did survive, and other, one-time counterculture members settled in places such as Vermont and Oregon, influencing the political and cultural dynamics of those states. But most of the communes and much of the counterculture dissipated during the 1970s and 1980s, absorbed into commercialized "New Age" settings or filtered through dominant institutions in limited, largely unchallenging ways. Despite those outcomes, both the New Left and the counterculture left important marks on the public discourse. For the counterculture, one crucial legacy was its largely undeveloped environmentalism: part cultural expression, part social dissatisfaction, part search for new environmental values. The counterculture survived not so much in the form of the "new settlements" that Paul Goodman had foreseen, but as a relatively diffuse environmental sentiment that differed in both language and focus from earlier movements. Representing an interlude between the old conservationism with its search for a managed or protected wilderness and a hidden urban and industrial environmentalism that had not fully cohered into an organized movement, the counterculture, along with the New Left, served as a transition to a new environmental politics in which the question of Nature could no longer be separated from the question of society itself.

Earth Day 1970: Between Two Eras

Earth Day started out, rather inauspiciously, as an idea for a teach-in, drawing on one of the early tactics of the New Left and anti–Vietnam War movements. It was an idea directly tied to the enormous surge of interest in quality-of-life and environmental issues in the late 1960s. This interest had influenced the development of new policy initiatives in the form of proposed pollution control legislation and discussions about how to craft an environmental policy on a national level. While these policies were designed to deal with such growing problems as air pollution and solid waste disposal, they were also directed at an impatient, emerging activism that some in the press began to call the new environmentalism or "ecoactivism."[52] In Congress, Wisconsin senator Gaylord Nelson and Maine senator Edmund Muskie became the most visible proponents of an environmental agenda. Nelson was the first to develop the idea for Earth Day, proposing a "National Teach-in on the Crisis of the Environment" at a Seattle, Washington, symposium in September 1969. Nelson argued that a teach-in might be an appropriate forum to help crystallize this new environmental constituency while also distancing it from more radical New Left and counterculture activists. The environmental crisis, Nelson asserted, was "the most critical issue facing mankind," making "Vietnam, nuclear war, hunger, decaying cities, and all the other major problems one could name . . . relatively insignificant by comparison."[53]

By late 1969, as the teach-in idea began to generate interest, a new organization, the Environmental Teach-In, Inc. (which soon became Environmental Action), was created to handle the numerous inquiries about the event coming into Nelson's office. The preparations for the teach-in, however, still seemed overshadowed by the rapid sequence of events influencing this budding new environmentalism. With the events at People's Park, with the alternative press issuing its calls for a revolutionary transformation based on ecological principles, and with several dozen new Ecology Action groups establishing centers, drawing up manifestos, holding demonstrations, and planning for direct action, the new environmentalism continued to be seen as an extension of the sixties movements. More moderate campus environmental groups were also active, pursuing letter-writing campaigns, conducting guerrilla theater–like actions, such as Berkeley's "Smog-free Locomotion Day," and undertaking their own teach-ins in advance of the national action planned for April 22, 1970. It was toward these more moderate activists that the Earth Day organization directed its efforts.

While the teach-in idea quickly began to take root, the actual planning for the proposed Earth Day, as the April 22 events came to be known, remained limited in scope. "Once I announced the teach-in," Nelson recalled, "it began to be carried by its own momentum. If we had actually been responsible for making the event happen, it might have taken several years and million of dollars to pull it off. In the end, Earth Day became its own event."[54]

Even as the event gathered momentum on its own accord, its definition remained problematic. The key figure Nelson recruited to help pull off Earth Day, twenty-five-year-old Harvard Law School student Denis Hayes, embodied the contradictory impulses associated with the event. On the one hand, Hayes assumed the stance of an activist and radical critic of existing government and industrial policies, not dissimilar to New Left and budding ecology action perspectives about the urban/industrial order. "Ecology is concerned with the total system—not just the way it disposes of its garbage," Hayes stated at a January press conference on the upcoming Earth Day event. On the other hand, Hayes wanted to project a less confrontational style and tactical approach than the New Left activists, to have the event "involve the whole society," as he put it at his press conference.[55]

As with the Vietnam Moratorium event of the previous October organized by Sam Brown, another law student who resembled Hayes in approach and demeanor, Hayes hoped to pull off what seemed to be an increasingly difficult task: to criticize the system through a peaceful mass mobilization at a time when polarization and confrontational events increasingly prevailed. "We didn't want to alienate the middle class; we didn't want to lose the 'silent majority' just because of style issues," Hayes later recalled. Like the Vietnam Moratorium, after which Earth Day was partially modeled, it was hoped that by decentralizing the event and allowing for a wide variety of local actions, as well as by obtaining as much official support as possible, Earth Day would somehow remove itself from the politics of confrontation. The event, its organizers urged, needed to be as much a celebratory as a critical happening.[56]

For many in the New Left, this approach to Earth Day caused concern. Despite contrary statements by Earth Day organizers, New Leftists feared the Earth Day events would be interpreted, as many in the media had already begun to do, as a rejection of New Left activism and a deflection of a growing environmental radicalism. The event was also perceived as providing a shift in focus from the critique of the urban/industrial order and its polluters to individual lifestyle issues. For both New Left and counterculture participants, the effort by Earth Day organizers to seek consensus by receiving the blessing of the press, government, and even industry seemed a betrayal of the search for alternatives. "When conservationists argue that everyone is in the same boat (or on the same raft), that everyone must work together, tempering their actions to suit the imperatives of coalition," one New Left environmentalist wrote of the Earth Day approach, "they are in fact arguing for the further consolidation of power and profit in the hands of those responsible for the present dilemma."[57]

While New Left and counterculture groups were wary of the Earth Day organizers and their proposed action, the older conservationist and protectionist groups, such as the Sierra Club, the National Wildlife Federation, and the Audubon Society, also seemed hesitant about these new activists and were generally absent from involvement in Earth Day. The characterization of a "new environmentalism" especially grated on the conservationists, who insisted on the importance of "the old values in their own right," as Edgar Wayburn, the Sierra Club's vice-president, put it. "We cannot afford to let up on the battles for old-fashioned Wilderness Areas, for more National Parks, for preservation of forests and streams and meadows and the earth's beautiful wild places," Wayburn warned.[58] Some leaders also feared this new environmental movement might be quite different than just "a relabeled Conservation movement," and that such a new movement might draw attention to "approaches other than those the traditional movement has pioneered and knows best," such as wilderness preservation. "We can hope our willingess to learn and to work with others will induce a spirit of cooperation rather than competition," the Sierra Club's Michael McCloskey said of the two movements. But McCloskey also warned that "either a better synthesis of philosophy must develop or hard choices will have to be made."[59]

There were some exceptions to the conservationist mistrust of the Earth Day organizers, most notably the leaders of The Wilderness Society and the Conservation Foundation. The Wilderness Society's executive director, Stewart Brandborg, sympathized with and was intrigued by the Earth Day activists and provided them with financial assistance and space in the group's publication. The Conservation Foundation's executive director, Sydney Howe, one of the original incorporators of Environmental Teach-In, Inc., was also an important sympathizer. Just a few months before April 22, Howe provided an under-the-table loan of $20,000 for the organization at a point when it had run out of money. Fearful that his conservative board would never sanction such a loan, Howe kept it off the books until Gaylord Nelson, through speaking engagements, was able to repay the money. The loan, which was never made public, also never came to the Conservation Foundation board's attention. Howe, an insightful but little-recognized figure among the traditional conservationists, continued to try to steer his organization toward urban environmental and social justice issues, and, for a time, succeeded in attracting staff and resources for this approach before he was finally fired in 1973.[60]

What became most disconcerting to the traditional conservationists was the intense media coverage of Earth Day and the sense of discovery, especially by the media, that a new issue and a new movement had emerged full-blown with little apparent connection to earlier conservationist and protectionist movements. This element of discovery was further reinforced by the increasing political significance attached to this new movement and its pollution-centered issues, most dramatically revealed by President Richard Nixon's State of the Union address three months prior to Earth Day. Proclaimed "Nixon's New Issue" by *Time* magazine, the president, in devoting the major portion of his speech to the environment, sought to elevate environmentalism as the new "selflessness." The great question of the 1970s, Nixon proclaimed, was whether "we surrender to our surroundings" or "make our peace with nature" through "reparations for the damage we have done to our air, to our land and to our water."

Concerned about the political inroads made by potential presidential challenger Edmund Muskie, Nixon sought to preempt the environmental issue by putting forth a technology-centered, pollution control approach to such concerns as water pollution, garbage disposal, sewage discharges, and air pollution. Designing and constructing waste treatment plants was at the heart of Nixon's new cause, with his cleanup of water pollution to be "the most comprehensive and costly program in America's history."[61]

Though Nixon's policy turned out to be short on specifics and funding mechanisms, the elevation of pollution control and technology-based solutions was especially made to appeal to Nixon's mainstream constituency, his "silent majority." These middle-class and working-class constituents, while hostile to the antiwar, black power, New Left, and counterculture movements, were nevertheless concerned about the environment, as polls at the time pointed out. By visibly identifying himself and his administration's policy with the environmental cause and linking those issues to the search for technology-based solutions, Nixon explicitly sought to distinguish between the antisystem New Left and counterculture activists and the consensus-seeking effort to fix the system. Nixon wanted to be seen as a new Theodore Roosevelt, a Republican champion of efficiency and technological improvement making peace with nature. He further cultivated this image by inviting traditional conservationist leaders to the White House for a pre–Earth Day meeting, which he characterized as equivalent to the May 1908 Governors' Conference.[62]

Nixon's avowed embrace of environmentalism, however, failed to convince even the more moderate of the Earth Day activists of the sincerity of his administration's intentions. A number of environment-related government agencies, for example, were instructed to support Earth Day through office displays, press releases, and speeches at Earth Day events, but then failed to address their own long-standing prodevelopment biases and actually increased security measures in anticipation of possible picketing or demonstrations at their offices. Nixon's secretary of the interior, Walter Hickel, was the most visible of the cabinet members wishing to assume the environmental mantle, but he was also the most vulnerable to criticism for his vigorous support of the Alaska pipeline. For the Nixon administration, the effort to coopt the issue, as various environmental critics characterized his new approach, seemed tentative at best and often backfired in terms of the reactions it generated.

While the Nixon administration sought to refashion its image and influence Earth Day, industry interests similarly adopted a stance aimed at incorporating the corporate point of view into a newly defined environmental consensus.[63] By 1969 and early 1970, a number of industry groups and individual companies had begun to reposition themselves around pollution and waste issues. They argued that they were making adequate, even innovative, responses to these environmental concerns. The packaging industry, paper companies, chemical manufacturers, automotive industry, and utilities were especially concerned that they might become targets of environmental action and legislation. In response, companies reluctantly began to associate themselves with the new pollution control technologies, such as mass-burn incinerators, air scrubbers, and catalytic converters for automobiles. Some companies, such as Monsanto, even heralded their readiness to become environmental leaders in the coming decade by applying pollution control technologies in their own facilities and then marketing such systems as a profitable line of business for the company.[64]

With Earth Day generating more and more media attention, a number of companies and industry groups decided to embrace the event through advertisements, displays at corporate headquarters, announcements about new demonstration technologies or other pollution control efforts, speakers made available for teach-ins and forums, and even financial and in-kind contributions, such as those received by the Earth Day steering committee at the University of Michigan from Dow Chemical and Ford Motor Company.[65]

The utility industry undertook perhaps the most visible campaign around Earth Day. Wary of the movement's possible influence on regulatory bodies, utilities made special efforts to counter the anticorporate sentiment within the new environmentalism. One utility, Chicago-based Commonwealth Edison, through its newly created environmental task force, planned to send 175 speakers to various Earth Day events at colleges and high schools. Other utilities mobilized as many as 1000 industry personnel to be made available on April 22. In a similar vein, New York's Consolidated Edison provided New York's mayor, John Lindsay, with an electric bus to ride around in during the day's festivities and displayed an electric car near its headquarters. These were rather disingenuous acts, since Con Ed had not pursued any plans to stimulate the development of electric vehicle technologies readily available for investment and support. One trade journal characterized the industry's public relations efforts "as a platform for outlining what it has done and what it is planning to do" to establish its environmental credentials.[66]

Not all industries, however, felt comfortable participating in Earth Day. Several companies feared the anticorporate rhetoric and the association with New Left and counterculture groups and ideas adopted by some of the Earth Day organizers. One industry trade journal warned that the ecological movement "carries with it the danger that it may be seized by the radical left as a broad attack on the entire industrial system." The article warned that "a lot of young idealists, waving banners that advocate love and a better quality of life, can be used by radicals for revolutionary purposes." For some industry groups, concerns about the radical nature of the event turned into paranoid fears stimulated by right-wing politicians who suggested a conspiratorial connection between the date selected for Earth Day and the centennial of Lenin's birthday. This coincidence of dates received a fair amount of media attention and was even raised by one Conservation Foundation trustee who worried that any support for the event would be tainted by the presumed association with Lenin.[67]

The actual events of Earth Day, as it turned out, reflected both the anti-establishment and consensus-seeking impulses of its organizers. Government and corporate speakers were subject to hostile attacks, frequent interruptions, and creative protests. At the University of Alaska, Secretary of the Interior Hickel was booed off the stage when he laid out administration support of the Alaska pipeline. In Denver, antinuclear activists presented the Colorado Environmental Rapist of the Year award to the Atomic Energy Commission. The use of guerrilla theater techniques was also prominent. At the University of Illinois, students came on stage to disrupt a Commonwealth Edison speaker by throwing soot on each other and coughing vigorously. In another instance, Florida activists presented a dead octopus at the headquarters of Florida Power & Light, a utility responsible for the thermal pollution of Biscayne Bay.[68]

Despite the desire of the Earth Day organizers to avoid confrontations and Nixon administration efforts to divorce environmental from Vietnam and race-related issues, a number of Earth Day actions occurred along those lines. At the University of Oregon, students demonstrated against the campus's Reserve Officer Training Corps (ROTC) headquarters throughout April. The protests culminated in a massive sit-in at the administration building on Earth Day. The sit-in was broken up by the police and National Guard that evening, with sixty-one students arrested and tear gas used against a crowd that had gathered to observe the events. During the next several weeks, Oregon students continued their protests, forcing the administration to suspend classes for several days in order to hold a series of teach-ins on campus. Subsequently, the University of Oregon administration agreed to utilize the consulting services of the Berkeley Center for

Environmental Structure, one of a handful of radical, environmentally oriented planning groups, to help reshape the campus environment and community. Working with activists and administrators, the Berkeley center helped put together the "Oregon Experiment," a highly acclaimed experimental design in environmental and community planning unveiled four years after the Earth Day events.[69]

Race and antiwar issues were also raised by Edmund Muskie in an Earth Day speech at the University of Pennsylvania. Muskie declared that "those who believe that we are talking about the Grand Canyon and the Catskills, but not Harlem and Watts, are wrong. And those who believe that we must do something about the SST and the automobile, but not ABMs and the Vietnam War, are [also] wrong." Similarly, Earth Day organizer Hayes, at the Washington, D.C., rally, called the Vietnam War an "ecological catastrophe" and suggested that "politicians and businessmen who are jumping on the environmental bandwagon don't have the slightest idea what they are getting into. They are talking about filters on smokestacks while we are challenging corporate irresponsibility." At the same Washington, D.C., rally, New Leftist Rennie Davis, one of the Chicago Seven on trial in regard to antiwar demonstrations at the 1968 Chicago Democratic Party convention, also spoke. Though Davis's talk showed little comprehension of the environmental issue, his presence provided another indication of the 1960s undercurrent to Earth Day.[70]

In contrast to the more radical protests, Earth Day also had a more moderate, symbolic side, consisting of semi-official, widely sanctioned happenings. In some areas, streets were closed off, and elected officials joined hands with celebrities and Earth Day organizers in parades held with police escorts, to be greeted by handmade posters held by demonstrators and slogans chalked on sidewalks or on the sides of buildings. Though much of the rhetoric in the speeches and slogans suggested imminent and potentially catastrophic outcomes, the mood of the day was mostly peaceful and even festive. There was a less confrontational quality to the demonstrations, an opportunity for those removed from, even hostile to, the cultural movements of the 1960s to mingle with their more expressive counterparts within the counterculture. For the Earth Day organizers, the strength of the event was defined by the number of participants. And although the largest of the rallies in New York and Washington failed to match the massive anti–Vietnam War mobilizations of the period, both the Earth Day organizers and the press emphasized the range and diversity of activities across the country, including those held at hundreds of campuses. Estimates of the overall number of participants were as high as 10 million.

What most distinguished Earth Day from other 1960s events, however, was not the spirit and character of the events or even the number of participants, but its symbolism, largely influenced by the extraordinary media coverage that ultimately gave the event its definition. In the weeks leading up to and following April 22, the media embraced environmentalism as the all-inclusive cause of the day. They combined such varied issues as population growth, air and water pollution, the loss of wilderness, and pesticide use. They also tended to emphasize the need for cleanup and science and technology rather than urban and industrial change. Environmental ills, in this context, represented "technical and mechanical problems that involve processes, flows, things, and the American genius [to attack such problems] seems to run that way," as a *Time* magazine cover story on the environment put it. The media further suggested that "water, air and green space know no class or color distinctions," as the same *Time* article put it. Consequently, it was implied, the environmental movement could avoid dealing with the intractable problems of race and class in society.[71]

Ultimately, the environmental issue and the newly defined environmental movement was afforded instant recognition by a media suddenly discovering the issue for the first time. Environmentalism became a movement without a history, with an amorphous social base, and with a clean slate on how best to proceed. Unlike the New Left and counterculture's search for alternatives and their differing vision of urban and industrial life, the interpretation of Earth Day as Day One of this new environmentalism also reflected the media's "fix-it" conception. This fix-it approach, even if huge and costly, would finally allow everyone, whether industry, government, or citizenry, to participate in helping the system work better.

In the months following Earth Day, a new environmentalism did begin to take shape. It evolved in relation to the series of laws and regulatory initiatives establishing a massive pollution control apparatus at the federal level. At the same time, a number of groups, including existing conservationist organizations, developed along more professional lines, creating a mainstream environmental movement in the process. This movement became a powerful force in extending and shaping the environmental policy system that so dominated legislative and administrative agendas during the 1970s.

At the outset of this "Environmental Decade," the mood of confrontation and rebellion that had characterized the oppositional movements of the late 1960s seemed to increase in intensity. It reached a climax almost immediately after Earth Day with the killing of two students at Jackson State University in Mississippi and four students at Kent State University in

Ohio during antiwar demonstrations. But the search to construct alternative institutions and utopian visions associated with the sixties rebellions and linked to the emergence of a new environmental activism was short-lived, a casualty of the narrowing political space that closed out the decade. Still, the activists of the 1960s—both New Leftists and counterculture participants—did not entirely disappear from the environmental scene. Many would become part of a newer generation of eco-activists helping to define key issues in this next period.

In this context, Earth Day can be seen as a transitional event, tied in part to the rebellions of the 1960s. But it also reflected the desire to fix things up, to conquer the beast of pollution by capping it, managing it, and ultimately controlling it. With the banners folded up, the street barricades taken down, and the activists dispersed, the legacy of Earth Day remained to be found in the continuing debates over what kind of environmental change was desired, how differing groups would seek to shape a movement, and how such a movement would define itself in relation to the contemporary urban and industrial order.

Part II

THE CONTEMPORARY MOVEMENTS

Chapter 4

Professionalization and Institutionalization: The Mainstream Groups

A CEO CULTURE: THE GROUP OF TEN

The meeting on January 21, 1981, at the Iron Grill Inn, a few blocks from the White House, promised an afternoon of good food, interesting discussion, and getting acquainted or renewing ties. Set for the day after the inauguration of Ronald Reagan as the fortieth president of the United States, the proposed three-hour luncheon meeting was also designed to be a strategic moment for the mainstream environmental movement. It involved a gathering of the leaders of nine (soon to be ten) of the largest national environmental groups, brought together by sympathetic funders whose participation would likely ensure a high turnout. The conveners of the meeting had already begun to warn that the new administration in Washington demanded a strong response, a "precise, enlightened coordination among envorgs (environmental organizations) when pursuing shared goals," as one internal memo put it.[1]

For the environmental organizations, the mood that winter of 1981 was somber. Uncertainty about the change in administrations was in the air. Through the winter, a series of meetings had taken place to assess the impact of the Reagan election, but no clear consensus had emerged among the organizations. Some groups thought the movement would need to send out emissaries to "establish diplomatic relations" with the Reagan team, as former Oregon governor Thomas McCall, a friend of the environmental groups, had put it earlier. Others despaired of maintaining the kinds of ties established during the Carter presidency and prepared instead for battle. Most everybody thought their actions would have to be defensive in nature, to ensure that the accomplishments of the Environmental Decade, with its "indisputable achievements" in environmental policy, as Conservation Foundation head William Reilly put it, might be preserved.[2]

Though it had been just a little more than ten years since the tumultuous days surrounding Earth Day 1970, the environmental organizations whose leaders had been invited to attend the Iron Grill meeting had experienced extraordinary changes. At the core of those changes were the vast array of environmental laws and regulations that had been passed and the huge, complex environmental bureaucracy the Iron Grill groups had helped establish. From the passage of the Clean Air Act in 1970 to the approval in the waning days of 1980 of the Comprehensive Emergency Response, Compensation, and Liability Act (CERCLA), better known as Superfund, this myriad of laws, regulatory agencies (particularly the rapidly expanding U.S. Environmental Protection Agency), new administrative bodies, and court decisions had developed into an environmental policy system. After Reagan's election that system seemed endangered, with the movement so identified with it a prospective casualty of the threatened Reagan administration onslaught against environmental regulations.

The nine environmental leaders who gathered that relatively mild day in January were in some ways a disparate group, reflecting the differing orientations and cultures of their own organizations. Traditional conservationist and protectionist groups were well represented. They included cautious and conservative Thomas Kimball of the National Wildlife Federation, Jack Lorenz of the Izaak Walton League, former Delaware governor and Council for Environmental Quality head Russell Peterson of the National Audubon Society, Michael McCloskey of the Sierra Club (who had overseen his group's transition from the David Brower era), and William Turnage of The Wilderness Society. These leaders were joined by the leaders of some of the newer, staff-based professional organizations that had been instrumental in reshaping the mainstream movement. Younger and more expertise-oriented, they included John Adams of the Natural Resources Defense Council (NRDC), Janet Brown of the Environmental Defense Fund (EDF), Louise Dunlap of the Environmental Policy Center (EPC), and Rafe Pomerance of Friends of the Earth (FOE).

These nine environmental heads were joined at the Iron Grill by several funders, who had come to the luncheon meeting more to listen and observe the potential interaction than to offer donor-related advice or opportunities. The key funder present was Robert Allen, vice-president of the Henry P. Kendall Foundation, a midsize foundation whose decisions were fully controlled by only three individuals: Allen and the two Kendall brothers, John and Henry. The Kendalls, one a businessman (John) and the other (Henry) a physicist (and founder of the Union of Concerned Scientists, a key environmental advocacy group of scientists), had decided to expand their family foundation after selling their family company to Colgate-Palmolive. The Kendalls had selected company employee Allen to run the foundation, partly for his environmental views. It was the environmental area that the brothers had decided would become the primary focus of the foundation.

It was a "heady moment" for Allen. "There I was," he recalled, "at age fifty, after fifteen years at the company, overseeing the distribution of two-and-a-quarter million dollars in grants a year, concerning issues I cared passionately about." Allen remembered ripping out articles in the Sunday *New York Times* about the latest environmental scandal and then bursting into his office Monday morning, prepared to call around to see which groups or environmental experts might be ready to litigate or otherwise challenge the status quo on that particular issue.[3]

The foundation world itself was in flux regarding environmental matters during the early and mid-1970s, when the revamped Kendall Foundation made its appearance. Old-line, moneyed interests, such as Laurance Rockefeller, and hunting, fishing, and bird-watching groups, such as the National Wildlife Federation, Izaak Walton League, and National Audubon Society, were being complemented and, in some cases, replaced by other foundations: Ford, the younger generation of Rockefellers, the Stern and Mott family funds, and wealthy individuals such as Marion Edey of the League of Conservation Voters. These funders were new players in the world of environmental politics. They were helping to create a new breed of environmental organization, with expert staff, especially lawyers and scientists, and a more sophisticated lobbying or political presence in Washington. As much as anyone else, the foundations had become part of the process of creating the environmental policy system of the 1970s.

Though other funders were involved in the planning or participated in the Iron Grill meeting, including David Hunter from the Stern Foundation, Sidney Shapiro from the Levinson Foundation, and Rob Scribner from the Rockefeller Family Fund, it was Allen who initially conceived and would ultimately become the driving force behind the group.[4] Through his role at

Kendall, Allen had become wary of some of the dynamics and interactions among the largest of the environmental groups. This included the incessant and fierce competition for funds, recognition, and overall political legitimacy, as well as the lack of colleagueship among the heads of the national groups. Even though groups had met together before, they had not developed any ongoing relationship or strategic division of labor concerning issues and organizational resources. Instead, competition often created pettiness and, sometimes, personal tension. Without a forum to create linkages, disputes between groups and leaders periodically developed and sometimes "festered," as one of the leaders put it.[5] Though increasingly professionalized in their composition and outlook, the groups had still failed to construct a coherent, movement-wide, institutional framework commensurate with their role in establishing a policy nexus in Washington. As leaders of disparate groups still searching for self-definition, some heads of mainstream organizations had become more willing to entertain the notion that their relationships needed restructuring, perhaps in ways similar to their industry antagonists.

One model available was the gathering of corporate chief executive officers that was established during the 1970s, in part because of environmental issues. At the national level, this gathering took place under the aegis of the Business Roundtable; at the regional level, it was structured through such groups as the California Business Roundtable and the Western Regional Council.[6] Audubon's Russell Peterson, a former official at DuPont, was especially attracted to the idea. Other environmental leaders, though less familiar with the Roundtable example, were nevertheless amenable to the idea, particularly since the foundations were promoting it. For Robert Allen, the "CEO gathering" concept was the best means to make the institutionalization process as efficient as possible. It provided a way to "rationalize the field" by avoiding reduplication of organizational resources and collectively targeting funds for issue development. The need to create a coherent environmental identity was no longer just an abstract question: it had to be forged in the midst of a Reaganite counterrevolution against environmental regulation. The groups needed to see themselves and their roles in new ways: as defenders of a system and the heads of multimillion-dollar operations who by coordinating their common interests and goals could forge a mainstream identity.[7]

As the environmental groups' "CEOs" and funders gathered at the Iron Grill, expectations about the luncheon meeting were rather vague. Allen had put together the initial list of invited participants, with advice from the leaders of some of the groups that Kendall had funded, such as the Environmental Policy Center's Louise Dunlap. There were a few who declined, most notably the Conservation Foundation's William Reilly, who remained wary of the staff-based, often adversarial groups such as the EDF, NRDC, and EPC. There were obvious omissions as well: by limiting the gathering to nine CEOs (soon to be ten, after Paul Pritchard of the National Parks and Conservation Association was invited to join), a number of groups, including several that matched the original nine in terms of membership and resources, would necessarily be excluded. But Allen and most of the CEOs thought the number of participants had to be limited to assure ease of conversation at the gathering and to maintain the capacity to follow through once consensus was reached. As with the Business Roundtable groups, the very existence of a mainstream environmental CEO gathering would signify an institutionalization process deemed necessary for environmentalism in the Reagan era.

Allen opened the gathering by laying out his premise and then posing a series of questions to the CEOs. With the new White House, Allen argued, there would be "powerful, increased efforts to subordinate environmental integrity to other national goals and private purposes." This, in turn, warranted improvement in the performance of the environmental movement through better coordination among the groups. How do you see the future? Allen asked the CEOs. Will you have difficulties working in the Reagan years? Would you as CEOs be willing to meet on a regular basis to establish this new form of coordination? Do you want your staff people involved?[8]

The CEOs responded briefly to the questions. They were happy to have met some of their counterparts, though disappointed in the low turnout of funders (and the implied promise of more funds for the groups). Although the Iron Grill gathering didn't prove conclusive in establishing new forms of coordination nor formal linkages among groups, the CEOs nevertheless decided to meet again on a quarterly basis, without their staffs, and with participants still limited to those initially invited. Agendas were to be set by the host CEO. Russell Peterson from Audubon would convene the next meeting, scheduled three weeks later for February 12. And although future meetings would be limited to the CEOs, Robert Allen was also invited to come back. He would serve as the glue for the group and a possible source of funds or fundraising for projects the group might collectively decide to undertake.

During the next months, the quarterly CEO gatherings continued, with agendas and chairs rotated among the ten participants. An air of informality prevailed, though discussions sometimes focused on particular topics, such as the Clean Air Act. While several of the CEOs had wanted the group to remain nameless (and without a specified agenda or set of projects), the name "Group of Ten" began to be used by both participants and staff members of the groups involved. By the end of the group's first year of meetings, the name had begun to stick, providing an important identity to this experiment in forming a coordinated institution. The exclusive nature of the meetings remained in effect, though the Ten also established a staff-based implementation group, which staff members derisively called the "B Team."[9]

Within the Ten's first year, two key projects emerged as possible bases for action. One centered around a series of proposed regional meetings, each to be hosted by one or more of the groups and designed to strengthen the lobbying capabilities of the Ten at the local level.[10] The second project involved the more ambitious task of identifying a common agenda. The details would be worked out by the B Team, with the completed product published in the name of the Ten. This book project, *An Environmental Agenda for the Future*, was published in 1985, when the Group of Ten was becoming a much more visible entity.[11]

The first years of the Ten also saw a shift from the defensive posture assumed at the Iron Grill meeting to an increasingly assertive, more demonstrative approach taken during the early Reagan years. This more aggressive stance was based on rapidly increasing membership, high rates of return on direct-mail fundraising efforts, and continuing foundation support. It appeared to place the mainstream environmental groups back on an adversarial track, even as the agenda of the Ten sought to extend rather than revise the environmental policies created during the 1970s. With the one significant exception of nuclear issues, the areas reviewed by *An Environmental Agenda for the Future* duplicated or reinforced earlier ad hoc agendas by "building upon the strategies of the past two decades," as the document put it.[12] What made the Ten's efforts most distinctive was the mainstream movement's dramatic clash with the Reagan administration that characterized much of Reagan's first term.

This conflict with the Reaganites, it turned out, also reinforced the movement's institutionalization process. On the one hand, it created a common focus for the Ten, with easily identifiable targets such as Secretary of the Interior James Watt and the EPA's director Anne Gorsuch Burford, dramatizing the conflict with the Reaganites and enhancing the organizational effectiveness of the Ten. Audubon, for example, with only limited success in direct-mail campaigns up to the Reagan period, shifted its appeal to a direct attack on the Reagan administration and raised more than ten times its largest previous total.[13] At the same time, while most groups benefited enormously from these symbols of confrontation, a few of the Ten, most notably the Izaak Walton League and the National Wildlife Federation (NWF), had substantial internal conflicts concerning the battles with the Reaganites. Watt especially targeted the NWF, hoping to separate hunting/sportsmen interests from both the traditional conservationists and the new environmental policy experts.[14] Though Watt himself largely failed to create such divisions, perhaps because of his temperament and divisive rhetoric, the NWF leadership under its new CEO, Jay Hair, maintained an ambivalent posture toward the Reaganite confrontation and the value of the Group of Ten process. This caused considerable consternation among other CEOs and B Team staff. In 1982, in anticipation of a "spring offensive" against Reagan's environmental policies, and again in 1983 and 1984 following the dismissal of Burford at the EPA and Watt at Interior, internal memos of the Ten and the B Team revealed strong fears about NWF positions. Ultimately, the NWF's Hair reluctantly continued to participate, both in the quarterly CEO gatherings and as a signatory to the environmental agenda document, modifying by his presence the adversarial stance taken by the Ten.[15]

By the mid-1980s, as confrontations with the Reaganites began to lessen in intensity, the Group of Ten became more secure in its defense of—and efforts to extend—existing environmental policies. Overtures were made to establish a dialogue with corporate leaders. This included a series of meetings with the heads of six chemical companies: DuPont, Exxon Chemical, Union Carbide, Dow, American Cyanamid, and Monsanto.[16] The Group of Ten process was also becoming more routinized, increasingly seen by its participants as a successful effort at establishing a common frame of reference for the mainstream environmental organizations. Despite Jay Hair's ambiguities and the reluctance of some staff, particularly in groups such as the Sierra Club and the Environmental Defense Fund, to give much credence or prominence to the Group of Ten idea, the "CEO gathering" concept continued to suggest ways in which the groups were tending to converge. By seeking consensus through its published agenda and related activities, such as press conferences and other media efforts, the Group of Ten pursued its search for a common denominator. The process of attempting to achieve such unity itself became one more basis for the institutionalization of the movement. Even when the formal organization of the Ten collapsed at the end of the decade, due to misgivings about possible negative associations for such groups as Friends of the Earth and the Sierra Club, the name "Group of Ten" continued to be used by its critics as *the* symbol of mainstream environmentalism. Strengthened initially by the Reaganite confrontation, forced to locate a common environmental identity amid potentially differing positions and constituencies, the Ten, in existence for less than a decade, had effectively redefined mainstream environmentalism less as a movement and more directly as an adjunct to the policy process.

THE ENVIRONMENTAL POLICY SYSTEM

The cornerstone for contemporary environmental policies, the National Environmental Policy Act (NEPA), was enacted, rather ironically, with little legislative debate and with little input from any of the mainstream groups who eventually constituted the Group of Ten. The brainchild of Washington senator Henry Jackson and his chief advisor on this legislation, Indiana University professor Lynton K. Caldwell, NEPA was, as Caldwell put it, "a political anomaly." A policy statement not rooted in any specific legislative context, NEPA, which was signed into law on January 1, 1970, referred only in a general way to environmental and quality-of-life concerns. Both Jackson and Caldwell were interested in removing environmental policy from the kind of political deal making exemplified by the Iron Triangle system in water projects. They hoped instead that environmental issues, through NEPA, would be debated and defined as technical issues, involving the application of science in environmental decision making in order to solve problems.[17]

For Jackson and other proconservationist office holders, the political basis for NEPA was the increase in leisure-time consumptive uses of Nature. These referred not simply to hunting and fishing interests of groups such as the National Wildlife Federation and the Izaak Walton League but also the growing middle-class interest in recreational opportunities. Beyond the interest in these environmental amenities were the quality-of-life concerns about pollution, whether in the form of smog, landfills, or drinking water contamination. Though important legislative and administrative action had occurred during the 1960s on issues such as air quality, water quality, and solid waste disposal, it was not until Earth Day and after the passage of

NEPA that a more fully developed environmental legislative agenda began to be introduced. These initiatives were influenced less by conservationist groups than by new consumer and public interest movements spearheaded by Ralph Nader and by budding staff-based groups of lawyers, scientists, and other environmental experts. Perhaps most significantly, the new legislative debate emerged also as a response to the relatively inchoate, yet large and disturbing, constituency of young activists and counterculturists who remained an undercurrent during the Earth Day events and who symbolized a general dissatisfaction with the established urban and industrial order that had worried policy makers during the eventful days of the late 1960s.[18]

For policy makers, those worries had enveloped the political debate in Washington by 1970. Paralleling the sense of system breakdown associated with the Vietnam War, inner-city riots, and turbulence on college campuses, the mood of environmental crisis seemed more and more overwhelming. This sense of crisis was already reflected in such disparate events of the late 1960s as the burning of the Cuyahoga River in the center of Cleveland, the eutrophication of Lake Erie, and the dying birds washed up on the oil-slicked shores of Santa Barbara. These were also events that referred both to the degradation of the natural environment and urban and industrial hazards. And while there was at first no organized lobby pressing for coordinated government action or intervention by the courts, the force of events combined with the unsettling times began to lay the groundwork for new forms of environmental policy.

In just four years, between 1970 and 1974, an extraordinary range of legislative initiatives, regulatory activities, and court action came to the fore. These established a broad and expansive environmental policy system centered around efforts to control the environmental by-products of the urban and industrial order. Through this system, a vast pollution control, or environmental protection, industry was created, including engineering companies, law firms, waste management operations, and consulting firms specializing in environmental review, standard setting, or other new environmental procedures. According to one industry estimate, the waste management business alone was transformed from a group of small businesses with revenues under $1 billion in 1970 into a huge, increasingly concentrated industry with more than $25 billion in revenues two decades later. Similar to the rise of the huge environment-related government bureaucracies, the fast-growing environmental industry of the 1970s and 1980s became an offshoot of and influential player in the development of environmental policies.[19]

As legislative and administrative environmental policies were put in place in the early 1970s, a revised and restructured mainstream environmental movement also began to take shape. While less instrumental in drafting early legislation, the mainstream groups became heavily involved in implementation, monitoring agencies and preparing to develop new initiatives in such areas as solid waste disposal, toxic substances, and cleanup of hazardous waste sites. Increasingly defined by the elaboration of these policies, the mainstream groups came to focus on four arenas: legislation, administrative and regulatory action, the courts, and the electoral sphere. Adopting an adversarial stance regarding much of the policy process in those first four crucial years, the groups also developed a more complex position, assuming simultaneously both insider and outsider status. As a result, by the end of the 1970s, mainstream environmental groups would become both a part of the environmental policy process and its watchdog.

Aside from NEPA, the first key legislative act helping to establish the contemporary environmental system was the 1970 Clean Air Act. By the late 1960s, air pollution had emerged as a potent political issue in a number of urban communities and industrial plants. Between 1966 and 1970, according to one poll, the number of people indicating "very serious" or "somewhat serious" concern over air pollution jumped from 28 percent to 69 percent. In response to these escalating concerns, Congress found itself increasingly embroiled in the issue. Air pollution legislation enacted in 1963 and again in 1967 only increased rather than decreased the pressure to eliminate or otherwise deal with the problem.[20]

This momentum for action was also spurred by the emergence of several grassroots or citizen-action organizations, such as the Chicago-based Campaign Against Pollution (CAP) and the Pittsburgh-based Group Against Smog and Pollution (GASP). Founded in September 1969 and led by an asthmatic housewife who became a creative and effective organizer, GASP immediately became a presence in the debate concerning air quality in the Pittsburgh area. There, air-related policies had long been dominated by a coalition of industry interests led by the steel industry. Allied with local steelworker activists, GASP lobbied for changes in state and regional regulations, awarded "Dirtie Gertie" certificates to polluting industries, sold cans of clean air to dramatize the state of local air quality, operated its own complaint department that forwarded citizens' complaints to regulating agencies, and engaged in a wide variety of other mobilizing tactics to pressure for change at legislative and administrative levels. To both industry opponents and public officials, groups such as GASP reflected what one industry representative called "an activist period, what with the racial dimension, colleges, and now the environment." But to policy makers in Washington, it was precisely the concerns about activism that pushed along the Clean Air debate.[21]

This debate was also influenced by the emergence of a new activist/lobbying presence in Washington tied to citizen-action advocate Ralph Nader. Nader's teams of graduate students with law, science, engineering, and public health backgrounds became immediate players in the legislative arena regarding pollution issues. These "Nader's Raiders," who identified a public interest rather than a conservationist perspective, combined intensive research efforts with direct advocacy work. In the air quality area, the Nader team published a key report, *Vanishing Air*, and helped form the Coalition for Clean Air, the major environmental lobbying effort. The coalition eventually included a few of the conservationist groups (who largely played a passive role during the Clean Air Act debates), some of the new expertise-oriented environmental groups (most notably the Natural Resources Defense Council), and Earth Day activists from Environmental Action (the group that inherited the structure originally established by Wisconsin senator Gaylord Nelson).[22]

These debates were framed in the context of national politics, especially the jockeying for the 1972 presidential election. In his 1970 State of the Union message and related speeches during the next several months, President Nixon declared that air pollution legislation, along with clean water initiatives, would be the cornerstone of his new environmentalist stance. During this intense 1969–1970 period, it appeared that the president's most likely challenger would be Senator Edmund Muskie, the chair of the Senate Committee on Air and Water Pollution and key to the passage of the 1967 Air Quality Act. Muskie, however, was a target of some of the new activists, particularly the Nader groups, for his unwillingness to challenge industry's position in the air quality debates and for his tendency to characterize the environmental issue as a problem of consumption rather than urban and industrial pollution.[23]

Despite their small numbers and the powerful array of industry interests mobilized around air pollution legislation, the new environmental coalition succeeded in getting the new Clean Air Act to extend the federal presence in the air quality area and establish more direct procedures for regulating air emissions at the national level. These procedures included standards for six specific criteria pollutants and a federal government role in the development of pollution control technologies as part of the act's cleanup emphasis. And, like other legislation that followed it, the new clean air legislation included target dates for such cleanup, raising expectations that pollution problems could be solved through back-end engineering controls.[24]

The Clean Air Act was signed into law in December 1970, a little more than seven months after Earth Day. It was seen as a pivotal event, defining a new environmental presence in Washington and providing the first block in the construction of a new policy framework. Subsequent legislation incorporated several key elements of the Clean Air Act approach: standard-setting mechanisms; an increased federal role in the development of control technologies; and ambitious target dates for eliminating or greatly reducing emissions or discharges, particularly more visible or treatable forms of pollution such as sulfur dioxide emissions or biological oxygen demand in water sources. These laws, including the 1972 Water Pollution Control Act Amendments (better known as the Clean Water Act), the 1970 Resource Recovery Act, and the 1972 Federal Insecticide, Fungicide, and Rodenticide Act Amendments (FIFRA), also helped to further develop the new forms of environmental coalition and organization with their focus on regulation, litigation, and implementation.

During the 1960s, most federal environmental policies had been handled through conservationist or resource management agencies, such as the Bureau of Reclamation, the U.S. Forest Service, or the Army Corps of Engineers. Efforts to consolidate those bureaucracies into a single environmental agency or even a single resource-based agency (e.g., by combining the Bureau of Reclamation and the Army Corps) were strongly resisted by both industry and conservationist or protectionist interest groups associated with those bureaucracies. The federal government's presence in the environmental area, whether in terms of pollution issues or resource policies, tended to be uncoordinated and weak due to uncertain mandates and ineffective regulatory powers for agencies that were also needing to respond to the mounting public concerns over the environment.

With the passage of NEPA and growing pressure to pass more expansive environmental legislation, the question of implementation, particularly the role of the Council for Environmental Quality (CEQ), mandated by NEPA, began to preoccupy the Nixon administration. Concerned about the CEQ's potential to "advocate an environmental position, and probably an extreme one," Nixon sought to counter its potential role. He hoped the answer lay in the proposal by his White House Commission on Executive Reorganization, which urged, shortly after Earth Day, that a single independent agency, separate from the CEQ, be established. Nixon endorsed the proposal and sent a reorganization plan to Congress to create a new and instantly massive bureaucracy that would essentially consolidate the functions of all existing entities dealing with pollution-oriented environmental issues. Borrowing from the language of systems analysis in his charge to this new U.S. Environmental Protection Agency, Nixon argued that it would need to treat "air pollution, water pollution and solid wastes as different forms of a single problem." The creation of the EPA also meant that pollution-based environmental policies would now have their first, fully developed administrative form.[25]

The EPA began operations on December 2, 1970. Its first administrator, William Ruckelshaus, initially sought to convey the impression that his agency would aggressively enforce the new policies, while partially adopting the systems approach by forming two primary program offices to handle the variety of issue areas and legislative mandates under its jurisdiction and several function-oriented divisions designed to be more responsive to White House concerns as well as fulfill certain agencywide objectives, such as enforcement and research. The new agency, however, was quickly overwhelmed by its rapidly expanding regulatory responsibilities, the conflicting signals from the Nixon and later Ford administrations on how aggressively it should pursue such regulations, and effective industry maneuvering, which used scientific uncertainty in the regulatory process to delay or counter the establishment and enforcement of standards.[26]

Faced with potential regulatory gridlock, a number of mainstream environmental groups, increasingly focused on both the implementation of the new environmental legislation and the environmental impact review process established under NEPA, turned to the courts to force the system to function as it was presumably designed to do. The focus on litigation highlighted the significant role that lawyers began to play in the movement. It also reinforced the adversarial posture several mainstream groups had assumed during the early 1970s.

During this period of intense lobbying and litigation, mainstream environmentalism came to be defined in terms of two sets of groups, each with a focus on federal government and congressionally based activities. One set included the new organizations, such as the NRDC and EDF, which were fully engaged in the policy process; the other involved traditional conservationist or protectionist organizations, such as the Sierra Club and the National Audubon Society, which were transforming themselves as well as being transformed by this same environmental policy process. Despite a convergence in focus, the groups remained distinct and often competitive and territorial in the issues they selected. But by the mid-1970s, the two kinds of groups had begun to resemble each other in how they structured their operations, set their agendas, and defined their constituencies and objectives. This mainstream environmentalism evolved rapidly, still primarily interested in issues concerning the natural environment and resource policy, but also, perhaps inevitably, addressing questions of pollution. With the EPA in charge of implementation of many of these policies, the mainstream groups found themselves focusing more and more on the new agency—and its pollution control–driven mandates—as central to their concerns and agendas.

By the mid- to late 1970s, the election of several new governors broadly sympathetic to environmentalist goals in places such as California, Colorado, Arizona, and Montana, as well as the 1976 victory of Jimmy Carter, who adopted a number of environmental themes during his campaign for the presidency, made the momentum to extend environmental policies throughout federal, state, and regional bureaucracies seem unstoppable. With Carter in the White House and environmental politics apparently on the ascendant, the mainstream groups found themselves intricately involved in the policy process as it continued to evolve and expand. These groups inherited the label "environmentalist," a term employed by the press to designate both the largest of the Washington-based mainstream groups (often excluding other locally based or direct-action movements and groups) as well as the environmental bureaucracies with which the groups were associated. Growing in terms of staff and financial resources, the mainstream organizations were increasingly absorbed by the operation and maintenance of the policy system itself. A revolving door between staff positions in the mainstream groups and government and industry positions cemented those connections, while the groups' advocacy role, focused especially in terms of crucial lobbying and litigation functions, became more and more centered on keeping the system intact.

During the late 1970s, this system maintenance function of environmentalism, tied especially to the prominent role of revolving-door "envirocrats," fully absorbed the mainstream groups. Key environmental leaders, such as Joseph Browder of the Environmental Policy Center; Katherine Fletcher, Harris Sherman, and Gerald Meral of the Environmental Defense Fund; and Gus Speth and John Bryson, cofounders of the Natural Resources Defense Council, secured positions in the Carter administration and at state resource agencies. They not only helped formulate policies, but at times found themselves in conflict with their former colleagues over specific issues. Browder, who headed a task force that selected sites for some controversial power plants in the Southwest, was a particular target of criticism, especially by local groups confronting those plans. Browder dismissed this more "provincial" perspective, arguing that while he would "rather be out watching birds than helping the utilities site power plants," the nature of his position required him to help the utilities find sites for their coal-fired plants. The revolving-door phenomenon also paralleled efforts by mainstream environmentalists to find common ground with industry antagonists concerning such key issues as strip-mining reclamation. Attempts at dialogue such as the National Coal Policy Project were pursued by those mainstream leaders, such as the Sierra Club's Michael McCloskey, who saw environmentalism as a cooperative effort of industry, government, and mainstream groups in identifying and making acceptable specific environmental policies to each of the parties involved.[27]

Still, several mainstream groups continued their more adversarial advocacy work, particularly in the courts, in response to the slow pace and effective industry countermaneuvers regarding legislative and regulatory activities. Implementation of the Clean Air and Clean Water Acts and, subsequently, the 1976 Resource Conservation and Recovery Act (RCRA) and Toxic Substances Control Act (TSCA) particularly preoccupied the newer, staff-based, and expertise-oriented groups such as the NRDC and EDF, which had carved out significant roles concerning these elaborate laws and regulations. Their court-related interventions, furthermore, underlined fears by both industry groups and the understaffed and overwhelmed agencies such as the EPA and OSHA that environmentalism was perhaps a more implacable, system-challenging movement.

But the advocacy work, despite the spirited and creative efforts of certain staff members within those groups, never became linked to the possibility of creating a new kind of social movement. The staff of the mainstream groups continued to limit their activities to their expertise function, whether drafting legislation or arguing in court about implementation. There was little effort to mobilize at the local level, despite a proliferation of local environmental groups throughout the 1970s. Several of these local groups at times cooperated and eagerly sought the expert advice of Washington-based mainstream groups concerning local hazards and environmentally destructive projects. But the process of interaction was largely one-way and failed to counter the growing impression that the mainstream groups were becoming indistinguishable from the environmental policies they had helped create and sustain.[28]

This perception was borne out by the political problems experienced by the Carter administration. Shortly after his inauguration, Carter announced new approaches for both water and energy policy. Through these initiatives, Carter hoped to demonstrate a kind of environmental populism, an outsider approach in keeping with some of his campaign themes. His attempt to scale down or eliminate certain federally subsidized water projects, dubbed the "hit list" by the press, was originally put together by several mainstream environmental group staff members, some of whom were on the Carter transition team and would subsequently be appointed to mid-level policy positions in the new administration. Similarly, in the energy area, Carter worked closely with environmental advisors to fashion environmentally oriented conservation and efficiency themes. But both Carter and the environmental groups failed to mobilize constituencies at the local level in advance of these new policies and allowed industry groups to assume a local, grassroots stance as part of their prodevelopment countermobilization. These debates further strengthened the new coalition of business groups that had already begun to systematically attack environmental regulations established by the EPA, OSHA, and a number of other agencies. Led by the Business Roundtable, these business coalitions effectively emphasized the costs of regulation and asserted that a kind of moral as well as economic decline had set in with the enactment and implementation of current environmental policies. Fearful of potential structural challenges to industry implicit in such legislation as TSCA, the industry lobbies, through their antiregulatory counterrevolution, put the mainstream groups and environmental bureaucracies increasingly on the defensive.

Politically unprepared to fight back, the Carter administration and its small group of envirocrats began a full-scale retreat regarding questions of efficiency, regulation, and resource development. Carter quickly backed down on the water-project hit list and proposed an Energy Mobilization Board to fast-track certain energy resource projects and an $88 billion Energy Security Corporation to underwrite the rapid development of a synfuels industry. With its extraordinarily hazardous processes and byproducts, synfuels development especially had the potential to transform parts of the West into an environmental sacrifice area.

Although the mainstream groups were concerned about Carter's shifts in energy and water policy, they remained loyal to his administration and endorsed him for reelection in the face of Ronald Reagan's nomination. Reagan, with his embrace of the western ranch and mining "sagebrush rebellion" designed to replace the vast federal ownership of land in the West, his rhetorical belligerence toward environmentalism and government regulation of industry, and his stated intention to deregulate in order to dismantle that system, represented the worst fears of the mainstream groups. But, like Carter, the mainstream groups were vulnerable to charges that their environmentalism represented a "special interest," an elite constituency whose call for individual sacrifice translated into lifestyle disruptions that forced people "to be hotter in the summer and colder in the winter," as Reagan put it. With Reagan's election in November 1980, the era of the first Environmental Decade came to a close, with environmentalism having been successfully linked to an overall decline in America's international standing and the daily life of its citizens. During that decade, an environmental policy system had been put in place and a mainstream movement had taken root. But the question now emerged: Would they both survive?

The Role of Expertise: New Organizational Forms

In the course of the debates about the 1970 Clean Air Act, participants in the environmental coalition that had formed to press for the bill's passage were immediately struck by the effective use of expertise by industry groups to amend and otherwise limit the legislation. Even more than earlier legislation, such as the Wilderness Act, which had required knowledge of the science involved in justifying certain provisions of the legislation, the pollution-centered bills debated during the late 1960s and early 1970s were defined at the outset in scientific and technical terms. Gaylord Nelson recalled that many of the traditional conservationist or protectionist groups felt removed from pollution issues and the maze of technical arguments that emerged during debates about those issues. "It wasn't just a matter of dealing with parks or wilderness or the aesthetics of the matter," Nelson remarked, "but dealing with complex subjects that required a different orientation, a level of sophistication and expertise that the groups had just not acquired."[29]

While pressure was developing at the legislative level to establish a new environmental expertise, advocates of quality-of-life issues associated with the New Left or counterculture remained deeply suspicious of the role of such expertise and the presumption that technical solutions were the best way to address such problems. Even the Nader groups, who prided themselves on careful, interdisciplinary research, shared much of the New Left critique of technical information, warning that "by unnecessarily technicizing and complicating the process, the law often creates insurmountable obstacles to public participation." At the same time, the Naderites warned, the understaffed, undertrained, and underfunded pollution control agencies "suffer as much from the complexity of the legislative scheme as do citizens." All too often "the putative expertise of industry prevails."[30]

Much of the lobbying and congressional debate about the Clean Air Act, such as the question of statutory auto emission standards, centered as much on technical details as on the larger political issues raised by the legislation. Confronting Rachel Carson's crucial axiom about industry's ability to purchase expertise, environmental lobbyists found themselves overwhelmed and outgunned by the technical experts mobilized by industry. While the final bill was shaped by the new focus on the environment, many of its details were modified during these technical debates. That use of technical expertise would become significantly magnified in the implementation process, where industry, through its "putative expertise," was able to heavily influence the agencies involved.[31]

The issue of where and how to direct organizational resources, particularly concerning this question of expertise, quickly preoccupied the new environmental groups emerging in this period. While there clearly remained, in the aftermath of Earth Day, the potential for further mobilization, most groups felt an even more pressing need to professionalize to better respond to and shape the new legislative and administrative structures rapidly being put in place. For a group such as Environmental Action, what direction to take remained a difficult question, never satisfactorily answered.

Following Earth Day, Environmental Action found itself a major player in a quickly evolving movement. It had benefited from the enormous publicity surrounding the Earth Day events, successfully marketed a couple of books that increased its prominence, and immediately attracted thousands of supporters and small contributors, more than for many of the traditional conservationist groups. By trying to maintain the political balancing act it had pursued for Earth Day, Environmental Action staff sought to combine advocacy and lobbying. As distinct from the traditional groups, the organization focused on urban issues, or what one staff member called the "ecology of the cities," including clean air, transportation and the role of the highway lobby, and social justice themes. Some of the staff were influenced by the 1960s movements and theorists such as Murray Bookchin and attempted to establish a collective form of organization, while seeking to encourage various kinds of direct action, including acts of what came to be called "eco-tage." Most of the Environmental Action staff, however, including Denis Hayes, also wanted to influence the policy process in Washington and the explosion of new environmental legislation under consideration on Capitol Hill. Like other groups wishing to cultivate a new environmental expertise, Environmental Action had, in fact, sought funding from the Ford Foundation to help develop its professional and lobbying capabilities. But after months of negotiation and the establishment of a separate Environmental Action Foundation to secure such funding, the group's proposal was ultimately turned down by the foundation. The loss of the Ford grant also meant the organization would not be able to translate its interest in policy development into a more high-powered, expertise-driven framework.[32]

While the organization had only limited resources for its lobbying, policy, and expertise-related work, it also discovered by the early 1970s that the climate for continuing mobilization and direct action was also changing. Though environmentalism as an idea continued to receive favorable attention from policy makers and the press, demonstrations and other forms of mobilization or direct action were becoming more difficult to pull off. At the same time, Environmental Action, as well as several of the other new environmental groups spawned by Earth Day, witnessed a decline in support, with the number of Environmental Action supporters only gradually increasing later in the decade.

With the loss of the Ford grant and declining media interest in the direct-action forms of environmental activity, the group experienced a protracted identity crisis throughout much of the 1970s. Denis Hayes, most focused on the group's policy emphasis, decided to leave the organization for a policy-oriented program set up by Douglas Costle (later President Carter's EPA head) through the Smithsonian. "I decided I needed to have more information to be more of an expert," Hayes later said of his departure. After Hayes left, Environmental Action continued to try to refocus its efforts and establish a viable identity, but without great success. The group was most successful in highlighting its media-based, electoral-targeting strategy known as the "Dirty Dozen," which first received significant attention in the 1972 elections. The campaigns against the "Dirty Dozen," twelve congressional incumbents selected for their poor environmental voting records, also helped reinforce a more adversarial political role for the organization. This independent role was also reflected in the pages of the group's main publication, *Environmental Action*, which served as a forum for a wide range of issues, such as occupational hazard issues, that received significantly less attention from other mainstream groups. But neither the group's publication nor its "Dirty Dozen" electoral campaigns were able to translate that oppositional politics or broader agenda focus into large-scale grassroots mobilization, partly reflecting the organization's decision to continue to function as a staff-based rather than membership-driven or community-based organization. With most of its small staff serving as lobbyists focused on the development of environmental policies, with sympathetic coverage in its biweekly (later monthly and then quarterly) publication of the themes of a developing alternative environmentalism, and with an electoral political strategy heavily dependent on its media visibility, Environmental Action remained a group without a clearly defined role, walking on the divide between the various movements.[33]

Unlike Environmental Action, which achieved instant national status through Earth Day but failed to carve out a clear role for itself, the ad hoc group of Long Island citizens and professionals who created the Environmental Defense Fund in 1967 found their group clearly situated in just a few years time as a major center of environmental expertise and policy action. The EDF began as a local environmental discussion group, the Brookhaven Town Natural Resources Coalition (BTNRC). It consisted of scientists from the Brookhaven National Laboratory and the State University of New York at nearby Stony Brook as well as concerned residents in the area. Similar to a number of other locally based antipollution or antidevelopment groups that sprang up in middle-class residential communities during the 1960s, the Long Island group addressed a wide range of environmental issues. These included pollution from duck farms, dredging, sewage pollution, groundwater protection, dump sites, wildlife and habitat preservation, and the use of DDT. The DDT issue was of particular importance for the group, given the widespread use of the pesticide throughout Long Island's Suffolk County.[34]

The BTNRC leaders were avid bird watchers, Rachel Carson fans, and frustrated DDT opponents unable to counter the county's high-profile spraying campaigns despite growing apprehensions about DDT use. The BTNRC's major break on the DDT issue came in May 1966, when the group joined forces with a local lawyer named Victor Yannacone. Yannacone had brought a class-action suit regarding a large DDT-related fish kill in Yaphank Lake on Long Island. With his energetic, combative, and powerful personality, the Long Island lawyer immediately enlisted the BTNRC group to provide scientific and technical support for his lawsuit. In later years, EDF leaders would see the Yannacone association with the BTNRC group as a new "partnership between scientists and lawyers." But in the period following their initial meeting, the association was most focused on stopping the DDT spraying. When a State Supreme Court judge unexpectedly signed a temporary injunction blocking the spraying, the new allies realized an extraordinary breakthrough had occurred: DDT could be stopped, and litigation, armed with scientific and technical arguments, could be the way to do so. The success of this initial DDT litigation also raised another question: How could the group expand its resources to take on other environmental issues and causes, particularly in the area of pesticide use?[35]

The opportunity to seek new resources presented itself in September 1967, when Yannacone was invited to address the annual convention of the National Audubon Society. The Yannacone/BTNRC group decided to try to get Audubon to establish a new "environmental legal defense fund" to initiate litigation concerning pesticides and other issues. Yannacone, it was hoped, would then become chief counsel for such an entity. With the Audubon Society torn between a conservative and cautious leadership that had played a largely passive role during the Rachel Carson controversy and a membership increasingly receptive to the argument that DDT needed to be banned from use, the Yannacone/BTNRC strategy stood a good chance of succeeding. Resolutions from the floor of the convention, orchestrated by the Yannacone/BTNRC group, called for the establishment of such a legal arm to be financed through Audubon's Rachel Carson Memorial Fund. This entity would initiate litigation to seek a national DDT ban. Though the resolutions were passed, the Audubon leadership, which (like the Sierra Club) included directors with ties to chemical companies, looked unfavorably on the establishment of a new litigation arm. Audubon chairman Gene Setzer successfully blocked implementation by having the matter referred to Audubon's general counsel, Wall Street lawyer Donald C. Hays. When Yannacone and others met with Hays two days after the convention, Hays indicated that any effort to establish a defense fund would not occur through Audubon. Audubon would only provide some financial backing through its Rachel Carson Memorial Fund, although Hays also made it clear that an activist-oriented litigation effort had no place within this traditional protectionist organization.[36]

Outmaneuvered by the Audubon leadership, the Yannacone/BTNRC group felt they had no choice but to establish their own organization, which they called the Environmental Defense Fund. Incorporated a few days after the Hays meeting, the new organization immediately sought local cases for which its budding legal and scientific resources could be utilized. The organization's first action involved support testimony for the Michigan Department of Conservation regarding the use of dieldrin, a chlorinated hydrocarbon far more toxic than DDT, in a Japanese beetle control program in Berrien County. This effort was followed by legal support work for several other local groups, most of whom were involved in pesticide issues.

In its first years, the organization was dominated by Yannacone through his forceful personality, effective use of cross-examination, willingness to go public and mobilize people as part of his legal strategy, and irreverent, adversarial stance that linked him in some ways to the protest culture of that period. Yannacone's trademark phrase "Sue the bastards"—also the title of his talk at the Washington, D.C., Earth Day rally in April 1970—reflected the sense of implacable opposition associated with direct action forms of environmental protest while placing such protest in the context of litigation. Such a stance, EDF leaders later commented, worked well in the pre-NEPA, pre–environmental policy system days of passion and advocacy, when the "crucible of the adversarial process" (i.e., cross-examination tactics) could burn away the "evasions, temporarizations and data tampering of the polluters and of the cozy-with-industry regulatory agencies."[37]

But by 1970 that approach was seen as increasingly problematic by key organization figures, such as the new executive director, Roderick Cameron, and EDF trustees and financial backers, including the Ford Foundation. Though a decision to sever relations with Yannacone derived in part from a personality dispute, the action helped refocus the EDF framework of linking litigation, science, and the policy process. The departure of Yannacone also served to moderate the rhetorical stance of the organization, shifting it away from a "sue the bastards" style of confrontation to a far more temperate tone and organizational image.[38]

By June 1970, several months after the break with Yannacone, the organization was ready to undergo another major transformation. The Ford Foundation, which had provided initial funds for the organization through the Audubon relationship, expressed an interest in providing a sizable grant of $285,000, matching the organization's existing annual budget. Ford, however, imposed a condition: the EDF would become subject to oversight by the "Gurus," a Ford-created committee of five past presidents of the American Bar Association who would review any proposed action by all environmental and public interest groups receiving funding from Ford. The EDF was also obliged to create a Litigation Review Committee comprised of "prominent members of both political parties who would have great political 'clout' and would have final say on any EDF litigation." Though the review committees were designed to partially shield Ford from EDF actions that might be deemed too radical, it was another step to further professionalize the organization, since any litigation now had to pass through a two-stage screening process, forcing the organization to justify each prospective action "legally, scientifically and practically." At the same time, the role of the Gurus and the establishment of the Litigation Review Committee, as well as the expansion of the EDF board to include such figures as Amyas Ames, chairman of the executive committee of the investment banking firm of Kidder, Peabody, further modified the adversarial approach. Ames, whose firm was a major underwriter of public utilities, later commented that his decision to join the board was a reflection of "EDF's very early policy of working in cooperation with business and industry that made this decision easy."[39]

Through the early and mid-1970s, the EDF became a major litigator in such areas as lead toxicity, the fight against the SST, the protection of sperm whales, and pesticide hazards, its preeminent issue since its founding days. The organization rapidly expanded staff resources, opening new offices in Washington, D.C., northern California, and Denver. It enlarged its financial base with support from foundations as well as from individuals who subscribed to the EDF's newsletter and were otherwise designated as members of the organization. Of the different staff-based professional groups, the EDF maintained through this period a strong reputation as an independent, adversarial organization, particularly in such areas as toxics policy, where regulatory activities provided a continuous arena for challenge and litigation. At the same time, the group kept its distance from grassroots activities and movements, continuing to emphasize its professional character and the leading role of its own staff in setting the organization's focus and agenda.[40]

During the late 1970s, the organization began to more systematically pursue the idea that Kidder, Peabody's Amyas Ames had first raised on joining the EDF board: cooperation with industry through the search for what the organization called "win-win" strategies. In the area of utilities policy, this was represented by an economic model put together by the organization's California office. The model sought to demonstrate to the utility industry that a conservation-oriented or demand-based approach, linked to certain incentives, would be more profitable in the long run than the volume-oriented, expansionary policies that had characterized the utility industry approach for more than sixty years.[41] This focus on a potential economic common ground between industry and environmentalists, encouraged by the growing numbers of staff economists within the organization, evolved during the Reagan and Bush years into an overarching EDF strategy to promote market incentives, replacing regulation as a primary tool for reshaping environmental policy. A key influence in the shift toward professionalization in the early 1970s, the EDF would ultimately become, by the end of the second Environmental Decade, the organization most wedded among the mainstream groups to the idea that environmental change was a matter of reinforcing rather than restructuring a market-driven urban and industrial order.

The Ford Foundation's support for the new environmental professionalism was underlined by its role in the emergence of another lawyer- and scientist-based staff organization, the Natural Resources Defense Council. The NRDC was formed as a result of two parallel developments: a career search for a public-interest law practice by several graduating Yale Law School students and the protracted Storm King power plant fight along the Hudson River. The merging of the two under the auspices of Ford created another trajectory for a professional environmental group: nationally focused at the outset, heavily weighted toward the practice of law, and an ad hoc and expertise-oriented approach toward issues.

The Yale student group, from the class of '69, first got together in the fall prior to their graduation. Several of the participants, including Gus Speth, who had been instrumental in bringing the group together, had been editors of the *Yale Law Journal*. All of them seemed destined to have successful careers in the worlds of law, business, or politics. Interested in policy matters, the group wanted to pursue alternatives, such as the budding field of environmental law, to more conventional career choices. They had no clear role models to guide them. The closest examples seemed to be the American Civil Liberties Union or the NAACP Legal Defense Fund, legal advocacy groups that act on behalf of particular causes, such as civil liberties or civil rights.

By the time of their graduation, the Yale group, calling itself the Legal Environmental Assistance Fund, or LEAF, had not yet secured funding to establish its own environmental law practice. As a result, group members became law clerks or took similar jobs in the interim, while continuing to talk with potential supporters, including the Ford Foundation. Ford, which had established a Natural Resources and Environment division a few years earlier, had already begun its own discussions with two prominent Republican attorneys from the Wall Street law firm of Simpson, Thatcher, and Bartlett. Along with a few associates, these lawyers, Stephen Duggan and Whitney North Seymour, Jr., had become interested in establishing a resource-oriented law practice based on their experiences with the Storm King case. This proposed water storage facility and power plant had generated strong local opposition in the Hudson River Valley, and litigation eventually caused the project to be abandoned. In the course of this fight, the Duggan/Seymour group had become familiar with similar local environmental and development-related campaigns across the country, but felt they tended to be "disorganized, one-shot deals," as Duggan later put it. "Instead of this finger-in-the-dike approach to protecting the environment," Duggan said of his group's thinking, "such fledgling environmental efforts" would benefit "from an organization staffed by lawyers and scientists who could provide ongoing professional help."[42]

In January 1970, the Duggan/Seymour group decided to hire John Adams, a lawyer formerly with the U.S. Attorney's office in New York, to become the first staff member for their new law practice. Incorporated a month later as the Natural Resources Defense Council, the new organization immediately approached Ford about start-up funding. The foundation, in turn, suggested that the organization join forces with the Yale group. Ford executives knew such a merger had to overcome generational differences in style, attitude, and organizational focus. The Yale group was suspicious of the orientation of the old-line Republican (and conservationist-oriented) NRDC group, while Duggan and Seymour remained wary about the new and possibly radical ideas of the former Yale students. The Ford executives were friends and associates of the NRDC board members, but were also impressed with the Yale graduates, and wanted to support both groups without duplicating efforts. To obtain foundation backing, Ford insisted that the Yale group work through the existing NRDC structure with its already established board of wealthy and powerful figures, such as Laurance Rockefeller as well as Duggan and Seymour. Duggan himself would be the group's first chairman.

Before the Ford funds could become available and the new group fully launched, a tax problem involving the IRS still had to be resolved. As a result, through the spring and summer of 1970, the organization experienced some tensions between the two groups exacerbated by the uncertain status of the funding. Yet both groups quickly agreed upon a common goal: to establish a new kind of law practice, a "law firm for the environment" capable of responding to the sudden explosion of environmental issues and legislation. In its effort to help mold a new kind of law and policy, the NRDC could help reconstitute a contemporary environmental movement by "bringing professionalism to the environmental scene," as one of the group's top leaders put it.[43]

Once the IRS allowed Ford to fund environmental and public-interest law activities, the NRDC, subject to Ford's review process (which included the five Gurus and an acceptable, corporate-oriented board), quickly became involved in several issues. "Our agenda was eclectic," Adams recalled, "but we felt like we were being dropped right into the middle of a war, since there were no other lawyers doing what we had decided to do." The group immediately involved itself in implementation of the Clean Air Act, became engaged in the coalition lobbying for the Clean Water Act, focused on strip-mining and stream channelization problems, helped bring suit regarding the Alaska lands bill, and tied into agriculture, forestry, and land use issues as well. It became best known for its monitoring and litigation on implementation issues: through Richard Ayres, the NRDC's air expert, the organization filed thirty-five of the first forty Clean Air Act–related lawsuits.[44]

By the mid-1970s, with the NRDC functioning on much of the same terrain as the EDF and the Sierra Club Legal Defense Fund, and with several of the traditional conservationist groups also hiring lawyers as well as scientists and economists, the question of division of labor and increased competition also emerged. The NRDC sought to distinguish itself from the other groups by presenting itself as a law service. "Instead of competing with the big groups like Audubon or The Wilderness Society," Adams recalled, "we were saying that we would be their mouthpiece, and in that way we wouldn't step on toes." An informal division of labor with groups such as the EDF also developed, though competition between the groups, the more directly they resembled each other, generally increased rather than decreased during the 1970s. Each of the professional groups also worried that their need for greater expertise was outpacing their capabilities in influencing the environmental policy process.

The NRDC decided it needed scientists and lobbyists; the EDF, with more scientists, added lawyers and economists. A number of staff members also became crucial participants in the policy process through the environmental revolving door. John Bryson, for example, one of the original Yale Law School group, helped establish a northern California office for the NRDC and soon parlayed his NRDC involvement in California resource issues into high-level appointments with the State Water Resources Control Board and then the Public Utilities Commission during Governor Jerry Brown's administration. Eventually Bryson secured a position at Southern California Edison, becoming its chief executive officer in 1990. In the process, Bryson's career demonstrated how one industry had come to recognize that the professional wing of environmentalism also provided training for high-level industry jobs.[45]

By the 1980s, the NRDC had become a dominant force in the professionalization of the movement. Though its agenda remained eclectic (it included, for example, nuclear arms race issues), it felt firmly implanted within the environmental policy system. At the same time, the NRDC, along with several other mainstream groups, became increasingly focused on global issues and helped promote certain domestic environmental approaches (such as energy efficiency and pollution control) within an international context. Over time, the NRDC became the environmental organization most identified with the technical expertise needed to draft legislation, issue reports, and use litigation as a tool in the policy process. By the end of the second Environmental Decade, it had come to symbolize the ascendancy of professionalism among the mainstream groups. Applauded for its continuing emphasis on expertise while simultaneously expressing a lack of interest in the tactics and strategies of mobilization, the organization had secured for itself a central place in the organizational culture of contemporary mainstream environmentalism.

One of the most forceful figures helping to reshape the environmental movement has been David Brower. A charismatic yet polarizing personality, Brower embodied the uncertainties and turmoil experienced by traditional conservationist and protectionist groups during the late 1960s. As his fight with the Sierra Club reached a climax, Brower made plans in 1969 to establish a new organization more in keeping with the new trends within the movement. He was aided in these efforts by the cult of personality that had built up around him. With his shock of white hair and rugged good looks, his evocative rhetoric, and his increasingly "no compromise" negotiating posture, Brower looked and acted the part of the "archdruid" so strikingly described by essayist John McPhee in his profile of Brower for *The New Yorker*. Aside from establishing greater latitude for his own interests, Brower wanted his new San Francisco–based organization, which he named Friends of the Earth (FOE), to pursue certain issues and strategies that the Sierra Club had not or would not pursue. These included a greater emphasis on international issues (including plans for non-U.S. chapters of the organization), a more direct ideological role through an expanded publishing effort, and a more expansive agenda, including but not limited to traditional wilderness and resource policy themes.

With his departure from the Sierra Club and incorporation of Friends of the Earth, Brower quickly established a board willing to back his ventures and selected a staff interested in more expertise-oriented approaches toward the new environmental issues. FOE's key staff figure was Joseph Browder, a former southern field staff representative of the Audubon Society who had been heavily involved in the fight to protect the Everglades. Browder and Brower, though, immediately clashed, with Browder emerging as a leader of the staff faction of FOE wishing to develop an interdisciplinary or integrative expertise concerning environmental policy. Brower, on the other hand, saw the staff, particularly those in Washington, as overly focused on the legislative, administrative, and political dimensions of the "beltway" process, the narrow locus of policy making centered in Washington. The bitterness of the dispute was magnified by the conflict over Brower's management style and personality, a key area of confrontation in the Sierra Club situation also.[46]

In 1972, most of the East Coast staff departed from FOE and created a new D.C.-based organization, the Environmental Policy Center. The EPC was structured as a quasi-lobbying, quasi-research and advocacy group of policy experts-in-the-making, complementing the parallel shift toward professionalization based on the use of law and science occurring within other parts of the movement. Though the EPC maintained a strong lobbying presence in Washington, it was one of the few groups to successfully establish ties with dozens of local organizations that had formed to tackle specific issues ranging from strip mining to water development.

Under first Browder and later Louise Dunlap's leadership, the EPC was able to define itself in professional terms with respect to both its lobbying and policy development functions. The organization became adept at working congressional committees, monitoring agencies such as the Army Corps of Engineers and Bureau of Reclamation and allying with a wide variety of interests to accomplish certain specific objectives. With respect to water policy, for example, the EPC (which changed its name to the Environmental Policy Institute) utilized cost-benefit techniques to enlist more conservative and less environmentally oriented opposition to water projects such as the Bureau of Reclamation's Central Utah Project and the Army Corps' Tennessee-Tombigbee project. During the 1970s, this approach helped delay authorizations of existing projects and limit new projects. Although successful at the policy level, the EPI and other environmental groups that had helped put together Carter's "hit list" were still unprepared to deal with the countermobilization organized by water industry interests.

The EPI achieved its most notable water policy successes during the Reagan administration years. Despite the harsh, anti-environmental rhetoric within the EPA and the Department of Interior, EPI staff discovered that administration officials in other agencies, especially the Office of Management and Budget, were willing to work with environmental groups to block or slow down specific water projects that they opposed for fiscal or ideological (reducing government's role) reasons. Ironically, the EPI's water policy successes eventually caused some of the organization's major foundation backers to eliminate funding in this area on the assumption that the issue was no longer critical, even though overall federal water policy remained uncertain and contentious. The dependence on foundations, which reinforced the tendency to address policy questions on an issue-by-issue basis rather than in more strategic ways or as part of a larger environmental vision, underlined the crisis management atmosphere of the beltway process and the ad hoc nature of the staff work within the EPI and other policy-oriented mainstream groups. Within those limits, the EPI had succeeded, by the end of the 1980s, in consolidating the group's role as lobbyists and policy analysts in the reconstitution of the contemporary mainstream movement.[47]

Shortly after its formation, but prior to the FOE/EPC split, Friends of the Earth played a role in encouraging the development of an environmental electoral arm for the mainstream movement. First conceived by Marion Edey, a young congressional staff aide, the idea was brought to the attention of Lloyd Tuppling, a long-time Sierra Club lobbyist who had stayed aloof from the internal fight that had recently paralyzed his organization. Tuppling suggested that Edey speak to David Brower, who was then forming Friends of the Earth. Edey, Brower, and several Brower allies subsequently met and decided that Edey should establish a separate group to be initially housed at FOE's Washington office.[48]

The new organization was named the League of Conservation Voters (LCV). It began in 1970 with few resources but only a small overhead. Supporting herself through her own inherited wealth, Edey became the LCV's first, unpaid staff person. She quickly began to raise funds, establish a board largely drawn from FOE, and search for candidates to support or incumbents to oppose. The group decided it could most effectively challenge potentially vulnerable anti-environmental Democratic Party incumbents in the primaries. Toward that end, its first major effort was an attempt to unseat George Fallon, the powerful head of the House Public Works Committee, who was also known as Mr. Highway Trust Fund for his support of highway construction and opposition to mass transit. Fallon was challenged in the primary by Paul Sarbanes, a young Baltimore attorney whom the LCV aided through its fundraising and volunteer support. The timing of the election proved to be fortuitous. A major air pollution alert in the Baltimore area shortly before the election, along with Earth Day events and debates on the Clean Air Act, served to highlight this and other pollution-related issues that were central to the LCV strategy. Fallon indeed proved to be vulnerable and lost the election. Two years later, an even more dramatic primary upset occurred in Colorado with the defeat of Wayne Aspinall, the powerful, prodevelopment chairman of the House Interior and Insular Affairs Committee, by a former regional director of the EPA, Alan Merson. In this race, the LCV raised more than $20,000, or about half of Merson's funds, and again provided volunteers and staff time. The 1972 Aspinall defeat, two years after George Fallon's loss, astounded many of the existing environmental groups and put the LCV, along with Environmental Action and its "Dirty Dozen" campaigns, at the center of the new environmental focus on the electoral process.[49]

During the early 1970s, when the LCV was establishing its role within the mainstream movement, campaigns were far less expensive to run than later in the decade, when limits on individual donations paved the way for corporate PACs and enormously expensive elections. As its ability to influence the outcome of specific elections diminished, the LCV focused efforts on its congressional ranking system, charting the votes of members of Congress on environmental legislation. The group saw itself as representing the interests of the entire mainstream movement by institutionalizing the movement's political and lobbying activities. One frequently employed tactic involved the purchase of a block of tickets to a fundraiser. Tickets were then distributed to various mainstream environmental lobbyists to allow them to be visible at the event, reinforcing their lobbyist profile.[50]

By the 1980s, the LCV's political capabilities had largely become dependent on its voting charts, reinforcing the perception that the group—and the mainstream environmental movement—were involved in a kind of interest group politics tied to the maintenance of the environmental policy system. Though the LCV never became an organization of professional lobbyists—many of the mainstream groups developed their own PACs and hired their own lobbyists instead—it contributed to the mainstream movement's conception of itself as a "career oriented and not cause oriented" group of staff professionals, as the LCV's founder Marion Edey put it.[51]

Ironically, the most conflicted of the new policy-oriented groups spun off or departed from FOE was Friends of the Earth itself. From its beginnings, the group was torn between David Brower's vision, evolving politics, and management style, and a reliance on a staff form of organization with its tendencies toward professionalization. During the 1970s, Friends of the Earth sought to explore new themes, new coalitions, and new environmental issues, such as the SST, nuclear energy, and urban development, even as it maintained a heavy emphasis on wilderness, population, and environmental protection concerns. Its internal organ *Not Man Apart*, its agenda documents, and its various other publications, such as *Progress As If Survival Really Mattered*, reflected this diversity of approach. FOE's efforts at establishing an international framework for its activities through its semiautonomous FOE chapters in other countries also contrasted with nearly all the mainstream groups, which maintained an exclusive domestic frame of reference in this period.[52]

Despite its interest in an active membership and local activities, FOE never fully resolved the tension between its organizational emphasis and its activist inclinations. During the 1970s and early 1980s, a number of FOE organizers saw themselves as the radicals of the mainstream movement, pushing the limits of the environmental agenda and seeking to broaden the movement's appeal. But FOE's resources were primarily available for its lobbying and policy-related professional functions, which clashed at times with the group's search for a more radical approach. This problem was significantly exacerbated by yet another conflict between some of the professional staff and the group's founder, David Brower, who had already been removed from the organization's daily operations during the late 1970s. By 1984, a bitter internal battle ultimately led to Brower's departure from the organization. While the FOE founder, anticipating his defeat, established Earth Island Institute, a more radical, think tank–like organization, the post-Brower FOE shifted its operations to Washington, D.C., further underlining its policy, rather than activist, orientation.

With a declining membership and an inability to compete in terms of professional resources and policy influence, Friends of the Earth finally decided to merge in 1989 with the Environmental Policy Institute. The merger integrated FOE's international presence, membership base, and "willingness to take risks" with "EPI's expertise in politics and key environmental issues," as the organization's new CEO, former EPI head Michael Clark, put it. While Clark suggested that by reaching out to new constituencies, such as minorities, FOE/EPI could conceivably expand its agenda to deal with "poverty, housing, income, food and shelter" as "legitimate environmental concerns," the merged organization nevertheless remained bound by its overwhelming Washington, D.C., beltway-process orientation, which precluded such a change. The Brower-inspired experiment in radical environmentalism, never fully escaping the professional impulse of the mainstream movement, had come full circle.[53]

The Restructuring of the Traditional Groups

With the emergence and growing prominence of expertise-oriented, staff-based professional groups, traditional conservationist and protectionist groups also experienced their own organizational metamorphoses. Long-standing debates about protection versus management and recreation values versus development values came to be modified by, and eventually gave way to, a new focus on environmental regulation and pollution control. Constrained by their own history and constituency ties, traditional groups nevertheless adapted to the conditions imposed by the development of an environmental policy system, embracing it as their own once its construction was well under way.

The key group in this shift was the post–David Brower Sierra Club. In the aftermath of the Brower conflict, the organization, with a new executive director and a board of directors wary of the sharp rhetoric and heightened visibility of the Brower era, worried about the political polarization in the period. Some feared the influence of the New Left and counterculture would push the organization toward addressing issues such as the Vietnam War. In a February 1970 report to the organization's board of directors, the club's president, Philip Berry, warned that the "growing popularity of our cause has attracted some whose motives must be questioned," including "anarchists voicing legitimate concerns about the environment for the ulterior purpose of attacking democratic institutions."[54] At the same time, with the election of Richard Nixon in 1968, the organization anticipated a retreat from conservationist themes and was unprepared for the rapid increase in interest and membership following Earth Day and the explosion of media coverage about pollution issues. The club, which maintained a modest-size staff at its San Francisco headquarters and a handful of staff at regional offices across the country, also failed to anticipate the rapid construction of new pollution-oriented environmental policies that would also force the traditional groups to reorient their activities.

The first major indication of change was the creation in 1971, through a Ford Foundation grant, of the Sierra Club Legal Defense Fund as a separate legal arm of the organization. Immediately plunging into both NEPA-related litigation and the implementation of new environmental legislation, the Legal Defense Fund, which both complemented and occasionally competed as an environmental legal advocacy group with the EDF and NRDC, had a major impact on the Sierra Club itself. It caused the club to rethink its staff functions, especially as Legal Defense Fund staff lawyers began to assert that the "real battle may not be in court, but long before, when EPA or whatever agency is involved arrives at its administrative position." This, in turn, required an ability to argue cases in "conference rooms rather than courtrooms," as the Legal Defense Fund's executive director concluded.[55]

While the Washington-based Legal Defense Fund, operating largely independent of the San Francisco–based club, helped stimulate a greater club emphasis on both litigation and lobbying, the organization still remained reactive and often wary of the developments reshaping the mainstream groups during the early 1970s. In the energy area, for example, the Sierra Club was influenced by the pronuclear position of its president, Laurence Moss, himself a nuclear engineer. At first, the organization took a cautious position on nuclear power, despite its rapid emergence as a pivotal local issue in areas where plants were being sited. The group's leadership also proceeded cautiously with respect to the question of coal-fired plants and related mining questions, seeking to separate landscape and natural environment questions from the pollution problems associated with the use of coal. During the same period that the Legal Defense Fund undertook litigation to stop the pollution caused by the coal-fired plants in the Four Corners area in the Southwest, the club became heavily involved in the National Coal Policy Project (NCPP). The NCPP, a group of industry executives and environmental staff initiated by Moss and Dow Chemical's corporate energy manager, Jerry Decker, sought to develop consensus on acceptable forms of mining and coal energy development. When Sierra Club executive director Michael McCloskey became cochair of the project's mining subcomittee during 1975 and 1976, he was forcefully attacked by several club activists, including the former chair of its energy committee. But McCloskey saw coal as a "bridge fuel" and worried that the mainstream movement's overall tendency toward adversarial actions made it difficult for organizations such as the Sierra Club to pursue "acceptable accommodations" as represented by the NCPP.[56]

Unlike expertise-oriented organizations such as the EDF and NRDC, which operated without any direct connection to—or constraints from—specific constituencies, the Sierra Club, with its legacy of chapters and membership and a post–Brower era reluctance to vest too much authority in staff leadership, at first resisted the powerful professionalizing tendencies within mainstream environmentalism. During the 1970s, disputes among and between chapters and the staff and the board occurred periodically. With senior staff creating their own cliques in alliance with separate board or volunteer factions, the club increasingly resembled a fractious organization without a clear definition or sense of purpose. This infighting weakened the shift toward centralization and professionalization. More than the other mainstream leaders, executive director McCloskey's participation in the Group of Ten process was influenced by these organizational conflicts, exacerbated by a mistrustful board and staff unwilling to grant him functioning CEO status.[57]

Nevertheless, the Sierra Club continued to grow and expand its reach through the 1970s and 1980s. It benefited enormously from the explosion of interest in environmental issues and its continuing reputation as a central player in the movement. By the mid-1980s, its membership had jumped to more than 400,000, with an organizational budget of $23 million and a staff of 225.[58] When Michael McCloskey decided to resign in 1985, a divided board sought to locate a new executive director capable of concentrating its professional, technical, and lobbying resources. After a bitter search process, the board selected former Nixon aide Douglas Wheeler. But Wheeler, who later became a top environmental aide to Republican California governor Pete Wilson, almost immediately became a magnet for organizational opposition at the chapter level and among various staff factions. After only sixteen months as executive director, Wheeler was fired on an eight to seven vote by the board, to be replaced five months later by Michael Fischer, a former California environmental official with numerous club ties who was himself hired on an eight to seven vote. Fischer sought to sustain the club's evolving, though still discordant, professionalization while maneuvering through its byzantine structures and organizational culture. Into the 1990s, the club continued to function as a less focused, more diffuse organization, with a relatively weak top leadership, a powerful, politically oriented staff group operating out of the club's national offices in San Francisco and lobbying headquarters in Washington, and a volunteer leadership that directed club activities at the local level.[59] While influenced by the reach of professionalization among the mainstream groups, the Sierra Club had also come to represent the movement's conflicting and diverging tendencies at the local level as well.

The Sierra Club's uncertain transition paralleled the uneasy shift toward contemporary mainstream environmentalism experienced by the oldest of the traditional groups, the National Audubon Society. Comfortably housed in its limestone-and-brick building on the Upper East Side of Manhattan, and later at its five-story, Georgian-style mansion just a few blocks south from its earlier site, Audubon had been able to grow modestly up through the 1960s, thanks to its bird-watching activities and sanctuaries, nature films and education programs, and conservation camps in places such as Maine, Connecticut, Wisconsin, and Wyoming. During the 1950s and 1960s, the organization also began its evolution from a more decentralized organization of state-based societies similar to the Izaak Walton League and National Wildlife Federation to a more centralized organization emphasizing individual membership and national programs and policies.

During the 1960s, under the leadership of Charles Buchheister and the organization's Washington, D.C., representative, Charles Callison, the Audubon Society followed the lead of the Sierra Club and Wilderness Society with involvement in issues such as the Wilderness Act and the fight against a jetport in the Everglades. The group was most absorbed by the pesticide issue, particularly in light of the controversies concerning *Silent Spring* and pressure within the organization to pursue a more forceful role against DDT. The anti-DDT efforts were led by the organization's vice-president, Roland Clement, who had worked closely with the EDF group to change the Audubon approach.

During the late 1960s and early 1970s, Audubon became one of the first of the traditional groups to try to adjust to the restructuring of the mainstream movement. Its new executive director, Elvis Stahr, a former secretary of the army and University of Indiana president, reflected board interest in establishing Audubon as a major player in the environmental policy arena. During Stahr's ten-year tenure from 1968 to 1978, Audubon hired a number of staff scientists, increased its lobbying presence in Washington, initiated direct-mail solicitation on specific environmental topics, and restructured its main publication, *Audubon*, to reflect a broader interest in environmental or ecology themes. It was in the pages of *Audubon*, especially, that membership unease with the transition was most forcefully expressed by those who felt the organization should remain "synonymous with birds and birding" and not environmental advocacy, as one member wrote. At the same time, the group's 1971 move to a midtown high-rise office building from its more reclusive setting on Manhattan's Upper East Side was another indication of the culture shift taking place within the organization.[60]

When Stahr retired in 1978, the Audubon board sought an executive director like Stahr who would have visibility and political influence. Though Audubon had grown more rapidly than other traditional organizations in the years shortly before and immediately following Earth Day, that growth had leveled off by the mid- to late 1970s, and the group's embrace of broader environmental themes had yet to translate into a clear organizational direction. After a lengthy search, the board selected Russell Peterson. They saw in him a credible environmental moderate whose background and interests as director of DuPont's research and development division, as Delaware governor, as chairman of the CEQ, and most recently as director of the Office of Technology Assessment would enable him to further broaden the Audubon agenda.[61]

Peterson immediately became a forceful figure at Audubon. His interest in global and population issues as well as politics conflicted at times with board, staff, and individual members "who considered the Society's mission to be historically that of wildlife protection."[62] During the confrontational years of the first Reagan administration when James Watt became a potent symbol of environmental polarization, Peterson helped initiate and hold together the Group of Ten while becoming the most visible spokesman for the mainstream movement's response to Reagan's approach. "As a Republican, a businessman involved with the chemical industry, a public official in both Republican and Democratic administrations, and an outsider to the movement prior to his appointment at Audubon," the EPI's

Louise Dunlap recalled the thinking among the Group of Ten, "he was the ideal person to get up at our press conferences and blast away at Reagan and Watt. Russell Peterson was our answer to charges that the environmental movement was a liberal cabal."[63]

For Peterson, politics, as much as professionalism, was critical to reshaping environmentalism. He sought to reinvigorate the chapter structure of Audubon as a potential lobbying force to accomplish political goals for the organization and the larger movement. In regional gatherings of the organization, in the pages of Audubon's publications and in its direct mail, and through Peterson's own public appearances, the Audubon CEO became the leading proponent for politicizing the movement.

Those political efforts culminated in the 1984 presidential election, when Peterson and Audubon and a number of other Group of Ten CEOs directly supported or provided resources for Reagan's challenger, Walter Mondale. But the Mondale campaign never effectively developed environmental themes, and the Reagan reelection effort was able to tag Mondale as the candidate of the "special interests," including the environmental groups. By the mid-1980s, with Reagan's sweeping reelection victory, his partial retreat with respect to deregulation, and a leveling off of support for the mainstream environmental organizations, a number of the mainstream groups began to reevaluate their public identities. This included rethinking the degree to which politics should be a primary frame of reference above and beyond the need to support and extend existing environmental policies.

By the time Russell Peterson reluctantly stepped down in 1985 (one year short of his seven-year contract), to be replaced by attorney Peter A. Berle, nearly all the organizations of the Ten had begun to focus on management style and how to better emphasize cooperation and solutions rather than the heightened adversarial stance of the early Reagan years. Each of the organizations, including Audubon, also strongly emphasized that CEOs had a fiduciary or corporate responsibility that measured group success by the bottom line, including staff resources.[64] At Audubon, Peterson's effort to elevate the chapters as a political force was deemphasized in favor of further strengthening the professional and corporate character of the organization.

For the conservative Audubon board, Berle seemed an appropriate choice for this new stage of development in the movement. The son of a well-known economist and a former state assemblyman and director of New York's Department of Environmental Conservation, Berle quickly sought to lower the Audubon profile, reemphasizing the centralized nature of the organization. He closed regional offices, eliminated staff positions, and reduced the share of membership dues available for local chapters. As a result of these moves, an organizational rebellion, led by a retired Charles Callison, challenged Berle's leadership and Berle's primary board supporter, Audubon chairman Donal C. O'Brien, Jr., a lawyer for the Rockefeller family. The clash between chapters and member activists and the professional staff became a critical part of the dispute. The Audubon leadership, as one of Berle's critics put it, worried that chapter "agitators" might elect directors "who wouldn't fit in socially with the rest of the Board, and whose manners could prove embarrassing in the presence of foundation executives and corporate officers."[65]

Though Berle and his backers were able to turn back the challenge from the Callison group, the question of organizational identity still remained prominent. Berle worried that Audubon, despite its large membership and big budget, had an image problem, a "bird watchers" reputation that it had never fully shaken off, even during the politically charged Peterson interlude. To deal with this image issue, the organization hired Landon Associates, an outside consultant. Landon reported that the "bird watchers" reputation still had deep roots, both among the membership and in terms of the perceptions of Audubon within the mainstream movement. Unless Audubon, with its legacy as an organization of birders, was accepted as a contemporary professional environmental group, it could conceivably fall behind in membership growth and access to financial and other organizational resources.

To accomplish that image shift, Berle sought to redefine the organization as "diverse and far flung, [where] there are unlimited opportunities to plug your energies into the Audubon machine." Recruitment, fundraising, and greater managerial skills continued to be given greatest priority. In a period extending from 1987 to Earth Day 1990, when all the mainstream groups were benefiting from a renewed popularity of the environmental issue, Audubon also aggressively sought to expand its budget and membership. By the new decade, the organization, like several other mainstream groups, had consolidated its own professional and corporate identity. For this one-time organization of birders, reconstituting itself also meant escaping its past.[66]

The question of organizational identity became particularly compelling for The Wilderness Society, one of the protectionist groups most bound up with its own history. During the 1960s and 1970s, the organization's publications frequently invoked the memories of its key founders and leaders, such as Benton MacKaye, Robert Marshall, Aldo Leopold, and Olas Murie. The death of Howard Zahniser, the group's long-time executive director, just a few months before final passage of the Wilderness Act in 1964, only reinforced that desire for continuity, based on the group's exclusive focus on wilderness issues. Unlike the Sierra Club and the Audubon Society, whose transitions to new executive directors paralleled their complex and often uneasy shift toward organizational restructuring, The Wilderness Society under Stewart Brandborg, Zahniser's assistant, who took over the position of executive director, maintained its earlier leadership style and wilderness focus.[67]

In the decade that Brandborg ran the organization, Wilderness Society resources became fully absorbed by implementation of the Wilderness Act and the group's other major area of concern, preservation of the Alaskan wilderness. For Brandborg, Alaska and the fate of the Wilderness Act were great causes. But his exhortative leadership style, his approach to "each battle as if it was the last battle," and his tendency to get absorbed in the details of staff work eventually caused him to clash both with staff and board members, who worried about the organizational debts he had incurred.[68] At the same time, Brandborg became personally sympathetic to the activism of the late 1960s and early 1970s, which caused additional clashes with his conservative and more narrowly focused board. Unlike the Sierra Club and the Audubon Society, which both grew faster and sought to expand agendas while professionalizing staff resources, The Wilderness Society was unable to make a clear transition to the new era of mainstream environmentalism, despite quadrupling its membership to 87,000 within a decade. When Brandborg was finally fired in January 1976, the organization seemed least prepared among the traditional groups to reconstitute itself along more professional lines.

During the next two-and-a-half years a crisis mood prevailed. The organization went through several leadership changes, rapid staff turnover, a 50-percent decline in membership, and an enormous budget deficit that came close to wiping out the principal from the endowment left by Robert Marshall that had long supported the organization. Many of the staff who remained were activist-oriented, poorly paid compared to their counterparts at other traditional groups, and locked in bitter conflict with the board. The board, in turn, sought a new executive director to "knock heads" in redirecting the organization. The eventual choice to head the organization, William Turnage, an aide of Ansel Adams who had worked at the State Department and studied at the Yale School of Forestry, promised to intervene forcefully in establishing a new framework for the organization.[69]

When Turnage took over in November 1978, the organization was suffering from uncertainty about its wilderness advocacy as well as from its internal organizational disarray. Many of the staff ("young, radical, crusader types," according to Turnage) had become frustrated and angered by the slow and complex Wilderness Act implementation process, including the proceedings involving Bureau of Land Management lands. By 1980, thirty-six of the thirty-seven pre-Turnage staff had departed, essentially forced out by the new Wilderness Society leader, who began to pursue a more aggressive professionalization process for the organization. Turnage wanted to carve out a specific role for the group within the mainstream movement, not by competing over the range of environmental policies being addressed by the different groups, but by limiting the organization's agenda to its own historic interest in public lands issues, including wilderness. Turnage hoped to concentrate organizational expertise, to have The Wilderness Society become a professional environmental group with a specific focus instead of a traditional advocacy organization with limited expertise. He was also interested in lobbying and public relations activities and linked the group's increased ability to analyze policy with its lobbying function. At the same time, the increased focus on public relations, relatively unique then among the mainstream groups, established greater visibility for the organization despite its narrower agenda.[70]

Similar to other mainstream groups, during the early 1980s The Wilderness Society was able to expand its financial base and substantially increased membership to 150,000 in 1985, when Turnage left the organization. His replacement, George Frampton, a Washington lawyer who had been an assistant special prosecutor on the Watergate Special Prosecution Task Force, reinforced the emphasis on public lands issues, although narrowing the focus even further by excluding any involvement in resource development questions on state or private lands. Frampton, seen by board members as more of an "analytic, rational manager" than the more intense Turnage, also decided to reduce the organization's high-profile public relations efforts and establish a greater emphasis on policy analysis and implementation and monitoring at the regional level.[71] Like other mainstream environmental executives, Frampton buttressed the group's corporate style of management, expanding the organization's finance, administration, and development departments, while also providing a think tank environment for the Resource Planning and Economics Department staff. With another spurt in membership and finances during the late 1980s, The Wilderness Society was able to complete its transition to a narrow but focused role within a mainstream movement that had itself become defined by its corporate/professional bureaucracies and focus on environmental policy making. Even when the growth subsided and membership actually began to decline after Earth Day 1990, the organization simply scaled back rather than refocused efforts, its strategic and organizational directions having already been set.

The most striking transition among the traditional groups was the evolution of the hunter-based National Wildlife Federation. Established in 1935 under the leadership of political cartoonist Jay "Ding" Darling, the NWF sought to fill a vacuum left by the declining fortunes of the most prominent conservationist organization of sportsmen, the Izaak Walton League. The IWL, or the Ikes, had exploded on the scene during the 1920s, particularly in the Upper Mississippi Valley area, where contaminated streams and other pollution problems had begun to have serious impacts on fish and wildlife populations. The IWL was especially critical of the evolution of the National Park Service, the Forest Service, and the Biological Survey, characterizing them as agencies "of destruction and not of preservation of outdoor America." Despite its initial popularity, especially in the Midwest, the organization began to experience bitter internal conflicts by the mid- to late 1920s. These eventually led to divisions between the state chapters and the central organization, as well as the forced resignation of its founder and most charismatic leader, Will H. Dilg.[72]

The origins of the NWF can be traced to a North American Wildlife Conference organized by IWL member Darling, who wanted to extend the approach of the Ikes by combining hunting and wildlife management through a single organization. Darling's plan was conceived in conjunction with and subsequently funded by the leading suppliers of ammunition to hunters, including DuPont, the Hercules Powder Company, and the Remington Arms Company. The new group was immediately forced to confront a key potential conflict: how to balance the needs and interests of hunters versus the growing pressure for a more absolute wildlife protection approach. This issue had been raging among organizations such as the IWL and Audubon and had preoccupied key wildlife advocates such as Aldo Leopold and William Hornaday. The NWF decided to side with the hunters' interests, defining the group in its constitution as a federation of hunter-oriented wildlife interests organized through state societies. Individuals and nonhunter groups who wanted to join—the most likely constituency for a more stringent approach to wildlife protection—were given only nonvoting membership. Thus, the new group established at its outset a structure that "reflected the reality of the sportsman-first philosophy," as the NWF's official biographer put it.[73]

Through its first several decades, the NWF, with a financial base tied to the organization's sale of wildlife stamps, continued to be dominated by hunting and fishing interests active through state affiliates. Many of the state groups functioned as pressure groups for the U.S. Fish and Wildlife Service. The NWF, however, was less action-oriented and more involved in its service functions. As a consequence, it tended to be less identified with some of the key wilderness and protectionist issues that emerged in the post–World War II period. Even when the group did become engaged in an issue, such as support for the Wilderness Act, it did so only after hunting activities were protected. During the 1960s, when the organization under executive director Thomas Kimball sought to broaden its agenda by involvement in the fight against the proposed dam in the Grand Canyon, it still faced conflict within its own ranks, particularly among the state societies that remained motivated by their hunting or recreational interests rather than a protectionist concern for the natural environment.[74]

During the 1960s, the NWF gradually shifted direction after individual voting membership was authorized to establish a broader constituency for the group regarding wildlife issues in general. By the early 1970s, the organization had made a modest transition toward a more professional staff and had expanded its mission. Still, the NWF remained one of the most conservative of the mainstream groups, reflected in its efforts to stay out of the courts, avoid adversarial relationships, and cultivate ties and personal relationships with policy makers. More than other mainstream groups, the NWF actively solicited relations with industry interests while decrying the "extremists and kooks" and "screamers and yellers" within the environmental movement. Nevertheless, the group's transition was still noteworthy, symbolized by its decision to change the slogan on the front page of its publication *National Wildlife* from "Dedicated to the Wise Use of our Natural Resources" to "Dedicated to Improving the Quality of Our Environment."[75]

In 1981, shortly after Ronald Reagan became president, Tom Kimball retired, the last of the old-line conservationist leaders to step down. After a two-year search, the NWF's board selected a new executive director: Jay Hair, a younger, professional-oriented leader from within the group's ranks. An associate professor of zoology and forestry at North Carolina State

University who had been president of the South Carolina state branch of the organization and a special assistant on fish and wildlife policy to Secretary of the Interior Cecil Andrus, Hair immediately sought to refashion the organization. He envisioned the NWF as an active, independent player in the mainstream movement, commensurate with its large membership, budget, and organizational staff. Once in office, Hair was immediately forced to contend with James Watt's efforts to actively court the NWF organization while attacking its Group of Ten counterparts. The interior secretary sought to appeal directly to hunters and recreational interests by distinguishing between the use and protection of wilderness and wildlife. At the same time, Watt's assistant secretary for fish and wildlife, G. Ray Arnett, tried to appeal to Hair's antagonists within the organization. Arnett, a former NWF president who cultivated a "great white hunter" image, had also been involved in NWF intrigues over the selection of Kimball's successor and the effort to have the organization "revert to its good-old-boy hunting club approach," as Hair put it.[76]

The response of Hair and his staff was ambiguous. On the one hand, Hair established his own direct link to the White House, the only Group of Ten CEO to develop a working relationship with top administration officials. Hair also expressed discontent with the Group of Ten process, suggesting that his organization was larger, wealthier, and more powerful than the other groups and that consequently he might not have the time nor inclination to participate as fully as the other CEOs. Though Hair never fully withdrew from the Group of Ten, his periodic distancing from the mainstream movement, at times in alliance with other conservative environmentalist figures such as William Reilly of the Conservation Foundation, revealed some of the movement's conflicting tendencies as well.[77]

One area of disagreement among the Ten was the issue of cooperation with industry groups. Shortly after assuming his NWF post, Hair decided to establish a Corporate Conservation Council in order to make a "fairly aggressive outreach to industry" to overcome the antagonisms that had developed during the previous decade. The council consisted of top NWF executives and representatives of oil and chemical companies and utilities. Its objectives were defined as identifying common points of agreement and ways to resolve disputes between industry interests and environmentalists. The council's quarterly publication, the *Conservation Exchange*, was sent out to Fortune 500 CEOs, major foundation executives, business schools, and mainstream environmentalists. This approach was further extended when both Hair and William Reilly, again with leading oil and chemical companies, created Clean Sites Inc. to promote privately sponsored cleanups of particular hazardous waste sites. The Clean Sites strategy, emphasizing voluntary action similar to William Reilly and George Bush's later concept of voluntarism as a substitute for regulation, broke new ground, according to Hair, by "enlisting the entrepreneurial zeal, the proven expertise, and the enlightened self-interest of America's private sector."[78]

By the late 1980s, Hair and the NWF, through its cooperation with industry, emerged as one of the two (the other was the Environmental Defense Fund) mainstream organizations in sharpest dispute with the grassroots and direct-action wings of the movement. The dispute extended not only to the content of issues (how, for example, cleanup of waste sites should be accomplished), but in the embrace of power associated with these organizations and their CEOs. One aspect of the professionalization of the movement—assuming the trappings of power such as higher salaries, chauffeur-driven cars, and elegant offices—stood out most dramatically with the NWF. Even more than the other traditional groups, this organization of hunters had made its transition by imprinting on the contemporary professionalized movement its concept of the necessity of relating to the dominant sources of power in the Reagan-Bush era.[79]

By the early 1990s, the professionalization and institutionalization of the mainstream groups seemed secure, though incomplete. The Group of Ten meetings, for one, had given way to larger gatherings that provided recognition for the continuing process of institutionalization, but these still fell short of fully defining the movement's institutional framework. The continuing debate about whether to assume an adversarial or negotiated approach toward industry and government also remained unresolved. George Bush's selection of William Reilly as EPA head in 1989 suggested one approach. As a leading advocate within mainstream environmentalism of a government/industry/environmental movement decision-making triad, Reilly sought to expand on his and Bush's voluntarism approach by creating policies independent of the regulatory system itself. This approach broadly corresponded to the growing interest in market arrangements, such as water transfers or marketable air pollution permits, that became prominent among key mainstream groups such as the EDF. These market and voluntarism approaches influenced the design of such new legislation as the 1990 Clean Air Act Amendments and such agency actions as the EPA's "33/50" voluntary reduction plan in the toxics area.[80] Both approaches were controversial, frequently attacked by direct-action and grassroots groups that were themselves situated outside the mainstream movement's arena of activity. These activists argued that such market approaches established a "license to pollute" and that both market and voluntary approaches decreased rather than increased the pressure to restructure industrial decision making in order to incorporate pollution prevention and other environmental values. Mainstream groups such as the EDF dismissed the criticisms, focusing instead on their broader objective of further influencing and refining the environmental policy system through its "win-win" solutions worked out with government and industry.

The mainstream groups today stand at a crossroads. On the one hand, they seem ready to pursue an even more comprehensive process of professionalization and institutionalization, their activities defined increasingly as career training for the initiation and management of environmental policies.[81] At the same time, having enlarged their agendas in response to the development of an environmental policy system, these groups remain vulnerable to criticism by those who have come to define environmentalism in broader social terms as a response to urban and industrial change. The conflicts and contrasts between mainstream and alternative environmental groups, central to this question of definitions, has ultimately become a crucial determinant in the reshaping of environmentalism in the contemporary period.

Chapter 5

Grassroots and Direct Action: Alternative Movements

On the Move with Penny Newman

Monday

On the face of it, the Golden Arches, a block from the ocean in Santa Monica, California, seemed an unlikely place for Penny Newman to start another week of organizing for what she and others called the Movement for Environmental Justice.[1] But McDonald's had become a key target for dozens of grassroots groups, several of them high school–based, that were involved in the McToxics Campaign. Launched in early 1987, the campaign was initially designed to pressure McDonald's to eliminate its trademark clamshell package, a polystyrene foam container produced with the use of chlorofluorocarbons (CFCs), an ozone-destroying compound.[2]

On this bright, brisk, Monday morning in October 1989, Penny Newman, one of the most dynamic organizers within the antitoxics movement, was being interviewed in front of McDonald's to explain the McToxics Campaign as part of a film for high school students. Just a couple of hours earlier, at daybreak, she had left her modest three-bedroom house sixty miles inland near the Stringfellow Acid Pits in the unincorporated town of Glen Avon to gear up for another week of politics, confrontation, and organizing.

As Newman described it to the film interviewer, the McToxics Campaign was a colorful and effective affair. It included boycotts, guerrilla theater (protesters dressed in styrofoam suits), hit-and-run tactics (hundreds of used clamshells left on countertops or mailed to McDonald's owner Joan Kroc), and related efforts at education and mobilization. Flustered by the campaign's tactics, McDonald's officials first denied there were problems. They argued that styrofoam was just "basically air" and performed a valuable function by "[aerating] the soil." Sensitive to issues of image, the company eventually eliminated the use of CFCs and, in a November 1990 agreement with the Environmental Defense Fund, shifted entirely away from polystyrene.[3]

Though the grassroots groups didn't receive public recognition for these company decisions, it had been the McToxics Campaign that had forced the issue for McDonald's and had placed pressure on mainstream groups such as the EDF to respond to the campaign's themes. For the activists, the real value of the campaign had been its ability to demonstrate how ordinary people could get involved and "make things happen," as Newman put it. The fast food industry as well as other toxics producers and users were coming under scrutiny for the first time from "average people" who were relaying a potentially radical message, Newman declared in her interview. "What we're saying," Newman continued, looking up at the Golden Arches, "is that you don't have to be an elected official or an industry executive to have an impact on waste policy."[4]

Tuesday

After checking the mail and the continuous flow of messages from the answering machine in her storefront office, Newman was off to a meeting in Riverside concerning the Stringfellow Acid Pits, the high-profile Superfund hazardous waste site that had consumed more than ten years of Newman's life. This meeting would plan for the next day's gathering of the Stringfellow Advisory Committee, on which Newman served as a representative of the community along with industry and government officials.

The industrial waste dump site at Stringfellow had opened in 1956 and was less than two miles from Newman's home. The first major dumper at the site was the U.S. Air Force, which deposited chemicals used for refurbishing missiles at nearby Norton Air Force Base. For more than sixteen years Stringfellow had received huge quantities of hazardous materials, including more than 34 million gallons of liquid wastes. The wastes included heavy metals, solvents, pesticides (including DDT), and large amounts of sulfuric, nitric, and hydrochloric acid (thus its name, the Acid

Pits). These wastes were deposited by chemical, aerospace, steel, and aluminum companies as well as by plating operations, agricultural concerns, and the U.S. Air Force. Fifteen of these companies would eventually be held liable for cleanup costs estimated then to exceed $600 million. The California Department of Health Services would also be held liable for negligently overseeing construction of the dump site and then delaying taking action to rehabilitate it.[5]

When Penny Newman, nineteen years old and pregnant, came to live in Glen Avon in 1966, she paid little attention to the dump site up the road. Glen Avon was quite rural at the time, though the urban edge of the fast-growing Inland Empire to the east of Los Angeles would subsequently stretch closer to her town. Newman liked the area. She had grown up about twenty miles east of Glen Avon and assumed she would raise her family in a peaceful, nonurban setting, where neighborliness and community values prevailed. She and her husband, a fireman, were well liked and soon became well-known community figures. A school teacher who specialized in speech pathology for special education classes, Newman also became involved in community activities, including the Jurupa Junior Women's Club and the local PTA, of which she eventually became president.

In 1969, ten years before the concept of a "toxic dump" entered the public's vocabulary, Glen Avon residents, as a result of heavy rains that caused a local earthen dam to overflow, experienced the first of several spills carrying wastes from the dump site into town. Community residents mobilized soon after, protesting the inaction of local agencies in dealing with the dump. Three years later, the residents got the dump operators to voluntarily close the site after it was discovered that several local wells had become contaminated.

During the 1970s, the Acid Pits became a political football. Various government agencies entered the picture, hiring consultants, issuing reports that included numerous health risk disclaimers, and focusing on containing the wastes rather than pursuing a more expensive cleanup process. In 1979 and 1980, heavy rains and floods caused another 5.5 million gallons of liquid hazardous wastes to spill from the site, an event that had a dramatic effect on Newman's family. One of Newman's sons, both during and after the discharges from the site, began to suffer serious asthma-related breathing problems, dizzy spells, headaches, and blurred vision. Many residents had similar complaints. Livid at the inaction of the agencies and their reassurances about no risk, the residents intensified their own efforts. A new organization, Concerned Neighbors in Action, was formed, and Newman emerged as one of its leaders.

Wednesday

When Newman arrived at the Jurupa Community Services District building a few minutes before the meeting of the Stringfellow Advisory Committee, the tension in the room was already apparent. Several members of Concerned Neighbors in Action had arrived to monitor the meeting, lend support to Newman, and alert the agency representatives that their plan to try once again to contain the wastes through a clay cap device would be challenged.

For Newman and her fellow residents, a clay cap was totally unacceptable, a red flag that brought back angry memories. During the early 1980s, after Stringfellow had become one of the first and most prominent sites put on the Superfund National Priorities List, the clay cap had come to symbolize all that had been wrong with previous containment efforts. The situation was exacerbated by the literal civil war taking place within the EPA during the first term of the Reagan administration. Anne Gorsuch ran the Agency then, and Rita Lavelle, a former employee of Stringfellow dumper Aerojet-General, ran the Superfund program. Newman, the best known and most knowledgeable of the local residents, had begun to receive revealing documents from an anonymous EPA whistleblower regarding Lavelle and the agency's actions concerning Stringfellow and Superfund. Newman passed these documents on to congressional committees investigating EPA actions.[6]

It was a tempestuous period in Newman's life, with the press camping out at her doorstep, government agencies wary of her activities, and industry interests marshaling their forces as they worried about liability. For a time, Newman felt that some EPA officials had become genuinely interested in a full-scale cleanup to be financed by industry dumpers. But cleaning up a hazardous waste dump site created numerous uncertainties about cost, technology, and health concerns. The EPA was then a novice in this area, easily manipulated by the industry players who became increasingly important to the cleanup process after the 1987 ruling that established their liability. The situation was compounded when George Bush became president and his new EPA head, William Reilly, devised a strategy that sought to emphasize results, even if such results failed to fully address the problems. Newman had met with Reilly about Stringfellow and had come away from the session convinced that Reilly's push for results and enforcement was "another way of saying that the Agency was ready to turn the solution over to the polluters."[7]

At the advisory committee meeting, industry representatives wanted to discuss technologies and technical solutions. Newman argued that these issues weren't technical but political. "For three years, we've come to the table and discussed technologies, and have even found technologies worth considering, only to have you return to the clay cap as the least-cost solution," Newman said to the committee members. Agency and industry representatives felt all political questions were subsumed under technical decisions. The industry representatives, who had formed their own "technical committee," would be meeting the next day and asked if Newman's technical consultant, but not Newman, would come. "Let's get beyond emotion and argue about technologies," appealed one of the industry representatives.

Thursday

When Penny Newman, uninvited, walked into the basement room at the Camelot Inn in Riverside where industry officials and invited agency representatives were about to meet, many of those standing and chatting with each other turned her way. A look of expectation and concern was on their faces, as they feared a confrontation rather than a negotiation on technical details. In the opinion of Newman and other community activists in the antitoxics movement, such technical details were continually used to intimidate "nonexpert" community residents, while disguising crucial political or economic choices. At the same time, a whole cottage industry of consultants and experts had emerged, hired by industry polluters and government agencies to convince communities that the risks they faced were insignificant. Meanwhile, mainstream environmental groups, themselves adept at the use of technical information, were uncomfortable at what they perceived to be the stridency and anti-expertise posture of the "not in my backyard" (NIMBY) community groups. But the NIMBYs, scorned by all the major players, had forced their way into the toxics debates, partly through their organizing successes and partly by insisting that they could develop sufficient expertise to influence decisions. In fact, NIMBYs were seeking to change the terms of the debate, both with respect to decisions about their particular waste sites as well as the overall approach toward toxics use and disposal. The groups proudly wore their NIMBY label as a kind of badge of honor. Their actions, as Newman and her fellow NIMBY activists liked to say, were a way to "plug up the toilet."[8]

After much maneuvering about whether to begin the meeting or eject Newman and thus force a confrontation, a key industry player summoned

Newman to a private meeting. For Newman, it seemed apparent that this was where the real power was located. Technical debates and government agencies were more incidental to the process. The industry figure wanted to know how much Newman and her community activists would push to force more expensive cleanup options. "Just what is your bottom line?" he asked.

Friday

Camped out in her paper-strewn movement office, with the pungent fumes from the next-door beauty parlor drifting in, Penny Newman settled in for a day of organizing by phone. The messages on the machine got her started. An activist/organizer who was a young mother with three children had called to say that she had discovered an illegal waste transfer station and that she was ready for action. Not all the messages, however, were as spirited. Newman needed to console many of those from her network of organizers who were women with families and whose husbands objected to their new roles. Newman herself had experienced a rocky personal period during the Lavelle affair, but had weathered it without giving up her new life as an organizer.

Newman had also been fortunate in this period to find a kindred spirit in Lois Gibbs, with whom Newman could share her personal thoughts and political growth. Gibbs, a housewife who became a community leader and high-powered organizer at Love Canal in New York, had also confronted polluters, regulators, cautious government bureaucrats, and high-priced corporate attorneys. As Gibbs's prominence brought her into contact with other community activists like Newman, she resolved to establish a more formal network of antitoxics organizers, to aid grassroots groups that had sprung up adjacent to waste sites around the country.[9]

Gibbs's group, the Citizen's Clearinghouse for Hazardous Wastes (CCHW), was formed in 1981 and immediately made its presence felt. Through its network, which eventually reached thousands of communities, the group confronted industry forces, government agencies, and occasionally mainstream environmentalists supportive of existing approaches. The CCHW also challenged the way mainstream groups used expertise and lobbying to frame waste and toxics issues. For the community groups, the use of expertise by others was frequently a disempowering experience, a method of defining issues that effectively eliminated participation by those most affected by such decisions. The use of expertise and related professional skills also established a hierarchy of participation in which those engaged in the minutia of specialization and technical details were given authority, while "nonexperts" were perceived as nonparticipants. Newman herself, at one meeting in the early 1980s, recalled being told directly by Blake Early, a top Sierra Club staff member and former lobbyist with Environmental Action, that community-based female activists were inappropriate participants in the toxics arena. "How come women like you aren't at home," Early had said, half-jokingly, to Newman's shock and dismay.

CCHW activists were perhaps most effective in their mocking criticism of the revolving-door relationships of industry, government, and mainstream groups. In one revealing episode, an activist in the CCHW network, a Vermont housewife, took on the National Wildlife Federation when she learned that Dean Buntrock, the chairman of Waste Management Inc., had been invited to serve on the NWF's board. Waste Management had become a particular bête noir for the CCHW activists, given the company's poor track record concerning its waste disposal activities (as the operator of some of the country's largest dump sites, including the notorious Emelle, Alabama, hazardous waste landfill) and its growing power in the waste field (by far the largest waste company in the country, with a strong penchant for monopolizing the collection and disposal of waste in a given area). The Vermont activist wrote a letter to NWF president Jay Hair, who defended the relationship and Waste Management's role. The CCHW, with its flair for the irreverent, had much fun publicizing the Waste Management/National Wildlife Federation ties and correspondence. The situation became even more explosive when antitoxics activists discovered a few years later that Hair had intervened with long-time associate and then EPA head William Reilly on behalf of Waste Management and other waste firms.[10]

Becoming part of the CCHW and its engaging spirit, growing network, and insistent politics had brought Newman into a new dimension of her life as organizer. With her children nearly grown and with encouragement from her family and neighbors (as long as she didn't stay away too long from Stringfellow matters), Newman became the CCHW's West Coast organizer. Her growing talents as organizer, researcher, and political analyst allowed Newman to become involved with a wide range of would-be activists, from rural, wheat-growing communities in eastern Washington to inner-city Los Angeles neighborhoods.

As Newman settled into her organizer's role, she became a prominent figure in the organization and a charismatic speaker who could inspire her audiences of like-minded activists. One such moment had been a keynote speech at CCHW's annual convention in 1989 that had brought delegates out of their seats and caused some to weep. "The lesson of the Movement is that we are the power," Newman had told the gathering. "We are the experts. We are the ones who have watched our community devastated. We are the ones who have watched our life's investment in our homes disappear. We are the ones who lie awake at night listening to our children struggling to breathe." This struggle, Newman continued, is "not just about environment but about basic issues of justice and fairness, of right and wrong, of the have nots and those with the economic power who would seek to exploit all of us. We are indeed People United for Environmental Justice and, by God, we will win."[11]

Saturday

On the road again, this time 100 miles west, to the Dunes area by the ocean between Oxnard and Ventura. A group had gathered at an apartment in a building constructed atop a waste dump. The organizer and host of the meeting, Linda Paxton, had called Newman two weeks earlier to ask how she could begin to organize her neighbors to take on the agencies that had been dragging their heels regarding the site. Paxton was the very profile of a potential Movement for Environmental Justice recruit: angry, passionate, rooted in her community, wanting something to happen. She had invited interested residents to meet with Newman, hoping the CCHW organizer could help get her started.

After introductions were made, Newman began a kind of "how to" organizing strategy session. There was no set agenda, no plan of action. "You have to rely on people, and people learn by their experiences," Newman said of her role, arguing that although she acted as a resource, if she were to become "the kind of organizer who goes in and sets it all up, then it won't work." Newman and the CCHW have used an organizing approach first developed by Saul Alinsky and his Chicago-based Industrial Areas Foundation. The object of the organizing has been to empower local citizens to do their own organizing and networking rather than for the CCHW to create an organized social movement. After the organizers' network is consolidated and expanded, only then might such a movement emerge. "I can go anywhere where our network has developed, in small towns, at the edge of the city, in the suburbs where the organizing has begun or is about to begin, and I feel at home," Newman said of this process.[12]

Resonating with some of the language and spirit of the 1960s movements, these networks have helped bring about specific victories by mobilizing communities, especially to block waste sites. Despite the fact that several of the community groups have disappeared once proposed facilities have been stopped, the network itself has continued to grow. More directly, the Movement for Environmental Justice exists primarily as an idea, a possibility more than a fully defined organizational approach. Dwarfed by the resources and media recognition of the mainstream environmental organizations, these grassroots groups, part of a potential social justice movement, have had to struggle to keep afloat and maintain their vision of community defense and environmental transformation. Still, as Penny Newman prepared that chilly October day to undertake her long drive back to her home by the Acid Pits, she knew that her week of organizing was continuing the process of constructing a new kind of environmentalism, a movement involved in issues of everyday life made up of advocates like herself.

A New Environmental Framework

Since the 1970s, there has emerged, distinct from the mainstream groups, a powerful current in contemporary environmentalism focused on issues of empowerment, environmental justice, equity, and urban and industrial restructuring. This current consists of alternative groups—alternative in their critical view of the environmental policy system and the role of the mainstream groups in helping to shape and sustain it. This alternative movement is predominantly local in nature, more participatory and focused on action, and critical of the roles of expertise and lobbying in defining environmental agendas. With a direct lineage to earlier urban and industrial movements, the alternative groups have sought to develop a new framework for environmental change, relying on constituencies often underrepresented or excluded from environmental decision making and drawing on such critical concepts as citizen empowerment and the prevention or reduction (rather than management or control) of pollution. In establishing this framework, they have also been influenced by a number of key analysts and advocates, including Ralph Nader, with his critique of corporate power and call for citizen action; Barry Commoner, with his critique of postwar technologies and production decisions; and Paul Connett and Peter Montague, with their critique of the use of expertise in policy development.

Ralph Nader's influence on the development of contemporary environmentalism still remains to be fully explored. Emerging as a public figure in the mid-1960s following the publication of his book *Unsafe at Any Speed* and his successful invasion-of-privacy suit against General Motors, Nader quickly developed an interest in industry decision-making issues and industrial sources of pollution. The General Motors episode also transformed Nader into a symbol of anticorporate resistance. Building successfully on his growing prominence, Nader was able to create and consolidate within just a few years a far-reaching public interest organizational network, which also became a major environmental force. Nader groups included the PIRGs (Public Interest Research Groups), the Center for the Study of Responsive Law, Public Citizen, legislative-oriented coalition spinoffs such as Clean Water Action, and various task forces that produced a remarkable series of studies and reports on more than twenty-five different topics, such as air pollution, the politics of land development, the role of DuPont in controlling the affairs of Delaware, and the structure of the paper industry and its environmental impacts. Nader's emergence as a public figure both reinforced and helped elaborate the development of this organizational network with its focus on corporate power and the need for citizen involvement.[13]

Among the network groups, the PIRGs have been most directly engaged in environmental issues. The PIRGs were first organized during the Earth Day period. This was also during the cataclysmic days of student protest against the Vietnam War when many of the organizational expressions of the New Left had largely disappeared from the campuses and opportunities to create new vehicles for student action were available. Shortly after Earth Day, Nader gave a speech at the University of Oregon, the scene of major antiwar and environmental protests. More than 500 students showed up for the talk, and Nader urged them to act on his citizen action message. The Oregon students responded by creating the Oregon Student Public Interest Research Group, which eventually included participation from all seven schools in Oregon's state college system. Encouraged by the Oregon experiment, Nader and one of his associates, Donald Ross, helped launch a similar effort at the University of Minnesota, where students, through a massive petition drive, created a Minnesota PIRG. Both state college organizations were successful in establishing a system of student fees to help finance their work and hired professional staff to undertake research and initiate specific campaigns. Based on this organizational model—professional staff plus student volunteers financed by student fees—individual PIRGs quickly sprang up in several different states and became the primary stimulus for a wide variety of local and statewide anticorporate and public interest campaigns. Many of these campaigns had environmental themes, ranging from support for bottle bill legislation for glass recycling to banning the use of asbestos in new construction. The PIRG message also sought to emphasize the idea that students—and citizens—could bring about change, with campaigns selected on the basis of their being "winnable."[14]

During the 1970s as many contemporary environmental policies took root, the Nader groups, unlike most mainstream environmental organizations, sought to develop a multi-issue approach and help coalesce different constituencies. For Nader and his network, questions of citizen empowerment and corporate behavior were directly related: without a strong basis for citizen action, whether in terms of consumer protest, the labor movement, or environmental activism, corporate power would remain unchecked. Consumer, labor, and environmental issues were themselves different expressions of a corporate-dominated political economy. The multinational corporation thus became the focus of attack, the engaged citizen the instrument for change. Paralleling the critique of the turn-of-the-century settlement workers who focused on the corruptions and hazards of the industrial city, Nader and his citizen network defined a form of environmental action that incorporated professional skills into a broader, anticorporate movement.

While Ralph Nader and his network emphasized the role of the individual corporation and its antisocial behavior, biologist Barry Commoner came to focus on the post–World War II industrial system and its need for fundamental change. Commoner first became prominent in the 1960s through his work with the St. Louis Committee on Nuclear Information, which he had helped found in 1958, and through the publication of his book *Science and Survival*, which focused on the uses of science and technology in relation to environmental hazards. These themes were also explored in *Scientist and Citizen*, an early environmental journal on whose board Commoner also served.[15]

In these forums, Commoner argued that the use of technical information associated with such issues as radiation emissions, lead in the ambient environment, and pollution from organic chemicals had created an "apparently insuperable barrier between the citizen, the legislator, the administrator and the major public issues of the day." At the same time, the technologies that had emerged since World War II represented a new and quite different environmental threat, requiring an informed citizenry and a more socially committed group of scientists.[16]

These concerns about the problems of science and the new technologies were more fully developed with the 1971 publication of *The Closing Circle*, Commoner's most influential book. Commoner, whose picture had already adorned the cover of *Time* magazine as one of the "new jeremiads," stated in his new book that the sudden eruption of environmental protest required an analysis of the problems of pollution and the "enveloping cloud of science and technology" that he had identified in *Science and Survival*. This analysis, in turn, needed to generate a "deeper public understanding of the origins of the environmental crisis and its possible cures."[17]

The Closing Circle identified a series of ecological laws: "everything is connected to everything else," "everything must go somewhere," "nature knows best," and "there is no such thing as a free lunch." These laws, Commoner argued, had long been rooted in human perceptions about the natural world. What created special concern in the contemporary period was "the rising miasma of pollution" tied to the development in the postwar period of synthetic organic chemicals and other petroleum-derived products such as pesticides, synthetic fibers, plastics, and detergents. These products, moreover, had displaced less environmentally destructive materials and products. Such a technology factor, according to Commoner, was driving the postwar environmental crisis. Though *The Closing Circle* didn't focus specifically on solutions to this crisis, it provided, according to a leading environmental activist, "a new coherency" for the environmental movement "which could be a revolutionary vision of an ecologically sane society."[18]

During the 1970s, Commoner's technology-centered ecology argument increasingly focused on energy issues as the problems of energy production and supply reached center stage. Drawing on the work of researchers associated with his Center for Biology and Natural Systems (CBNS), Commoner extended the notion of an interconnected ecosystem and its relation to production choices to the structure of the economic system and the way all three systems were influenced by the decisive role of energy. His analysis, which first appeared as a series of articles in *The New Yorker* and was subsequently published as a book in 1976 under the title *The Poverty of Power*, defined the energy choices of the post–World War II era as inherently inefficient and thus directly contributing to the contemporary environmental crisis.[19]

Commoner's solutions, spelled out more directly in *The Poverty of Power* than in earlier writings, linked his advocacy of renewable energy technologies to the need for different economic decision making (which Commoner called the social governance of production decisions) and a new political force outside the two electoral parties.

Commoner's interest in economic restructuring and political action made him a more controversial figure for the mainstream groups and a magnet for the developing alternative movement. *The Poverty of Power* was especially attacked by mainstream environmentalists for advocating an environmentally grounded socialism and a third political party.[20] At the same time, Commoner became a much-sought-after speaker, from unions contending with occupational hazards to anti–nuclear power activists attracted to Commoner's arguments about the structural inefficiencies of nuclear power. The growth and success of the antinuclear campaigns suggested to Commoner that a third political party could be a viable strategy, a U.S. version of the new Green parties of Europe. By 1979, Commoner, in conjunction with other activists, decided to pursue this electoral route. Hoping to wed environmental concern to his radical critique of the post–World War II industrial system, Commoner would become the 1980 presidential candidate of the new Citizens Party.

Despite the hopes of a new Green politics, the Citizens Party campaign never effectively escaped the marginalizing process for third-party efforts in this country. Unlike the European Green parties, such as Die Grünen in West Germany, which could successfully draw on environmental protests and tap into a politically oriented labor movement and a well-developed social democratic tradition, the Citizens Party in the United States became limited to an amalgam of small protest groups. It was also a victim of the numerous impediments, from campaign financing to onerous electoral laws, that made most third-party efforts problematic at the outset. Nor did the Citizens Party, despite Commoner's prominence as a major environmental analyst, develop an environmental identity, as Die Grünen had been able to secure in West Germany. This was partly a result of the nearly unanimous support for Jimmy Carter among the mainstream groups and the lack of cohesion and political focus of the grassroots and direct-action groups that might have otherwise constituted an initial base of support for the new party. By 1984, Commoner would detach himself from the Citizens Party and its third-party hopes and become an unofficial environmental advisor to Jesse Jackson's campaign for president.

During the late 1980s and into the 1990s, Commoner shifted his focus again, this time to solid waste and toxics issues. In a 1987 article for *The New Yorker* and in a subsequent book, *Making Peace with the Planet*, Commoner argued that the failure of the environmental policy system was embedded in its pollution control laws and regulations. This contrasted with the system's occasional successes resulting from the direct prohibition of specific toxic substances such as leaded gasoline. Commoner also criticized prevailing environmental approaches to solid waste, particularly the search for high-tech disposal options such as incineration technology. He raised as an alternative to incineration the concept of intensive recycling, which aimed to divert, through recycling, a substantial portion of the solid waste stream.[21]

Commoner's proposals for these new recycling and pollution prevention strategies heightened his profile within the environmental movement at a point when toxics and waste issues were forcing a reevaluation of the dominant pollution control approach. Like Ralph Nader, who used his celebrity status to further his goal of citizen action, Barry Commoner continued to use his prominence to popularize an alternative explanation of the roots of the environmental crisis, while simultaneously exploring ways to turn his analysis into a guide for action.

The development of an alternative environmental approach has also been linked to efforts to make technical information more accessible. A network of activist experts has emerged as part of this attempt to democratize information. Two of its most influential participants have been Peter Montague and Paul Connett, both editors of widely read newsletters and frequent speakers at forums and events of the alternative groups.

A one-time SDS activist, Peter Montague became strongly attracted to the ideas of Barry Commoner and his associates at *Environment* magazine (formerly *Scientist and Citizen*) in the late 1960s and early 1970s. Montague, who became involved with a new Environmental Studies program at the University of New Mexico, also linked up with the Nader network in 1971 through the creation of the Southwest Research and Information Center (SRIC) in Albuquerque. In contrast to the approach of mainstream environmental professionals, the SRIC staff sought to work directly with community groups by integrating their research and professional skills within a framework of community activism. In the process, the SRIC became deeply involved in such issues as uranium mining and development in the Four Corners region, linking up with constituencies such as the Navajo and northern New Mexican Mexicanos, whom the SRIC staff considered essential participants within a broader environmental movement.[22]

In 1979, Montague left New Mexico for Princeton, where he established a newsletter, *Hazardous Waste News*,[23] to monitor the activities of the waste industry on behalf of the growing network of antitoxics activists in New Jersey. Through his newsletter, which quickly became national in scope, Montague became a popular activist-expert advisor to community groups and assorted NIMBYs confronting the waste industry and their government allies. He subsequently moved his publication to Washington, D.C., and continued to issue it on a weekly basis while briefly working with Greenpeace on toxics-related issues.

Similar to Peter Montague, Paul Connett also achieved informal advisor status with local groups, primarily in the area of solid waste, though he had no previous activist background. Connett, who received his Ph.D. in chemistry from Dartmouth in 1983, had taken a job at St. Lawrence University in upstate New York and, with his wife Ellen, a librarian, and their three children, prepared to lead a quiet life at the rural campus. In 1985, however, the Connetts became aware of plans to build a 250-ton-per-day solid waste incinerator about twenty miles from the campus within a food-producing area. Although Connett knew nothing at the time about the technology, he quickly became concerned about the incinerator's potential for contamination.

As a resident NIMBYite, Connett immediately familiarized himself with incineration technology issues and related questions of solid waste generation and disposal. Concerned about the relative absence of alternative information sources for the grassroots groups, the two Connetts decided to produce their own newsletter, *Waste Not*, a solid waste counterpart to Montague's *Hazardous Waste News*. Like Montague, the Connetts also began to use the newsletter as a resource and information clearinghouse, with Paul constantly on call for community groups seeking specific information and counsel.

By the late 1980s, both the Connetts and Montague had become important adjuncts to the grassroots groups and antitoxics networks, such as Lois Gibbs and Penny Newman's Citizen's Clearinghouse for Hazardous Waste, the Boston-based National Toxics Campaign, and groups such as Greenpeace, the PIRGs, and Clean Water Action. Their publications became essential reading for community groups. They made obscure documents and reports accessible, covered project battles, and revealed information the waste industry would have rather kept removed from public view. In presentations to community groups, they shared information sources and made accessibility to technical information an empowering and confidence-building experience. They constantly reminded the community groups that "you can learn it too!" As activist experts directly confronting the disempowering experience of how expertise was used, Montague and the Connetts extended the concept of environmental justice to the arena of environmental democracy, complementing the citizen-action philosophy of Ralph Nader and the industrial critique of Barry Commoner.[24]

The Antinuclear Movements

The most important and far-reaching of the alternative movements to emerge in the period after Earth Day were those focusing on the nuclear issue. During the 1950s and early 1960s, the nuclear power and electric utility industries, in conjunction with the Atomic Energy Commission (AEC), enjoyed unprecedented support and success in launching this new technology. Opposition to the siting and construction of new nuclear power plants was often weak, particularly after the issue of radiation emissions from nuclear weapons testing, a major environmental concern of the late 1950s, began to fade. Antinuclear protests, even when successful, reflected the strength of local opposition groups rather than indicating the presence of a national movement. And while several local groups focused on natural environment impacts, including thermal pollution, the nuclear power issue remained removed from nearly all conservationist agendas.[25]

During the late 1960s, the nuclear issue reemerged on several levels at once. Unresolved problems of thermal pollution were complemented by major debates about the health effects of low-level radiation emissions as well as plant safety considerations. In this period, pro-industry advocates, including the AEC, dismissed new research that suggested that low-level radiation emissions had far more serious health impacts than previously assumed. This issue became particularly compelling when two scientists, John Gofman and Arthur Tamplin, from the Lawrence National Laboratory in Livermore, California, argued that the AEC was seriously underestimating these risks by as much as a factor of ten. The AEC and its allies responded harshly to the Gofman-Tamplin findings, attacking their credibility and eventually undermining their positions at the Lawrence Laboratory. But the problems for the AEC generated by the claims of Gofman and Tamplin were reinforced by internal disputes about altered information regarding plant safety, which raised issues not only about the health and safety of nuclear technology but the credibility of the regulatory process as well.[26]

These episodes had an important impact on the growing antinuclear movement. They intensified the mistrust of an officially sanctioned and politically motivated science used to buttress the position of the nuclear power industry and minimize concerns about health, safety, and environmental impacts. The growing controversy concerning the conflict of interest of the AEC, the chief sponsor and funder of much of that science, also created divisions within the scientific and engineering community, as increasing numbers of scientists and engineers began to participate in antinuclear activities and campaigns. At the same time, local citizen activists who had become interveners in the permit process were themselves becoming knowledgeable critics regarding the complex technical information involved.[27]

By the early 1970s, a substantial number of proposed nuclear facilities were being challenged, and a major debate about restructuring the nuclear regulatory process also began to occur. The focus of the antinuclear activists on permit and administrative restructuring, however, was lengthy, complex, costly, and successful only in delaying rather than actually stopping the construction and completion of new plants. And while the antinuclear movement grew in numbers and reach and strengthened an increasingly sophisticated counterexpertise, a great deal of frustration remained about the ultimate outcome of these campaigns. This was further heightened by the mid-1970s energy crisis, which the nuclear industry immediately sought to exploit by raising concerns about national security.

By 1974, a standoff had developed between the antinuclear groups and the industry. In that year, however, a series of events shifted the focus from technical arguments and regulatory procedures, where the industry remained strongest, to new arenas of protest and political mobilization, where the industry turned out to be most vulnerable. In early 1974, the Ralph Nader network, in alliance with Friends of the Earth and the Union of Concerned Scientists, sponsored a conference for antinuclear groups to establish a new network called Critical Mass. The frustration level of the conference participants testified to the difficulties of organizing around primarily technical issues. But the large numbers of activists in attendance at the conference and its Nader-inspired focus on citizen action also suggested that a powerful national movement could emerge, capable of combining the self-taught expertise of local interveners with the citizen movement appeal of the Nader groups.

While Critical Mass was being organized, a number of mainstream groups, including the Sierra Club and the Environmental Policy Center, also began, in response to the energy crisis, to confront the nuclear power issue more directly, albeit not as forcefully or with as militant a posture as Nader's Critical Mass organization or Friends of the Earth. The positions of these groups, such as the Sierra Club's decision to support a conditional moratorium on new plant construction until some safety, waste disposal, and potential theft and sabotage issues were successfully addressed, further broadened the growing political opposition to nuclear technology.[28]

What most directly and dramatically transformed the nuclear power issue, however, was the sudden emergence of direct-action antinuclear protest groups. The 1960s-style direct-action tactics within the antinuclear movement were first employed in February 1974, when Sam Lovejoy, a member of the Montague Farm rural commune in northwestern Massachusetts, used a crowbar to knock down a local utility's tower in his protest against a proposed nuclear power plant. Lovejoy's action found support among an eclectic constituency of rural commune members, former anti–Vietnam War activists, and other one-time counterculture participants in and around northern New England. This antinuclear activist core was further inspired the following year by a massive direct-action protest in Whyl, West Germany, where more than 28,000 activists sat in at the site of a proposed nuclear power plant and successfully blocked construction. The New England activists, increasingly focused on the proposed Seabrook nuclear power plant in New Hampshire, were joined in their opposition by the New England Coalition on Nuclear Pollution (NECNP), the major antinuclear coalition in southern Vermont and western Massachusetts. As plans for construction at Seabrook began to be formulated, NECNP expanded its technical critique of possible environmental impacts from nuclear plants to question the technology itself. By the time the Seabrook plant received its permit in 1976, the local, environmentally oriented antinuclear participants in NECNP were ready to link up with the direct-action advocates to create a new coalition of antinuclear forces under the name of the Clamshell Alliance.[29]

On August 1, 1976, eighteen Clamshell activists, adopting the strategy first developed at Whyl, occupied the site at Seabrook as a form of civil disobedience. Three weeks later, on August 22, another 180 activists undertook what was then being called "the occupation of Seabrook." These events electrified the antinuclear movement, with the occupation capturing the imagination of antinuclear groups around the country. The action of the Clams, as they were called, also helped coalesce with respect to the nuclear issue constituencies with roots in the New Left and the counterculture who had been largely dormant since the 1960s.

Through late 1976 and 1977, the Clamshell Alliance became a powerful but unstable group bringing together a wide range of activists. It sought to create a form of organization and protest that maximized participation, decentralized decision making, and relied on the tactics of mobilization rather than lobbying or the use of expertise to draw attention to its position to halt construction of the Seabrook plant. On April 30, 1977, another Seabrook occupation by 2400 activists took place, drawing an enormous amount of press attention in the process. For some protesters, the coverage, with its "one paragraph about nuclear power in thousands of inches of coverage," failed to address the content of the actions. Instead, as Dick Bell, the managing editor of the alternative Boston weekly *The Real Paper* put it, the Seabrook occupation became "a tremendous spectacle, the people moving onto the site, digging the latrines, being taken to the armories, the legal process. It could not have been a better media event."[30]

Despite the attention generated by the direct-action protests, the occupations also revealed conflicts within the coalition that were never successfully resolved. On the one hand, "seacoast people" (local residents who lived near the plant) felt they had a special stake in the outcome of the protests and how the Clams might proceed. Other activists, including those from the Boston area as well as those associated with Murray Bookchin's Vermont-based Institute for Social Ecology, felt that Seabrook was a regional issue requiring a regional response based on the broader geographic risks associated with a catastrophic accident. The debate about the roles of different constituencies paralleled the debates about protest tactics, including whether more confrontational approaches, such as fence-cutting to occupy the site (i.e., destruction of property), were warranted. Mirroring the posture and approach of late-1960s direct actionists, the more militant of the civil disobedience protesters, led by a Boston affinity group that called itself Hard Rain, didn't hesitate to block consensus, even as a minority within the larger coalition. It was these groups especially, as analyst Barbara Epstein put it, who hoped "to revive the confrontational style of the anti-war movement in an organization that had been formed in the hope of finding a different approach to protest."[31]

The major impact of the Clamshell protests was in stimulating similar direct-action groups, such as the Abalone Alliance in California. Abalone, formed in the late 1970s to protest the proposed Diablo Canyon nuclear power plant along that state's central coast, grew out of local resident protests against the siting of the plant. The most important of the protest groups, the San Luis Obispo–based Mothers for Peace, which had earlier been active in opposition to the Vietnam War, helped coalesce initial opposition to Diablo by focusing on the discovery of an earthquake fault offshore from the plant. A few years later, at the time of the first Clamshell occupations, a national peace walk, sponsored by a northern California organization, passed through San Luis Obispo and decided to commit civil disobedience at the Diablo site. The peace activists subsequently hooked up with Mothers for Peace and other anti-Diablo activists to develop a new, more expansive campaign against the proposed nuclear facility. As with the Clamshell Alliance, a new organization was created in 1976 through an affiliation of local groups, including affinity groups[32] of friends and associates prepared to engage in acts of civil disobedience. And, like the Clams, the Abalone Alliance operated without formal leadership, with decision making tied to the rotation of group representatives, or "spokes."

Like the Clamshell Alliance, the Abalone Alliance was able to attract enormous attention for its protests, particularly in the wake of the Three Mile Island accident in March 1979 and the release of the film *The China Syndrome*, which portrayed a fictional accident that eerily approximated the Three Mile Island event. Between April and June 1979, several major Abalone protests took place. These included a demonstration involving more than 25,000 people in San Francisco, the picketing of ninety-three different offices of PG&E (the utility constructing the facility) throughout the state, and a local demonstration in San Luis Obispo that started out at the movie theater playing *The China Syndrome* and ended at the local PG&E office. Those events culminated in the largest anti–nuclear power protest at the time, as 40,000 people gathered in San Luis Obispo on June 30 to hear rock music, proclaim their support for alternative energy, and vow to stop Diablo.[33]

The San Luis Obispo demonstration drew on several different currents within the antinuclear movement: counterculture activists attracted to the vision of a decentralist "Solar America"; local residents concerned about safety and health issues; direct-action advocates wanting to connect issues of nuclear power with the nuclear arms race, as well as those who saw the Diablo issue as a way to critique capitalism and militarism in the society; and an assortment of local and statewide environmental activists who hoped to raise concerns about the environmental impacts associated with nuclear power. The environmental issues, however, were "the occasion, rather than the impetus for a movement that was fundamentally about social, communal, and personal transformation," as Barbara Epstein has argued. Nor were the Diablo protesters articulating a politics about the natural environment except as it tied into the symbolism of a technology perceived as out of control. In that context, when David Brower, one of the speakers at the Diablo demonstration, recalled the battles of the late 1960s within the Sierra Club over the siting of the Diablo plant, his remarks generated little interest among an audience that had little connection to that earlier environmental history. Similarly, the enthusiastic response to Barry Commoner's talk about the inefficiencies of nuclear power and its relation to the political and economic system didn't signify any direct interest in an organized political effort, as revealed by Commoner's Citizens Party debacle a year later. As potent a force as the antinuclear movement had become, it remained a movement of distinct constituencies and contrasting perspectives held together by the single overriding objective of stopping nuclear power.[34]

The Diablo demonstration also occurred at a time when the possibility of actually stopping nuclear power seemed greater than ever. While nearly all statewide antinuclear ballot measures between 1972 and 1976 went down in defeat, the one striking success was a referendum in Missouri that prevented utilities from recovering through the rate base the costs of construction while a nuclear plant was being built. The elimination of this obscure financial practice, known as Construction Works in Progress (CWIP), became a potent accomplishment for the antinuclear protesters by seriously undercutting a source of capital for new plants. The attacks against CWIP also linked antinuclear efforts to the rapidly growing movement for utility rate reform inspired by huge energy crisis–induced rate increases. The utility rate reform movement—defined more as a consumer rather than an environmental movement—shared with the antinuclear movement concerns about access to information and the need to establish a citizen-based expertise, criticism of a regulatory process that reinforced an anticonservation approach, and a reliance on the politics of protest and mobilization as integral to the movement's campaign for change. By the end of the 1970s, both the financial issues and the growing public mistrust fueled by both protest movements had placed nuclear energy in its most precarious position ever, with no new plants on order and a number of proposed plants indefinitely postponed or abandoned. By the 1979 Diablo demonstration, the chant of the protesters had shifted from a moratorium on new plants to a call to shut them down.[35]

During this period, a number of participants within the antinuclear movement began to explicitly link the question of nuclear power to nuclear weapons and the arms race. One key organization, the Mobilization for Survival, was formed in 1977 to draw together the two sets of issues, while several direct-action groups organized protests against nuclear weapons–related installations and research facilities, such as the Hanford complex in the state of Washington and the University of California's Lawrence National Laboratory. The strong success of the anti–nuclear power forces by the late 1970s and early 1980s also helped to bring about this shift in emphasis, with the decline of nuclear power contrasting with an escalating arms race. Between 1979 and July 1982, when anti–arms race efforts culminated in a massive demonstration of more than a million people in front of the United Nations in New York City, a diverse set of anti–arms race protest movements and protest actions exploded on the scene. These ranged from the lobbying-oriented Nuclear Freeze movement to direct-action and counterculture-based activities, including a Women's Pentagon March, where the form of the protests was linked to the search for a new politics or a new spirituality.

The success of the demonstrators in mobilizing the public and helping identify the fears and concerns about the arms race eventually forced a change in the political discourse concerning the arms issue. By the mid-1980s, the Reagan administration had begun to shift toward a less confrontational rhetoric, with the changes in the Soviet Union under Gorbachev further eroding the language of the Cold War. The possibility that the Cold War could be diminished, while welcomed by the protest movements, reduced and eventually eliminated the sense of urgency associated with the threat of a nuclear conflict. Even the Chernobyl accident, which refocused attention on nuclear power plants, also appeared to reinforce the prevailing view that nuclear arms reduction rather than any further arms race escalation would become the major policy focus of the two superpowers. With the collapse of the Eastern European regimes and the erosion of central power within the Soviet Union between 1989 and 1991, the anti–nuclear arms race and anti–nuclear power movements found themselves lacking a focus and diminishing in numbers and activities.

Despite the decline of both sets of antinuclear protests, the nuclear issue in the United States remained a potent local concern during the 1980s and into the 1990s, especially in communities where nuclear facilities were located. Questions of waste disposal, polluted groundwater basins, toxic air emissions, and contaminated soil were increasingly linked with the hazards associated with nuclear facilities, including nuclear weapons plants. The community-based groups that emerged in this period to confront these nuclear/military toxics issues were becoming an important part of a larger movement seeking to redefine the nature of environmental politics. For while the anti–nuclear power movement had earlier helped situate a locally based and direct-action form of environmentalism, the movement against toxics, including its antinuclear wing, moved center stage to address the question of pollution in its contemporary forms.

COMMUNITIES AT RISK: ANTITOXICS MOVEMENTS

In 1980, as part of a broad survey on environmental issues, Resources for the Future (RFF) asked respondents a series of questions about locating certain types of facilities in their neighborhoods. RFF was interested in calculating a way to determine a "residential avoidance curve" associated with such facilities as nuclear power plants and waste disposal operations. Though the popularity of nuclear power had plummeted in the period prior to the survey, and hazardous waste was still an emerging issue, the residential avoidance curve for hazardous waste facilities was nearly the same as for nuclear plants, with more than 50 percent of the respondents unwilling to accept a facility within a fifty-mile radius of their home. The expression of discontent noted in the RFF survey, also reflected in the failure of the waste industry and local governments to convince residents that facilities sited in their backyard represented a negligible risk, would ultimately become the source for new forms of environmental protest and action, reconfiguring the shape and agenda of the environmental movement.[36]

Up until the 1970s, hazardous wastes were treated in much the same manner as other kinds of wastes, whether construction debris or residential household wastes. These wastes were dumped into pits, landfills, or waste ponds or discharged into surface waters, deep wells, or the ocean. The landfills of the 1950s and 1960s, even those defined as "sanitary landfills," were filled with numerous toxic substances due to the lack of information and limited controls on the types of wastes being disposed and the kinds of impacts they represented.

During the 1970s, as many of the new environmental policies were set in place, a number of mainstream environmental organizations, particularly the expertise-oriented professional groups, began to take an interest in the question of waste disposal. Much of the legislation drafted in the 1970–1974 period introduced pollution control and waste management regulations on a medium-by-medium basis (air, water, and land disposal), but failed to deal directly with polluting substances or wastes at their point of introduction. The number of new and possibly toxic chemicals entering the market each year was creating enormous stresses on this regulatory system, given its inability to manage the range of hazards represented by these new chemicals. Concerns about an uncontrolled toxics crisis were already surfacing by the early to middle 1970s, concerns magnified by such episodes as the dumping of the hazardous chemical kepone in the James River in Virginia and the widespread presence of polychlorinated biphenols (PCBs) in various products and wastes.[37]

During the early 1970s, two key pieces of toxics-related legislation were introduced. Mainstream environmental groups especially focused on the Resource Conservation and Recovery Act (RCRA), whose provisions for waste recycling had become a major objective. Prior to these debates, the recycling of wastes had emerged as a compelling issue during the late 1960s and early 1970s, particularly for many of the Ecology Action groups associated with the counterculture. Thousands of new recycling centers were established, with a mission of personal transformation and environmental consciousness-raising rather than the development of a viable recycling business. The interest in recycling as a cultural expression extended to some mainstream groups and was particularly influential at the chapter level for such organizations as the Sierra Club, whose members were heavily involved in campaigns to extend recycling's reach. The most prominent of these campaigns, first successfully pursued in the states of Vermont in 1971 and Oregon in 1973, involved the passage of what were called "bottle bills," mandating a deposit or fee for recycled glass containers.[38]

The bottle bill fights, which extended to nearly two dozen states by 1976, pitted glass, retail food industry, and labor interests against both mainstream environmental groups and dozens of local organizations and recycling centers for whom recycling remained the dominant metaphor for environmental change. Issues such as job loss stemming from environmental measures, voluntary programs versus government regulations, environmental programs that emphasized public participation as opposed to those that focused on technological solutions all became part of the various bottle bill fights. These also extended to the debates in relation to the RCRA legislation, in which the bottle bill concept proved to be a main area of conflict between environmental groups and industry interests. Ultimately, environmental lobbyists were not able to keep the bottle bill provisions in RCRA, but they were able to establish the principle that recovery and management of wastes needed to be developed more systematically at the national level.

While RCRA was being debated, mainstream environmental lobbyists were locked in a bitter battle with chemical industry interests over the Toxic Substances Control Act (TSCA), which provided for the review of new chemicals prior to their introduction. As with RCRA, mainstream groups were most interested in management strategies, such as procedures for notification and self-testing by industry. Industry interests, however, feared that TSCA had the potential to allow government agencies to decide which chemicals could be produced on the basis of how toxic they were. Chemical companies, most notably Dow, fought hard against TSCA but were less resistant to the development of the hazardous waste provisions in RCRA, which focused on issues of disposal rather than review at an earlier stage of production. RCRA's Subtitle C provisions for hazardous wastes required an extensive record-keeping and manifest system to track from cradle to grave the generation, handling, and disposal of those wastes. Similar to its Subtitle D provisions regarding nonhazardous solid waste, Subtitle C sought to allow the EPA to better regulate the management of hazardous wastes at the back end of the production cycle.[39]

The passage of RCRA and TSCA in late 1976 extended environmental policy into the wastes and hazards area. The legislation's intent, particularly RCRA with its extensive waste management and disposal provisions, was similar to earlier media-specific legislation such as the Clean Air Act and the Clean Water Act in helping stimulate a pollution control industry. Though sometimes in conflict over implementation issues, the triad of mainstream environmental group, government regulator, and pollution control or waste management industry promoted these hazardous waste management laws and regulations as an effective way to deal with the growing problem of toxics in society. But the problem of cleanup, it turned out, was not so easily addressed—a problem made far more visible when the new community-based groups entered the picture.

The first significant indication of the limits of the new waste policies emerged at Love Canal shortly after the passage of RCRA and TSCA. The dramatic events at Love Canal (which involved such issues as the nature and evaluation of risk and the role and responsibility of industry and government regulators) became a staging ground for the development of a new movement that was primarily about community empowerment. The sequence of events that took place—the casual dumping of highly toxic industrial chemicals by Hooker Chemical over several decades, the leasing of the waste site for one dollar for construction of a school on the site, the discovery of widespread health impacts on local residents and their continuous battle with agencies such as the State Health Department to acknowledge such impacts, and the residents' ability to mobilize effectively and use certain forms of confrontation to force public officials to respond—has since become part of the folklore of the antitoxics movement. A number of the tactics employed, such as the community-initiated health survey, the willingness to challenge rather than just lobby politicians, and a flair for the dramatic symbolized by the hostage-taking of EPA officials, have also become the signature of the antitoxics groups.[40]

The Love Canal events were also significant in demonstrating that new forms of environmental leadership were capable of addressing complex technical issues related to the nature of the contamination and assessment of the risks involved. The Love Canal Homeowners' Association—consisting of nonprofessional lower-middle-class and middle-class family members, especially women—and its leaders, including Lois Gibbs, who emerged as a kind of Everywoman in her appearance and style, belied the traditional profile of environmental organization and advocacy. The Homeowners' Association and Gibbs did effectively use outside scientists and even environmental groups such as the EDF to help situate and buttress their arguments and familiarize themselves with the language and application of technical information. But it was the community residents and their leaders who devised the tactics, framed the strategies, and established their bottom line—in the case of Love Canal, the group's insistence that homes on or adjacent to the dump site had to be purchased by the government, since the contamination had become too extensive and intrusive in its impacts on daily life.

The emergence of this new, alternative movement at Love Canal also differed in certain ways from the anti–nuclear power movements of the 1970s. While many of the antinuclear protests were locally grounded, they also included professional and academic participants who kept a focus on the technical debates about safety and waste disposal issues. The direct-action wing of the antinuclear movement, which drew on different kinds of constituencies, including those associated with the counterculture and New Left, represented a search for a new politics or new forms of spirituality rather than a defense of community or neighborhood that preoccupied the antitoxics groups. In terms of the people who became involved and the spontaneous agendas they developed, Love Canal ultimately prefigured a new way of defining what it meant to be an environmentalist.

Yet the label "environmentalist" did not come easy to this new movement. Many of the community groups quickly became wary of the language, style, and agenda of the mainstream groups. At the same time, the mainstream groups responded to the Love Canal events not by seeking to restructure their approach toward toxics policy but by extending existing environmental policy to deal with the problem of contaminated hazardous waste sites. But it was the threat that additional Love Canals were waiting to be discovered and that loud and intractable community groups were waiting to be formed, more than the lobbying of the mainstream groups, that pushed Congress to fast-track legislation to establish new cleanup procedures. This resulted in the December 1980 passage of the Comprehensive Emergency Response, Compensation, and Liability Act (CERCLA), better known as Superfund.

Instead of drawing community-based groups and mainstream organizations closer together in terms of its stated intent to clean up waste sites, Superfund only served to accentuate their differences. Like RCRA, CERCLA established an elaborate waste management system. This included complex mechanisms for identifying a site, placing it on a National Priorities List, creating mechanisms to name the parties responsible for the contamination, attempting to secure payments to help pay for the effort, and eventually selecting the cleanup technology to be used. The legislation revealed the environmental policy system at its most technically complex and unsuccessful. Each step of the process was subject to challenge, overly bureaucratized, slow to develop, and unable to meet its objectives. Superfund also became vulnerable to political maneuvering due to its high visibility, high costs, and high-stakes outcomes. And while many of the mainstream environmental groups sought to monitor and influence the course of Superfund and RCRA and other toxics and waste-related legislation passed during the 1980s, a significant faction of the movement, led by the National Wildlife Federation and the Conservation Foundation, called for cooperation with industry and the substitution of voluntary initiatives for the unwieldy regulatory framework that had been established.[41]

For this new community-based antitoxics movement, the focus by the mainstream groups on the regulatory system and the push for voluntarism reinforced their perception that they were a movement apart from these kinds of environmental activities and agendas. As groups sprang up in places such as Jacksonville, Arkansas, or Nanotches, Louisiana, their focus on the plight of their own communities and the sense of urgency they brought to their actions led to immediate—and often dramatic—

confrontations that rivaled the Love Canal events. Lois Gibbs herself became a central figure in giving shape and definition to that process. Having relocated to Arlington, Virginia, after her house was bought out, and after achieving even greater visibility with the airing of a television movie about the Love Canal events that focused on her role, Gibbs found herself constantly sought after by local groups for both advice and inspiration. With a few of her allies and advisors from the Love Canal fight and an organizer who had been involved with the Saul Alinksy–initiated Industrial Areas Foundation, Gibbs established the Citizen's Clearinghouse for Hazardous Waste in 1981 to consolidate and extend her rapidly expanding networking activities. Other antitoxics networking groups formed in this period, most prominently the National Toxics Campaign Fund, a spinoff of Massachusetts Fair Share, a Nader-like citizen-action organization. Together, the groups involved with the toxics issue, including Citizens for a Better Environment (CBE), Clean Water Action, the PIRG groups, Greenpeace, and hundreds of community-based associations, helped shape the beginnings of a new social movement.

By the mid-1980s, the number of community groups dealing with toxics and waste-related issues had increased dramatically. Their focus extended beyond existing hazardous waste sites to deal with a range of other waste issues, such as solid waste landfills and new waste-management facilities, including solid waste, medical waste, and hazardous waste incinerators. Many of the sites or facilities were a direct outgrowth of the existing environmental policy system's focus on managing wastes and disposing of them in a more environmentally acceptable manner. But the community groups learned to be skeptical of the claims of that system, particularly with incineration strategies, which had emerged as the preferred high-tech waste management option in an era of growing conflict concerning landfilling.

The incineration fights especially established the potency of this new movement. Although some initial battles of the antitoxics groups had focused on existing garbage dumps and hazardous waste sites and their impacts and hazards, the battles centered on proposed incinerator facilities, which escalated in number during the mid- to late 1980s, were predominantly fights about potential impacts. These battles thus raised questions about how to determine and allocate risk, who appropriately should be making decisions, and about how best to confront the waste crisis. With local governments, the EPA, the increasingly powerful waste industry, and "more responsible" environmentalists (as one consultant characterized certain mainstream groups) touting incineration as one solution to the waste crisis, the antitoxics groups found themselves challenging an important new component of environmental policy.[42]

This challenge took as its starting point the unacceptability of community hazards and a deep mistrust of negotiated waste management solutions that failed to account for such hazards. As a result, antitoxics groups, characterized as NIMBYs, became a focus of attack and efforts at persuasion. This much-discussed "NIMBY syndrome" filled the pages of the waste industry's trade press and generated numerous suggestions and commentaries by consultants schooled in the art of "risk communication" (convincing communities that certain risks were acceptable). Recommendations even included the creation of "environmental flying squads" of friendly scientists and community relations experts who could enter resistant communities to "explain the truth about an environmental hazard," as former New Jersey governor Thomas Kean put it. But instead of isolating facility opponents, the NIMBY focus only seemed to reinforce the message of the community groups, which became even more insistent in their demands to stop the disputed projects. Instead of simply saying no to incineration and landfilling, the antitoxics movement became the single most potent force calling for a new framework for waste policy and a new, overall environmental approach.[43]

The two most important antitoxics groups to emerge were the CCHW and the National Toxics Campaign. Both pointed to new directions in strategy and approach for an alternative environmental movement. The CCHW's action-oriented style of organizing and networking was feisty, humorous, and hostile to mainstream organizational approaches such as lobbying. Though the organization defined itself as a network and sought to emphasize the need to develop community leadership, it engaged in several national campaigns, such as its wide-ranging attack against Waste Management Inc., the largest waste industry company in the country. The CCHW's forte was organizing: spreading its network of organizers—many of them women like Lois Gibbs and Penny Newman—to go community by community in order to spread the gospel of their grassroots Movement for Environmental Justice.

The Boston-based National Toxics Campaign (NTC), which also established a strong grassroots network of antitoxics community organizations, was more focused on developing strategy for the movement, especially in the design of an alternative approach to the toxics issue. During the mid-1980s, NTC organizers launched their own community-by-community campaign, the Superdrive for Superfund, aimed at influencing the debate concerning reauthorization of CERCLA. The final passage of the Superfund Reauthorization Act (SARA) included a number of features, such as right-to-know provisions, technical assistance grants, and emergency planning requirements, that were a direct result of this antitoxics campaign.[44]

In the wake of Superdrive for Superfund, NTC organizers and their allies, critical of environmental policies that emphasized managing wastes at the disposal end, began to formulate a new approach defined by the group as "toxics use reduction." This approach addressed the generation and use of hazardous materials at the front end of production through such interventions as process design changes and product substitutions. Similar in intent to Barry Commoner's concept of the "social governance of production decisions," toxics use reduction helped focus debate on how certain environmental policies failed to account for occupational hazards (the use of toxics that impacted workers) and/or allowed the transfer of emissions or discharges from one medium to another. The antitoxics movement's elaboration of the concept created significant pressure on environmental policy makers to frame a more inclusive approach. By 1990, the EPA began to use the term "pollution prevention," which, despite its ambiguities and unaddressed definitions, suggested a necessary revision of the dominant environmental policy framework.[45]

By the 1990s, the antitoxics movement had succeeded in altering the terms of the policy framework for toxics. The emergence of new constituencies and the intensity of their opposition to environmental policy solutions such as incineration had shifted the focus of policy making increasingly toward the front-end issues related to the structure and functions of the post–World War II industrial order itself. At the same time, antitoxics groups presented an unprecedented challenge to the mainstream environmental groups, which found themselves on the defensive regarding their (lack of) community ties and the related issue of the limits of existing environmental policy. Still, despite the strength of their individual campaigns and the potency of their arguments concerning strategy, antitoxics groups remained a network in search of a movement, a social force within environmentalism not entirely comfortable with the environmental label. With toxics issues continuing to be central to the debates about the future direction of environmental policy, the existence of these groups suggests that the differing tendencies within environmentalism will continue to be a dynamic question in the years to come.

NEW FORMS OF ACTION AND NEW PARADIGMS

While the antinuclear movements were able to coalesce around a unifying concern and the antitoxics groups were organized to protect and empower their communities, a number of other alternative environmental groups came together during the 1970s and 1980s. These groups were defined less by specific issues or rootedness in particular communities than by their need to protest, to bear witness, or to articulate a new set of ideas about social and environmental change. Many such groups expressed a kind of prefigurative or transformative politics, seeking to demonstrate by their actions and ideas new ways of defining environmental concern.

The most successful of these groups has been Greenpeace, the multinational, direct action–oriented organization that has grown exponentially in numbers and visibility during its two decades of existence. Greenpeace was formed in 1971, its name derived from both its antimilitarist (or antinuclear) and environmental focus. "I'm not a Red; I'm a Green," one of the group's early participants liked to say, and the distinction was crucial. Though the various participants in the first years of the new organization saw their roots in the New Left, the focus and direction of their actions was less a political statement than a desire to invent new forms of protest.

The organization's original home base was Vancouver, British Columbia, and although most of its early participants were Canadian, its focus from the outset was transnational. Its first major activity involved an attempt to sail Greenpeace vessels into the nuclear testing zone off the Aleutian island of Amchitka in the northern Pacific Ocean. Though the sailing mission failed to penetrate the zone and halt the tests, the publicity, generated in part by the key role of journalists in the new organization, created enormous visibility for Greenpeace. The Amchitka action also spurred an antinuclear demonstration at the U.S.-Canadian border that drew more than 20,000 participants and immediately elevated the organization into a major new player in the antinuclear and environmental movements.

During the next decade, Greenpeace came to be associated with the danger, excitement, and witness-bearing aspects of its actions, particularly its several voyages on the seas to interfere with nuclear testing, whaling, and the killing of seal pups. In the process, the new organization attracted a number of former New Left and counterculture activists as well as those wishing to find for themselves a new identity tied to the high-profile protests. "The oldest urge of all—to go to sea," as the group's eventual chairman, David McTaggart, put it, was a major part of the Greenpeace identity. Its actions on the seas, where small Greenpeace vessels would seek to confront big whaling ships or the French and U.S. military, gave the organization its individualized, high-risk, macho-style identity. The action itself became the message, the media its transmittal point. The objective, according to key participants, was nothing less than a change in mass consciousness. "If crazy stunts were required in order to draw the focus of the cameras that led back into millions and millions of brains, then crazy stunts were what we would do," Greenpeace's first president, Robert Hunter, said of his organization's approach. "Mass media," Hunter asserted, "is a way of making millions bear witness at a time."[46]

Through the 1970s, a number of Greenpeace groups were formed, primarily in Canada. Sympathizers worked with the group in the United States as well as in France, which had been a focus of anti–nuclear testing protests led by McTaggart. The organization essentially divided its efforts between various sea-based expeditions and the media efforts and organizational support work oriented toward underwriting and highlighting those protests. Each of the expeditions sought to make visible an identifiable "evil," such as the harpooning of whales by the Russians and Japanese or the clubbing of baby seals in Newfoundland. This latter protest, which also witnessed the organizational expulsion of Paul Watson, one of the more defiant of the early Greenpeacers who went on to found the Sea Shepherd Society, was particularly controversial since the action was aimed at centuries-old native practices in a part of Canada that was poor and fiercely independent.

Although the organization grew in numbers and influence during this period, it also suffered from uncertainties about its future agenda and from serious internal divisions that were often personality-based. Watson, for example, who celebrated the idea that action involving a high-risk defiance of authority was crucial to the organization's definition, warned against the group getting "bogged down in the details of organization."[47] On the other hand, one of the group's founders, Robert Hunter, who had joined several of the expeditions, suffered a case of direct-action burnout from the continuous drain of the high-intensity actions and the tensions of sustaining an organization that survived from one expedition to the next. The organization's ultimate strength was its own visibility and international character, which had been strengthened by the success of McTaggart's actions in France (related to his lawsuit concerning the antitesting protests) and the organization's incorporation in the Netherlands, providing it with a European base of operations.

During the 1980s, Greenpeace began to diversify its actions to include campaigns on toxics and energy issues to complement its continuing focus on antinuclear and oceans-related concerns. The group grew considerably in this period, in both membership support as well as staff, most of whom were attracted to its direct-action approach. The organization's growth accelerated especially after the 1985 *Rainbow Warrior* episode. This Greenpeace vessel, which had been docked at a New Zealand port, was blown up by French security agents, and one of the boat's crew, a photographer, was killed.[48]

The *Rainbow Warrior* explosion proved to be a watershed event for the organization. While its emphasis on direct action had ultimately led to a major international incident, the group's subsequent phenomenal growth in membership, staff, and field offices, surpassing in size every mainstream environmental group, turned it into a kind of organizational hybrid. This hybrid was part mainstream (particularly with its growing emphasis on research and publication of expert reports and its dependence on direct-mail solicitation for membership and financial support), part campaign-oriented (with its toxics, nuclear, energy, and oceans-related campaign structures operating as semi-autonomous units), and part direct action (with a continuing reliance on stunts, guerrilla theater, and imaginative forms of civil disobedience).

The organization was still best recognized for its flamboyance—one *New York Times Magazine* profile called the group "Daredevils for the Environment"—and for its international character, with offices in more than 150 countries by 1990, including Amsterdam, the headquarters for Greenpeace International. Greenpeace affiliates were like franchise operations, with several of the groups outside the United States, especially in places such as Germany, the Netherlands, and Austria, being influential forces in their own countries. The organization was particularly effective in its use of communication technologies, which strengthened its international role by linking the different field offices. It also enabled the group to function as a kind of international sleuth, tracking and targeting the actions of its antagonists while preparing to expose their impacts on the environment. At the same time, the organization's dependence on direct-mail appeals meant that its huge membership was "essentially soft," as one former Greenpeace president put it, and subject to dramatic swings in finances. In this regard, major staff layoffs that occurred during the early 1990s were primarily due to a drop in direct-mail income and membership.[49]

Along with its extraordinary growth during the 1980s, Greenpeace still suffered from the absence of any clear sense of its organizational and strategic role within the environmental movement. One Greenpeace organizer recalled the attempt by senior Greenpeace U.S.A. officials from the different campaigns to hammer out a single mission statement, an effort that ended "disastrously," without any basic agreement. As a staff-based organization, Greenpeace lacked the rootedness of the community-based movements, although its decision to elevate the toxics issue as a primary focal point for action often placed it in alliance with community organizations. As its budget and operations grew, particularly in the years after the *Rainbow Warrior* episode, the organization suffered from a range of management problems, exacerbated in part by its hybrid nature. Despite its size and reach, Greenpeace remained an organization most defined by the connection between action and media visibility, a group that still saw itself as bearing individual witness to dramatize what it urgently portrayed as the rapid deterioration of the Earth's environment.

While Greenpeace actions, particularly in the organization's early years, sought to focus attention on the plight of the seas and its inhabitants, such as sperm whales, it never entirely defined itself through a Nature-based perspective, similar in this respect to the antitoxics and antinuclear groups. During the 1970s and 1980s, another set of alternative groups and perspectives emerged that advocated the absolute defense of Nature as their starting point. The key group in this context was the action-oriented Earth First! organization, with the primary conceptual framework for this approach known as deep ecology.

The term *deep ecology* was first used in the early 1970s by the Norwegian philosopher Arne Naess. Naess sought to distinguish between a radical, biocentric view about ecosystems and the need to bring humans into harmony with the natural environment and a shallow, anthropocentric view that placed humans at the center of the human-environment relationship. Naess and other deep ecology analysts felt that the shallow ecology approach, associated with the ascendancy of a professionalized environmentalism, failed to protect the natural environment and confront the kinds of questions posed by the deep ecology philosophy. For the deep ecologists, all life, nonhuman as well as human, has intrinsic value and should not become subordinate to other needs, such as economic development or growth in general. This essential principle, the primacy of all living things, has led deep ecologists to emphasize issues of wilderness, population, and industrialization. They argue that a new kind of philosophy and new forms of social action are now required to reverse the course of the urban and industrial order and to challenge and ultimately eliminate or unmake a technology-based industrial civilization.[50]

The concepts of deep ecology were popularized during the late 1970s and early 1980s through the publication of books and articles that sought to elaborate on Naess's original writings. These ideas influenced the development of a range of different conceptual approaches and organizations, such as bioregionalism (the notion that human societies—and environmental organizations—should be established according to patterns set by the natural environment)[51] and the various kinds of spiritual and animal liberation groups that had direct lineage with the counterculture. Though many of these groups were small in terms of participants and often sectarian in their organization style, they nevertheless helped define an emerging Nature-centered movement, differing from the older-line protectionist groups in their search for prefigurative forms of organization and their rejection of professionalization. For groups such as Earth First!, the most prominent and controversial of these groups, the form of the organization and its actions described the kind of Nature-centered society they wanted to create. To succeed in that effort, to learn how best to defend the Earth, meant becoming not just a tree hugger but a tree spiker as well.

The idea for Earth First! emerged out of the conflicts within The Wilderness Society in the mid- to late 1970s prior to the restructuring of the organization by William Turnage. Several Wilderness Society staff, including Dave Foreman from the group's Washington, D.C., office and Bart Koehler from the organization's regional office in Wyoming, had become disillusioned with the complex bureaucratic process associated with RARE II, the U.S. Forest Service's 1979 Roadless Area and Review Evaluation, which had preoccupied the organization in that period. Environmental groups had been badly outmaneuvered by such corporate lobbying organizations as the Western Regional Council, which had been successful in making available, from the 80 million acres under Forest Service jurisdiction, more than 65 million acres for possible development, including logging, mineral exploration, and recreational activities. Turnage, who forced almost the entire staff involved in the RARE II process out of the organization, defined this failure as a lack of professional skills. But Foreman and his colleagues decided that the problem was one of focus and tactics. "We clearly needed," Foreman later commented, "a harder-nosed approach."[52]

In the spring of 1980, former Wilderness Society staff members Foreman and Koehler, along with Howie Wolke from the Wyoming chapter of Friends of the Earth and two other associates, Mike Roselle and Ron Kezar, went on a hiking, boozing, and camping spree in the Pinacate Desert near a tiny Mexican border town. There they thought up the idea of a new kind of pro-wilderness organization, at once militant and uncompromising in its defense of the Earth. Seeking to recreate the ambience and philosophy of Edward Abbey's eco-marauders from his 1976 novel *The Monkey Wrench Gang*, the new group set out to combine the tactics of civil disobedience and an implicit advocacy of "monkeywrenching" (the use of illegal tactics such as tree spiking or disabling road-building vehicles) with a flair for theater and a kind of rhetorical rejection of industrial society. Earth First! members, who scorned the mainstream groups for their "soft" membership and compromising approaches, would chain themselves in front of Park Service facilities or perch themselves on the tops of old-growth trees slated for logging. Foreman, the group's charismatic leader, who published a how-to manual on monkeywrenching to make clear his support for this "practice against the destruction of the wild and the spread of urban cancer," also played a central role in seeking to situate this new movement outside the traditional bounds of the environmental discourse. "We aren't an environmental group," Foreman said of his organization. "Environmental groups worry about health hazards to human beings, they worry about clean air and water for the benefit of people and ask us why we're so wrapped up in something as irrelevant and tangential and elitist as wilderness. Well, I can tell you a wolf or a redwood or a grizzly bear doesn't think that wilderness is elitist. Wilderness is the essence of everything. It's the real world."[53]

By the late 1980s, Earth First! had become an increasingly visible advocate for this new Nature-centered movement. At meetings, through actions, and at their annual Round River Rendezvous gatherings, Earth First! figures, particularly Foreman, would preach the group's anti-industrial, pro-wilderness, absolutist message. Avoid attempts to clean up the group's "bad image," Foreman warned Earth First! participants about the organization's growing recognition. Monkeywrenching continued to be supported; "just don't get caught," Edward Abbey facetiously told would-be tree spikers. Most strikingly, Foreman and such key Earth First! figures as Christopher Manes argued that putting the Earth first was required in "all decisions, even ahead of human welfare if necessary."[54]

During this period, the organization found itself subject to enormous stresses stemming from its positions and actions. On the one hand, the group's pro-wilderness/anti-industrial posture began to be criticized for its antipeople implications, as exemplified by talk of "eco-wars" against countries such as Brazil (for their deforestation activities) and rhetorical arguments about the value of AIDS, Third World starvation, and even nuclear war as a form of population control or an anti-industrial cleansing action. Earth First! was sharply criticized by other alternative movement figures such as Murray Bookchin for its "eco-fascism," and the debates increasingly absorbed the organization, particularly whether Earth First! needed to develop a social agenda or "stick to the original concepts," as Foreman argued. At the same time, the organization's flirtation with monkeywrenching eventually led to an FBI-conceived and -orchestrated plot to blow up some power lines, which in turn resulted in the filing of conspiracy charges against Foreman and three other Arizona Earth Firsters.[55]

By the early 1990s, the organization had divided along these stress lines. A majority faction of the organization, many of whom were associated with the group's most ambitious project, the 1990 Redwood Summer campaign against old-growth logging in northern California, rejected the anti–human welfare perspective associated with cofounder Foreman. They opted instead to create a more expansive social agenda by exploring potential alliances with community groups and even groups of workers such as those within the timber industry. Foreman and several of his allies, meanwhile, left the organization, complaining that Earth First! was becoming like the Greens in the former West Germany, where an ecological group had been transformed "into a leftist group [with] a more overtly counter culture/anti-establishment style and the abandonment of biocentrism in favor of humanism." The split in the organization also reflected differences in style and political temperament. With Foreman and his Arizona associates plea-bargaining for a reduced sentence in their monkeywrenching trial and the organization's split eventually reducing its media visibility, Earth First! ultimately decided to step back from its earlier, irreductionist stance while remaining an uncertain, though still defiant, part of a loosely connected alternative environmental approach.[56]

The search for alternative forms of environmental theory and practice in the United States was also significantly influenced by the organizational and electoral successes of Western European Green parties in the early to mid-1980s. The Greens sought to transcend traditional political categories by combining a radical critique of the industrial culture, a strong rejection of the arms race and Cold War–dominated perspective on international relations, and a deep commitment to the need for environmental change at a time when most European Christian Democratic and Social Democratic parties were not identified with the issue. The Greens were also associated with what were called in Europe "the new social movements," ranging from anti–nuclear power protesters to housing squatter groups. These movements had developed outside the traditional labor, social democratic, and communist movements and functioned primarily as an extraparliamentary force. And although Western European countries had initiated their own regulations and pollution control measures as extensive (if not more far-reaching) in certain areas as the environmental policy system in the United States, environmental politics in these countries had less to do with the professionalization of the movement than as an arena of action for the new social movements, including the Greens. To be an environmentalist in Western Europe defined one as a Green, while the definition of a Green encompassed a broader social agenda concerning the need for restructuring the urban and industrial order.[57]

The event that most captured interest in the United States was the 1983 West German elections, when Die Grünen, the Green Party, received more than 5 percent of the vote and thus obtained parliamentary representation. The seating of the representatives from Die Grünen, with their jeans and long hair, talk of empowering social movements and new forms of democratic accountability, and advocacy of postindustrial utopian and radical approaches, seemed to symbolize the possibility that a new kind of environmental politics could take root in the industrialized countries. The 1984 publication of *Green Politics*, Fritjof Capra and Charlene Spretnak's book on Die Grünen, became the occasion for suggesting the forms through which a green movement could emerge in the United States.[58]

During the mid-1980s, a variety of groups and organizing networks set out to claim and appropriate the green political label as their own. These ranged from the Minneapolis-based Committees of Correspondence, which emphasized a populist approach; the Vermont-based Institute for Social Ecology, whose main figure, Murray Bookchin, sought to fuse the traditions and concepts of anarchism, the New Left, and ecology; third-party remnants such as the Citizens Party and California's Peace and Freedom Party; Nature-centered approaches, including bioregionalism, deep ecology, and the different kinds of eco-spiritualism associated with writers such as Spretnak, Capra, Bill Devall, and Kirpatrick Sale; and direct-action advocates such as Earth First! and some of the antinuclear activists. Most of these groups and networks identified green politics as constituting a new political tradition, "neither left nor right but up front," as Spretnak and Capra emphasized, borrowing from an early slogan of Die Grünen that had largely disappeared within that movement by the late 1980s. The U.S. Greens, the different groups asserted, would offer a new vision for transforming the post–World War II urban and industrial order, a new way of thinking about issues of technology, industry, and the natural environment.

The effort to construct a green politics in the United States, however, was fraught with conflict from the outset. It suffered from contending theoretical and strategic approaches, different organizing styles, and an inability to translate the green idea into a successful method of identifying constituencies to embrace this new politics. There were significant differences concerning whether the Greens should be primarily or even exclusively Nature-centered, as the deep ecologists and bioregionalists argued, or oriented toward an anticapitalist position, as those associated with the Left Green Network, a caucus within the various green groups, insisted. Divisions also erupted about whether to define the Greens as an electoral party or a social movement. Most crucially, Green groups lacked a firm constituent base and set of dominant issues, so crucial for the identities of the Western European (and Eastern European) movements. As a result, the U.S. Greens turned into a set of factions unable to cohere as a major social movement.[59]

Unlike their European counterparts, the U.S. Greens were also not able to draw upon and contrast themselves to a well-organized and politicized labor movement or social democratic tradition. Electoral campaigns in the European countries, with their proportional representation criteria, were far more accessible to minority parties. Similar to the 1980 Citizens Party experience, U.S. Greens at the local and especially the national level found themselves largely shut out from any effective participation in electoral arenas.

Perhaps most strikingly, the green movement in the United States tended to be divorced from the country's "cultural, ethnic, and geographic diversity," as Carl Anthony, president of Earth Island Institute, put it. Anthony argued that U.S. Greens had largely failed to incorporate the concerns and participation of communities of color into the new definition of environmentalism. "Cancer Alley along the lower Mississippi is not only a toxic wasteland," Anthony insisted, "it is also a cultural hearth, poisoned by poverty as well as toxic waste, of Louisiana's African-American community." An alternative environmental vision and strategy, Anthony concluded, would best emerge from those kinds of communities rather than in deference to the European experience, which tended to "screen out strong emotional attachment to place." Such an alternative environmental strategy necessarily had to address questions of class and race, community and the work setting, and gender and place, each of which were central to the urban and industrial experience in this country.[60]

Earth Day Revisited

It was the ultimate signature event. The day after, the *New York Times* suggested that although "it was not quite *The Day the Earth Stood Still*," it was nevertheless "the largest grass-roots demonstration in history," a "renewed call to arms for an endangered planet" that demonstrated how one could "change the world." With 200 million people participating in 140 countries, Earth Day 1990, according to event organizers, was a fitting successor to its forerunner twenty years earlier, uniting "more people concerned about a single cause than any other global event in history."[61]

For the mainstream groups, Earth Day 1990 was an event made to order. When Denis Hayes, then in his forties and a private attorney, began to explore the idea with several Group of Ten CEOs in late 1988, support seemed assured. The mainstream groups were entering another period of rapid expansion and further institutionalization of the movement. Global issues, such as the hole in the ozone layer and escalating global warming concerns triggered by the severe drought that summer, had brought environmental issues back to the top of policy agendas and further secured the expert role of the mainstream groups. During that fall, moreover, George Bush, in sharp contrast to his predecessor, had suggested he would become the "environment president," a position he later sought to underline by naming William Reilly of the Conservation Foundation to be director of the EPA. Major new environmental legislation, most particularly amendments to the Clean Air Act, was being vigorously debated in Congress. Unlike 1970, when the Ralph Nader–inspired Clean Air Coalition had had to force its way into the legislative debates, the environmental lobby, consisting of yet another coalition of staff professionals, was at the center of this policy process and would ultimately craft some of its key provisions, such as the marketable permit concept proposed by the EDF and endorsed by William Reilly and George Bush.

In this heady period for the mainstream groups, Denis Hayes's proposal for another extravagant Earth Day event, this one global in nature, seemed an appropriate way to close out the Second Environmental Decade. Even more than the first Earth Day, with its lack of centralized direction and its complex relationship with the movements of the 1960s, Earth Day 1990 was to be organized as an explicit media event. Certain themes, such as the need for individual action, would frame the activities and the message. While some groups fretted about the effort to define most of the Earth Day 1990 activities in media terms—particular activities were frequently evaluated by their ability to generate "media bites"—the heavy reliance on professional media skills central to planning the event resembled the professionalization of mainstream environmentalism. The Earth Day 1990 emphasis on individual responsibility, from planting trees to recycling one's garbage, was also directly incorporated into media coverage ranging from television specials to news and editorials. Even more than the efforts to secure consensus in 1970, such themes provided an opening for industry and public officials to embrace the event and shift the focus away from issues of urban and industrial restructuring to the claim of collective responsibility. At the same time, key rallies to celebrate the event became not only star-studded media extravaganzas, but demonstrations without focus, protests without a clear target for protest. Earth Day 1990 might well have been the largest single mobilization of people ever, but it was also the most porous in terms of objectives and outcomes.

The broad and unfocused nature of the event also underscored the failure to address the gap between alternative and mainstream groups, a gap that had become more pronounced than ever. During the 1970s, even as the mainstream groups concentrated their efforts within the national arenas of lobbying, litigation, and the use of expertise, they had still maintained cordial working relationships with myriad local groups involved in such issues as nuclear power and strip mining. By the 1980s, however, with the growing trends toward professionalization and institutionalization and an expanded interest in international affairs and global concerns, the mainstream groups had become far more aloof and separated from environmental organizing at the local level. In turn, grassroots groups, particularly those associated with the antitoxics movement, became sharp critics of what they perceived to be the preemptory role of the mainstream groups in seeking certain legislative solutions that conflicted with the objectives of the alternative groups. Examples included efforts to classify solid waste incinerator ash, a crucial issue for anti-incinerator groups, and an industry-initiated attempt to preempt tighter state pesticide regulations with weaker federal laws, a position some mainstream groups had been willing to accept in order to achieve new pesticide legislation. In these and other instances, alternative groups became sharply critical of how the professionalization process (e.g., the use of professional lobbyists) undermined the organizing and mobilizing efforts so central to the community-based groups.[62]

When the plans for Earth Day 1990 first took shape, event organizers had been concerned that the divisions among groups could potentially influence the competition among those seeking to organize Earth Day events. Key figures such as Ralph Nader, Barry Commoner, and Jesse Jackson had not fully committed themselves in terms of which, if any, Earth Day organizational efforts they would support. Jackson, who had increasingly sought to address environmental issues during the course of his 1988 presidential campaign, was of special concern, since he had joined the boards of the two main competing Earth Day groups, Earth Day 1990 (the Denis Hayes group) and Earth Day 20. Through 1988 and the first months of 1989, this territorial conflict over which Earth Day organization would prevail influenced some of the discussions about how to broaden the event itself. Most prominently, the Jesse Jackson dilemma was resolved when a five-city, pre–Earth Day tour was arranged to include Jackson, Denis Hayes, and John O'Connor of the National Toxics Campaign. The tour was designed to address issues of environmental racism, highlighting the hazards and risks faced by communities of color. Similarly, the Hayes group worked out arrangements with various community-based groups and activists to help coordinate Earth Day events in their communities and thus function as officially sanctioned Earth Day offices. In Indiana, for example, People Against Hazardous Waste Landfill Sites (PAHLS), a key affiliate of the Citizen's Clearinghouse for Hazardous Waste, helped put together a number of Earth Day events in three counties in the state. "It gave us credibility—and resources —that we often lacked outside our own communities," PAHLS organizer Sue Greer recalled, though Greer also questioned the event's long-term impact.[63]

As Earth Day 1990 approached, it seemed possible that at least some modest level of unity within the movement could be achieved. The timing of the event, the twentieth anniversary of the original Earth Day, also seemed to suggest that a moment of summation had arrived. But this review process, particularly by the media, came to be focused primarily on the strengths and unfinished agenda of the environmental policy system, a primer on laws and regulations and what they had achieved and still needed to accomplish. This process of summation and review was further extended to the area of personal responsibility, the idea that people acting in their individual capacities as consumers and waste generators could bring about needed change, complementing the lobbying, litigation, and use of expertise considered necessary to identify problems and solutions. But what was missing from this twenty-year review was any focus on social movements, the idea that people, by organizing and mobilizing through groups, could bring about change. The absence of this idea of group empowerment in defining the event ultimately situated Earth Day 1990 more as an episodic media extravaganza than as a way to measure the evolution of the overall environmental movement and its impact on the larger society.[64]

Earth Day 1990 also became a vehicle, even more than the first Earth Day, for industry interests to try to appropriate the environmental message as consistent with their own. A massive corporate advertising blitz in advance of the event was undertaken with the theme that environmental responsibility had become a corporate trademark. Toward that end, several large industrial polluters, such as Polaroid, Monsanto, British Petroleum, and Honeywell, successfully persuaded some local and regional Earth Day organizations (though not the central group based in Palo Alto, California) to accept their contributions to help pay for events. The most striking example was Polaroid, the biggest polluter of Boston Harbor, becoming the largest funder of the Boston events![65]

The passing of Earth Day 1990 at once closed out an era and foreshadowed events and issues of the future. The twenty-year process of constructing an environmental policy system, now a fixture of the urban and industrial order, seemed to demonstrate more the breadth, extent, and costs of environmental hazards than any effective mechanisms to reduce or eliminate them. The dominant policy framework offered a containment strategy at best, with mainstream groups becoming, in effect, the policy system's managers and caretakers. The alternative groups that had emerged to battle specific hazards, and that challenged the way the containment strategy addressed such hazards, formed a counterpoint to the mainstream groups. They asserted that environmentalism had to be redefined in terms of both mission and strategies. And with Earth Day 1990 giving way to less visible events and activities, and the media-inspired interest in environmentalism beginning to fade once again, the conflicts and uncertain relationships among the different segments of the movement remain the most difficult, unresolved questions facing environmentalism in the years to come.

Part III

ISSUES OF GENDER, ETHNICITY, AND CLASS

Chapter 6

Gender and Place: Women and Environmentalism

A Movement of Housewives

In many ways, the gathering of the forty-three women at the 1987 Women in Toxics Organizing conference at the West Park Hotel in Arlington, Virginia, represented a defining point for the antitoxics movement and the role of women within it. Several of those attending the meeting had already emerged as key figures in their own communities and regions. They had led the fights against unwanted facilities by creating new organizations, by helping recruit new organizers (many of them women and mothers like themselves), and by seeking to address such crucial dynamics as race and class, which were influencing and setting apart the antitoxics movement. But it was the purpose of the conference—reflecting on ways the participants could empower themselves as women—that gave this gathering its special significance.[1]

The tone of the conference was set early on when the West Coast delegation, led by Penny Newman, arrived, wearing their bright new shocking-pink tee shirts. On one side of the shirt was stenciled a figure of a woman flexing her muscles, a slight smile on her lips, with the words "Tough Women Against Toxics"; on the other side was the slogan "WE CAN DO IT." But it was the conference's opening speech by Cora Tucker that seemed to best place the tee shirt's message in perspective.

A mother of seven children and grandmother of six who had lived her entire life in Halifax County, a rural area of Virginia near the North Carolina border, Tucker was one of a handful of powerful African-American women who had become central to this new movement. A civil rights activist since she was a teenager ("if you're black, you're dealing with these things whether you like it or not," she would later say), Tucker first became involved in environmental issues in the early 1980s through the fight to stop a uranium mining project. The more she became involved in the uranium conflict and related issues (such as a proposed high-level nuclear waste repository to be located in the area), Tucker found herself confronting long-standing assumptions that environmental issues were white issues. She recalled her first encounter with mainstream groups, when she was asked to speak at a Sierra Club gathering on the uranium issue. She angrily discovered after arriving at the meeting that the club's leaders were unprepared for a local black activist and even considered postponing the event. Tucker had also been struck time and again by how public officials and even some mainstream groups had trouble with her style and focus of activity as a woman, which emphasized "the special knowledge and talents that women could bring to the movement," as she put it.[2]

For Tucker, as for many of those gathered at the West Park Hotel, toxics issues were invariably intertwined with questions of gender and place. "We know more about our community than anybody else," Tucker told conference participants. "None of your politicians know more than you do. And when you go to the local government bodies . . . they pat you on the back like you do a child when she's complaining and say, 'O.K., Sugar, we're going to look into it.' I find that the only people who are going to do things for us is going to be us."[3]

Tucker's opening remarks at the West Park meeting were noteworthy for highlighting a major underlying theme of the conference: the involvement of women in the antitoxics movement was a function of *who they were*, related to their sense of family and community. This involvement was also shaped by how women dealt with information, asked questions, and shared insights about such information with others. For the participants, how the toxics issue was experienced and how the movement was organized provided a common thread, a basis for identifying a new kind of movement identity.

To begin with, community-based movements against toxics have been led primarily by women. Many of these women leaders are new to social action movements. They often find themselves forced to deal with highly public situations despite having previously led private lives. These new antitoxics leaders have also discovered their talents as organizers, based on their ease with one-on-one discussions, their connectedness with those they are trying to organize, and their interest in communicating and sharing what they are learning as leaders. While men might, as Cora Tucker remarked in her speech, "be lost for a week and not ask nobody for directions," women, in contrast, know how to ask in order to figure out the things they need to know. Women became leaders in the antitoxics movement because they are often adept at a range of community leadership skills, a set of talents given little political recognition. These skills, derived from women's experiences in managing their homes, engaging in activities concerning their children's schools, and in holding their families together, have been particularly relevant in the context of a movement of neighbors and residents characteristic of the antitoxics groups.

Similarly, the experience of mothering has also become important in establishing women's leadership roles in this movement. "I think women bring so much more to an organization because we go at it, most of the time, from the point of view of how it affects our children," Cora Tucker said in her speech. Concerns about children have been central to the antitoxics movement. Children are seen as particularly vulnerable and often powerless constituencies in terms of the traditional avenues of environmental protest and power. "Women are the first to know when their kids are at risk," as Cora Tucker put it. They respond accordingly, demanding that hazardous sites or facilities be cleaned up or shut down. One of the most powerful protest images that emerged out of the Love Canal protests was a Mother's Day Die-In, focusing on children's exposure to hazardous wastes and issues of reproductive health. Yet mainstream environmental groups, with their reliance on expertise and professionalism, often define these same environmental and toxics issues as issues of science and policy, not questions of personal experience and pain. Community-based groups, on the other hand, tend to define such issues in terms of experience. In this context, mothers especially become more assertive in their claims that they know when something is not right and that something has to be done about it. As caretakers and as nurturers, mothers bring to the movement a sense of immediacy and passion and inclusiveness related to the task at hand. "Everything is a women's issue," Cora Tucker declared at the conference, "because every child that's born, some woman had it."

In this setting, toxics issues as women's issues are crucially issues of place. "We become fighters," Tucker remarked, "when something threatens our home." Threats to the home from waste sites or hazardous facilities are seen as daily life threats to health and property, to where and how one lives. The frequent charge that the antitoxics groups are selfish NIMBYs just wanting to maintain property values fails to address how property represents, within so many of the communities impacted by toxics issues, the equivalent of home or place. The threat to property is not simply a financial issue (a not insignificant concern when one's major asset is one's home) but a question of rootedness and survival.[4]

The importance of place has become a dominant and powerful metaphor within the antitoxics movement, particularly for its women participants. In one striking episode at an antitoxics convention, many of whose delegates were new to any form of activism, the gathering decided to endorse and participate in a major demonstration about the issue of homelessness scheduled to take place while the conference was occurring. As part of its endorsement, the convention issued a statement linking the experience of homelessness to the violation of place and possible loss of one's home stemming from the presence of toxics or hazardous wastes in a community. Despite the cultural gap between movements ("we had no idea whether any of our people would actually show up for the march, but almost everybody did, and it became an extraordinary, even empowering experience for them," Penny Newman later commented about the event), this "movement of housewives," as some of the antitoxics activists also refer to themselves, was able to make a connection not ordinarily made within the traditional domain of environmentalism.[5]

The significant role of women within the antitoxics movement has also created its own set of stresses tied to gender issues. Given their identity as housewives, or "home *makers*," as Cora Tucker characterizes herself and others, many women in the movement prior to their involvement in antitoxics activities tended to be seen by husbands, family, and neighbors as more deferential or passive in their attitudes and relationships. Engagement in the movement, however, has not only allowed women to actively participate and shape their organizations, but has often transformed how they perceive their own identities as well. As organizers and leaders who become capable of questioning and challenging various sources of institutional power, many of the women in the antitoxics groups are transformed, in the eyes of others, into different people, a gender-based form of empowerment.

This experience of personal as well as public transformation, parallel in some ways with the experience of consciousness raising described in feminist literature, also has the potential of unsettling relationships within the family, particularly between husbands and wives. Some antitoxics groups continually witness troubled and broken marriages caused by husbands unable to adjust to this process of transformation. Women organizers with young children also have time conflicts and pressures from their role as primary caretakers, particularly when husbands are unable or unwilling to adjust their own lives to help out accordingly. Once the commitment to the movement is made, many of these same women are able to become full-time participants as their children get older and time pressures are reduced. Thus, the community-based antitoxics movement relates to family issues as well, with changing gender roles an integral issue for the movement.[6]

The combination of this personal, transformative experience and the crucial importance of place has also been bound up with the powerful image of community often evoked by antitoxics movement participants. There emerges a kind of before-and-after concept of what a particular community is all about, tied to the events of each antitoxics struggle. On the one hand, the presence of toxics can represent a real threat to a specific area in terms of health risks and a community's livability. At the same time, the development of an antitoxics movement, particularly in small towns or rural areas, helps define a new sense of community established through the struggle. In this setting, new personal relationships and identities develop that parallel the emergence of new social roles and relationships, producing new community leadership. The fact that many of these leaders turn out to be women previously excluded from the sources of local, institutional power adds the dimension of community governance to the movement's framework—that is, the antitoxics movement also becomes a community development movement.

In Kenosha, Wisconsin, for example, a midsize industrial city long dependent on automobile manufacturing, an antitoxics conflict ultimately raised these questions of community development and governance associated with gender. During the late 1980s, Chrysler made plans to shut down its large plant in Kenosha. After significant community protests became bound up with the 1988 presidential elections, the company arranged to link the phasing out of the plant with support for new development in the community. The funds made available for such development, however, became tied to plans to build a large medical waste incinerator in Kenosha, which led to the spontaneous creation of a group to oppose the facility. This group, Kenoshans Against Medical Waste Incinerators, led by the wives of laid-off auto workers, quickly emerged as a powerful force in the community. It not only challenged the arguments of the incinerator advocates but successfully undertook a recall election of local elected officials, asserting that future development decisions had to become more participatory and sensitive to the women's concerns over the livability of their community.[7]

By the early 1990s, this "movement of housewives" had become a major force, not just in individual communities such as Kenosha or concerning issues of toxics and wastes, but in relation to the environmental movement as a whole. The focus on community development, on place, on children, and on participation and access to power has provided a gender framework for the antitoxics movement and has created a different kind of language for environmental protest. Historically, the environmental idea, long associated with issues of management and/or protection of Nature, had evolved as part of a set of male concerns. The toxics issue, on the other hand, particularly as it emerged during the 1980s and 1990s, not only challenged the substance and framing of this environmental idea, but fundamentally raised the gender-related question of who should be part of the process in defining and laying claim to that idea.

A Male Preserve

Through much of the late nineteenth and twentieth centuries, mainstream environmental groups functioned as male-dominated organizations, both in terms of leadership and conceptual framework. The early protectionist and conservationist approaches derived largely from the nostalgia regarding the passing of the frontier and the response to urban and industrial forces that transformed the frontier and restructured work, family, and community as well. These traditional interpretations of the roots of environmentalism contain powerful gender implications, manifested by the reaction to the shift from a frontier culture to an urban and industrial age.

The frontier experience was strongly associated with images of mastery over the natural environment, based on the ability to transform a "virgin continent" filled with wilderness by exploring it, seizing it, and making it available for economic and recreational activity. These male notions of frontier activity contrasted sharply with how women "came to know and act upon the westward-moving frontier," as Annette Kolodny has argued. In her historical narrative about women's fantasies regarding wilderness and the frontier, Kolodny distinguished between male images of conquest and alteration of the natural environment with the female desire to locate "a home and a familial human community within a cultivated garden."[8]

These contending images—one central to the creation of the frontier myth, the other eventually subsumed under the rise of domesticity and the division between work and home as male and female domains—also help situate the contrasting social movements that emerged during the Progressive Era, in the period between the 1880s and World War I. Exclusive male groups, such as the Boone and Crockett Club, established themselves as a kind of male bonding society dependent on reenacting frontier and wilderness experiences. Even the Sierra Club, which organized itself less around hunting issues than on mountaineering and an appreciation of and communion with wilderness, still remained overwhelmingly a male organization with a male perspective on the natural environment as rugged, adventuresome, and monumental rather than diverse, interactive, and holistic, concepts of the environment that emerged in later periods.[9]

The need and desire to experience wilderness, particularly through hunting and exploring as well as mountaineering, skiing, and fishing, were also associated with images of "manliness," as Theodore Roosevelt, the foremost champion of the masculine definition of the wilderness experience, often put it. During the Progressive Era, Roosevelt, through his writings and actions, became the embodiment of this highly gendered concept of wilderness. "Hunting big game in the wilderness is, above all things, a sport for a vigorous and masterful people," Roosevelt and George Bird Grinnell wrote in the Boone and Crockett Club book *American Big-Game Hunting*. The club's own constitution, reflecting Roosevelt's influence, stipulated that the object of the club was to "promote manly sport with the rifle." Roosevelt extended this frontier-related manliness argument with his descriptions of the new nationalism for a robust, imperial America. The ability to maintain frontier and wilderness values ultimately became central to the Rooseveltian view of the country's mission and its future greatness and national virility.[10]

The issue of the character-building importance of the natural environment combined with a growing nostalgia about the loss of the frontier helps explain the brief but extraordinary popularity of the "Nature Man," Joseph Knowles, who decided to recreate a two-month, back-to-nature wilderness experience in the Maine woods in the summer of 1913. Knowles, a Boston-based artist in his forties, had himself grown up in northeastern Maine not far from where he decided to undertake, naked and without food or weapons, his much-publicized foray. Only through his instincts, wits, and desire to appreciate wilderness, he argued, would he be able to survive.

Knowles's descriptions of his experiences, chronicled in his best-selling *Alone in the Wilderness* (complete with finely realized drawings that he claimed were drawn in the woods with charred sticks on birch bark), sought to capitalize directly on concerns about what he called the "artificial life" of the cities in contrast to the "wild rugged life of the outdoors." As opposed to the materialistic "accumulation" images found in *Swiss Family Robinson* associated with the development of a mercantile class 200 years earlier, Knowles's tale of his wilderness encounter, from locating food to trapping a bear and then bashing it to death with a club, is filled with Rooseveltian images of "resolution, manliness, self-reliance, and capacity for hardy self-help." Knowles sought to convey the impression that what he accomplished was essential for character building: the experience of Nature allowed men to learn to live the way they were "meant to live."[11]

Knowles's wilderness tale caricatured the nostalgia about the loss of the primitive frontier. Despite his presumed life in the wilds cut off from any human interaction, the dispatches he wrote, using the same instruments employed in his drawings, were somehow "discovered" and reproduced in one of Boston's papers, creating an instant following for his wilderness encounter. Comparing the book's frontispiece photo of the bearded, bear-skinned, vigorous-looking frontier man holding his staff while posing for photographs on the day he came out of the woods, with the picture of the somewhat flabby, awkward-looking, middle-aged man posing with reporters and friends before entering the woods two months earlier, seemed only to reinforce the staged nature of the event.

Nevertheless, Knowles was still celebrated for his exploits. Immediately undertaking a speaking tour, he was greeted by huge crowds in Boston and other cities, and more than 300,000 copies of his book were sold the following year despite an exposé in a rival Boston paper that pointed to evidence that Knowles had spent his time in an enclosed cabin. The attention Knowles received turned out to be fleeting. His proposal to establish a wilderness colony "where every lover of nature may live as he wants to," on thousands of acres of government-owned lands, which, "if not utilized, would remain a waste for hundreds of years," was, for example, never seriously considered. Still, the sensationalism generated by the episode underlined the efforts that had emerged to recapture Nature and wilderness as part of men's lives.[12]

The most striking example of such efforts could be found in the meteoric rise of the Boy Scouts movement in the period prior to World War I. The Scouts organization, which formally incorporated in 1906, quickly became the largest male youth organization in U.S. history, with more than 350,000 scouts and 15,000 scoutmasters by the end of its first decade. Scouting advertised itself, similar to Joseph Knowles's wilderness encounter, as an "environmental surrogate" for the frontier. "The Wilderness is gone," lamented Scout movement leader Daniel Carter Beard, who declared in a Scout publication in 1914 that the "hardships and privations of pioneer life which did so much to develop sterling manhood are now but a legend in history, and we must depend upon the Boy Scout movement to produce the MEN of the future." A writer for *Colliers* declared during this same period that "the Scout movement is most potent" because of its "firm base deep in the natural outdoor instincts of boyhood."[13]

The rise of the Boy Scouts, with its emphasis on reestablishing a male encounter with Nature and wilderness, could also be seen as a response to what one analyst has described as a "masculinity crisis" in the Progressive Era that related to both the diminishing of wilderness and the changing roles of work and family. Although frontier and wilderness nostalgia extolled manliness and rugged individualism, the farm and frontier life prior to the Progressive Era had in fact relied heavily on shared experiences of work and survival for men and women alike. The separation of work and home that became dominant with the first industrial revolution and the rise of the industrial city, as well as the small yet significant increase in the numbers of women entering the urban and industrial work force, created a false nostalgia about gender roles that paralleled the search for recreating the wilderness experience. The development of such organizations as the Camp Fire Girls and the Girl Scouts in 1912 reinforced gender dichotomies concerning relationships to the natural environment and the home by the groups' insistence that "the girls of this generation must have a chance to learn the household tasks of cooking and sewing and care of the babies in the same spirit in which the boys rehearse the life and activities of men." One Girl Scout advocate proclaimed that "service in the home is emphasized because the home is the unit of society, and the good home-maker makes the good citizen."[14]

While women's domain was defined as the home and women as homekeepers, men's roles were increasingly defined by work separated from family life. The promotion of outdoor activities such as scouting and male-oriented activities such as athletics, with which Roosevelt was associated, sought to reinvigorate a masculinity displaced by an urban and industrial rather than a rural or Nature-based definition of work. The outdoors, or Nature, increasingly separated from concepts of home, work, and community, was thus situated as a male preserve, with the preservation of Nature and the conservation of resources distinctive male goals. "My own hope for the conservation of American forests and waters is to plant into every

American father these queries," Zane Grey wrote in an Izaak Walton League publication. "Do you want to preserve something of America for your son? Do you want him to inherit something of the love of outdoors that made our pioneers such great men?"[15]

If men were to be defined by their work and the effort to recreate manly roles related to Nature and vigorous sport, a woman's role was to be defined primarily as "keeping the home and bearing and rearing her children," as Roosevelt insisted.[16] Yet the home could also serve as an environmental surrogate, tied directly to the environmental hazards of urban and industrial life, as a number of female-led urban reform and public health groups discovered. While these movements explored issues of sanitation and workplace hazards, another group of primarily female environmental advocates, led by scientist and nutritionist Ellen Swallow, founded a new kind of gender-inspired, home-based environmental improvement movement.

Born on a New England farm in 1842, Ellen Henrietta Swallow, a graduate of Vassar College and the first woman to attend the Massachusetts Institute of Technology, was the first American to apply the German concept of *oekologie*, or ecology, to her research and advocacy. Trained in the fields of sanitary chemistry and nutrition, Swallow—who began to write under her married name, Ellen H. Richards—effectively integrated Progressive Era concerns about efficiency and rational management into her analysis of the home environment as a central arena for environmental activity. The emphasis on home-keeping for a new urban middle class had also produced a level of waste, according to Swallow, reflected in the "garbage-pail and overfurnished rooms." The public welfare thus demanded that home life be "governed by the best knowledge which science has been able to gather with reference to health and efficiency," while women needed to "become imbued with the scientific spirit of the age."[17]

In pursuing this study of the sanitary science of the home, Swallow organized a Woman's Laboratory at MIT to help train women for science careers and helped establish or direct several organizations and associations that dealt with home environment issues. Some of her early works, such as *The Chemistry of Cooking and Cleaning* and *Food Materials and Their Adulteration* focused on what Swallow considered the special problems of women with respect to food and nutrition as well as house-cleaning issues. By the early 1890s, Swallow began to explore larger issues of the urban environment, such as air and water contamination, and home environment concerns as part of a more integrative science she called euthenics, defined as the "science of controllable environment."[18]

Swallow was a prolific writer whose books and lectures covered a wide range of issues, and Swallow's concept of euthenics became integral to what Carolyn Merchant has called her "principles of ecology of earth and home." Defining human ecology as the study of the effects of human environments on people's lives, Swallow sought to differentiate between a natural environment and an "artificial environment" produced by human activities that created the "noise, dust, poisonous vapors, vitiated air, dirty water, and unclean food" found in everyday life. "Fresh, clean air, free of the pollutants of offensive factories, was necessary to human health," Merchant wrote of Swallow's approach, while "fouling a stream caused injury to one's neighbor below; and fertile soil was required to grow nutritious food. Any individual who selfishly used these life-sustaining elements squandered the human inheritance. Each family and city were points in the Earth's larger cycle of water flow and vaporization, of soil dissolution and deposit, and of plant scavenging, cleansing, and purification."[19]

Swallow's writings and advocacy work helped inspire several types of movements among middle-class women, including a consumer nutrition movement related to the hazards of industrialized food production and an environment education (or municipal housekeeping) movement tied to contemporary sanitary reform movements. She was also instrumental in founding the home economics movement, linked primarily to the environmental issues of the household. It was through the home economics movement that Swallow popularized her concepts of *oekologie* and euthenics, with different environments—whether urban or Nature-based—defined as interactive rather than separate and discrete.[20]

While Swallow identified poor and working-class issues such as dust emissions and water contamination as an appropriate field for the home economics teacher, home economics evolved into more of a middle-class movement tied to Progressive Era changes separating male and female domains at work and home. As a middle-class concept, home economics became increasingly associated with efforts to restrict women's activities in the workplace, even as growing numbers of women entered the work force.

The urban and industrial implications of Swallow's human ecology science also contrasted with Aldo Leopold's land science and the soil science that Charles Clements began to write about during the 1930s. As opposed to Swallow's urban orientation, these new sciences of land and soil ecology came to focus primarily on the pressures caused by resource development, such as timber cutting, grazing, water development, and agriculture. With advocacy about the protection of Nature seen as more directly constituting male concerns, and urban and especially home environments perceived to be part of a woman's domain, conservationist and preservationist groups tended to define their work through such gender distinctions, often limiting female roles to the kinds of upper- and middle-class concerns of garden clubs and similar women's organizations.[21]

Reinforcing these distinctions has been the male dominance of these groups. With the important exception of Rosalie Edge's persistent campaign against the leadership of the Audubon Society during the 1920s and 1930s, few women have participated within traditional conservationist and preservationist organizations, and even fewer have assumed leadership roles. A review of the boards of directors of the Sierra Club, Wilderness Society, Audubon Society, and National Wildlife Federation up through the 1980s reveals an almost exclusive male presence historically, with only a gradual improvement over time. Even the prominence of Rachel Carson, whose *Silent Spring* generated its own set of gender-based controversies (the prevailing conception of science, for example, was considered a male sphere, distinct from Carson's own notion of a more accessible science focused on "all living things"), still failed to ensure wider participation of women in the movement at that time.[22]

Only during the 1970s and 1980s, coinciding with the rise of the contemporary feminist movement, did gender issues more directly come into play regarding certain environmental themes, a process that culminated with the predominance of women and "women's issues" within the antitoxics and antinuclear movements. The major changes of the post–World War II order, tied to the new technologies of production, a restructured urban environment, and the emergence of a wide range of new and protracted environmental problems, also had their own gender implications. These were tied to the changing nature of work and home, community, and family. The concept of the environment, long associated with male images of and concerns about Nature and its loss and with male-dominated organizations that focused on those concerns, had come to be challenged by new movements and new concepts that linked the question of the environment to daily life concerns and the issues of gender bound up with them.

Women in the Workplace

It has become something of a sociological truism today to state that increasing numbers of women work two continuous jobs—at work and at home. Since the mid-1950s, the proportion of women in the work force has continued to grow dramatically, including women with young children or of child-bearing age. This growth has occurred at the same time as an increasing range of chemical hazards and environmental exposures experienced at work and in the home have come to be identified. Concerns about occupational and environmental health for women have touched off a critical debate regarding how best to protect women as future mothers and how industrial issues also reflect on issues of women's roles in society as a whole.

The question of workplace hazards for women has long been debated at both a movement and policy level. With the nineteenth-century rise of the industrial city and new manufacturing industries such as textiles, where women were first employed in large numbers, occupational hazards for women workers became a significant and contentious area of dispute and intervention. As early as 1908, Alice Hamilton would write that women workers were particularly susceptible to exposure to industrial hazards. This occurred because women often worked within hazardous workplaces such as the lead industries or were employed by industries "not in themselves dangerous," but that had become so because they were "habitually carried on under unhealthful conditions which could perfectly well be done away with." As one example, Hamilton spoke of the widespread occurrence of "phossy jaw" due to the use of white phosphorus within the match industry, in which significant numbers of women were employed and thus exposed. Focused on these kinds of situations, women worker advocates such as Hamilton, Florence Kelley, and union organizer Rose Schneiderman sought to raise questions about the hazards of the workplace for women and promote organizing efforts to address them. Heavily involved with a small number of female-based trade unions and support organizations, such as the Women's Trade Union League and the National Consumers League, these advocates sought to empower working women, who were especially vulnerable as a largely unorganized constituency, given the mistrust and, sometimes, hostility of male-dominated unions.[23]

This empowerment strategy took on several different forms. In the period prior to World War I, a number of women workers, aided by the advocacy groups, began to mobilize over their poor working conditions. These workers included most prominently thousands of women garment and apparel workers in New York and Chicago who eventually went on strike to protest their deplorable and hazardous working conditions. These events were capped by the most notorious occupational disaster of the period, the 1912 New York City Triangle Shirtwaist factory fire, in which 154 women and young girls burned to death or died jumping from windows because there were no fire escapes and doors were locked from the outside.

This tragedy was a pivotal moment for the growing, female-based trade union movement, elevating the question of working women to major prominence in the discussion of occupational hazards that preoccupied both policy makers and advocacy groups at the time.[24]

Aside from these heroic, though often limited, organizing efforts, some women worker advocates began to express an interest in developing protective or exclusionary legislation to distinguish between female and child labor and the male work force. Protective legislation advocates such as Alice Hamilton based their arguments on the difficulties involved in overcoming the traps experienced by women employed in hazardous industries. "We must remember that we are dealing with a class which is not really free, and which is compelled to a great extent to follow certain trades, and to work in certain places, and has very little choice," Hamilton wrote of these women workers. For Hamilton, these were matters of both workplace discrimination and public health. By World War I, reform groups such as the National Consumers League had become strong proponents of specific protective legislation, including the eight-hour day for women and children, frequently opposed by industry interests. But it was the changing conditions of work and the new kinds of industrial activity during and after World War I that especially influenced those, including Hamilton, who saw the need to develop a new approach concerning women's workplace hazards. These hazards were becoming more extensive and complex, due in part to the "great many new and more or less unfamiliar industrial poisons [that] have come into use since the war," Hamilton wrote in a 1926 *Bulletin of the Women's Bureau* of the U.S. Department of Labor. For Hamilton, Florence Kelley, and other worker advocates, protective legislation for women offered a kind of halfway measure of protection in the face of the continuing vulnerabilities experienced by women workers.[25]

Gender-based protective legislation, however, also reinforced the contention that women were less capable of pursuing certain kinds of work, whether jobs requiring physical labor or those affecting their potential role as mothers. In the first major court case on the issue, the U.S. Supreme Court in 1908 ruled that a woman's "physical structure and a proper discharge of her maternal function—having in view not merely her own health, but the well-being of the race—justify legislation to protect her from the greed as well as the passion of man." Although Hamilton argued that protective legislation could extend efforts to address occupational hazards as a whole, by the 1920s protective legislation had also become synonymous with attempts to keep women workers out of certain jobs not necessarily related to occupational hazards.[26]

Thus, the discriminatory nature of protective legislation became a major concern in the 1920s for advocates of an equal rights amendment (ERA) for women, including the Woman's Party and the Women's League for Equal Opportunity. These groups felt that "restrictions on the conditions of labor should be based upon the nature of the industry, not on the sex of the workers." Protective legislation advocates such as Hamilton and Kelley, however, saw a class distinction between middle-class proponents of an ERA-mandated repeal of protective laws and unorganized working women, who potentially benefited from such laws. "The practical effect of the efforts of the Woman's Party," Hamilton angrily wrote of the ERA advocates, "is to hold down in their present condition of overexploitation large classes of working women and to do nothing to alleviate their lot." This division concerning how to best confront occupational hazards for women extended throughout the 1920s campaign for the ERA, with Hamilton becoming an ERA supporter only in the early 1950s as a result of the changing position of women in the work force that had occurred during and after World War II.[27]

While the employment of women increased from 18 percent of the work force in 1900 to more than 25 percent in 1940, the kinds of jobs that became available to women as well as the nature of the female work force changed considerably during World War II. As a result of war-induced labor shortages, as many as 19 million women—or more than one-third of all women over the age of fourteen—were employed by 1945, many of them in manufacturing industries such as auto, steel, and rubber, which had higher-paying, unionized jobs. But these jobs also involved hazards not previously experienced to any significant degree by women workers. The profile of women employed in these and other jobs had changed as well. By 1940, 32 percent of all employed women under the age of forty-five were married. Those figures increased significantly during the war, thus adding pregnancy to the range of issues confronting women workers. By the end of the war, major debates began to unfold about whether new restrictive legislation should be instituted, and, more generally, whether women workers hired during the war should be displaced by returning male veterans. This latter issue was also tied to the deeper cultural question of women's roles and whether married women—working class or middle class—were obliged to define themselves exclusively as homemakers.

Unlike the debates about the ERA during the 1920s, nearly all protective legislation advocates in the post–World War II years were associated with these "back to the home" campaigns. Much of the literature on occupational hazards for women, including but not limited to pregnancy issues, reflected that bias. It emphasized that differences between men and women, whether physical strength, tolerance of fatigue, or higher susceptibility to occupational disease, demonstrated the need to eliminate women from certain jobs. Many of these arguments were based on incomplete and even inaccurate research as well as widespread assumptions in the culture that sought to underline differences between the sexes when it came to industrial hazards.

Many of these assumptions were challenged by a comprehensive 1946 study undertaken by Dr. Anna Baetjer on behalf of the Army Industrial Hygiene Laboratory, which cautioned against applying exclusive gender criteria to the analysis of industrial hazards and sought to review systematically the range of specific problems associated with women in industry. Baetjer's conclusions were complex and varied. She argued that there was "no reliable evidence to support the generally accepted view that women [are] more susceptible than men to occupational diseases," and that restrictive policies limiting the kinds of jobs available to women, especially those involving the use of toxic chemical substances, were misplaced and discriminatory. "Under all circumstances," Baetjer asserted, "the exposures [to such chemicals] should be kept below harmful levels." Women, in fact, appeared to be "more critical of working conditions than men," reinforcing the appeal for reduced hazards and more adequate sanitary facilities on the job. "If a working environment is safe for men," Baetjer concluded, "it is equally safe for women."[28]

On the other hand, Baetjer also argued strongly that occupational exposures were a factor in pregnancy and that pregnant women should not work in such occupations as the lead industries, where the permitted concentrations of harmful substances would not be considered safe for pregnant women. But Baetjer also warned against policies that caused women workers to be fired once it became known that they were pregnant. Citing the U.S. Department of Labor's "A Maternity Policy for Industry," Baetjer underlined the need for such policies as extended maternity leave to better protect the status of working women rather than reduce their rights or narrow their job options. While acknowledging that the relationship between pregnancy and industrial hazards required intervention, Baetjer felt that the potential for discrimination had unfortunately framed much of the discussion about how to intervene. To counter that approach, Baetjer recommended an "equal pay and equal opportunity for advancement" policy as most conducive to protecting women workers.[29] Despite its contemporary significance, this recommendation failed to generate interest or controversy when Baetjer's 1946 study was published.

Although women remained in the work force in significant numbers during the first couple of decades after the war, surpassing the wartime employment peak for women by 1955, Baetjer's concerns about the potential for discrimination were reinforced by the powerful cultural images about the working father and stay-at-home mother that predominated during this period. A number of the writings associated with the emergence of contemporary feminism during the early to middle 1960s, such as Betty Friedan's *The Feminine Mystique* and Alice Rossi's 1964 essay on women and work in *Daedalus*, were responding primarily to the cultural stereotyping and economic discrimination regarding women in the professions, but ignored the issue of occupational hazards for women. The early women's movement thus functioned primarily as a civil rights movement dealing with discrimination issues and as a cultural rebellion challenging the assumptions about women's roles. The industrial issues—what kinds of jobs were available to women and the kinds of hazards women would encounter on those jobs—would reemerge during the mid-1970s through the development of a separate movement sympathetic to, yet distinct from, feminism.

Two key studies helped inspire this later movement: a 1975 study, *Occupational Health Problems of Pregnant Women*, undertaken by Vilma Hunt for the U.S. Department of Health, Education, and Welfare and a 1976 guidebook on women's job health hazards called *Working for Your Life*, written by Andrea Hricko on behalf of the Nader-affiliated Health Research Group in conjunction with the Labor Occupational Health Program at the University of California at Berkeley. Hunt's study raised again the wide range of occupational hazards, some well documented but many still poorly understood, that jeopardized the health of pregnant women and their fetuses. Hunt was particularly critical of the absence of gender-based research on work hazards since Baetjer's 1946 study, even though women workers tended to be concentrated in relatively few industries, allowing for more comprehensive research on this subject. Hunt's study was complemented by Hricko's guidebook, which not only explored the kinds of hazards present in jobs where large numbers of women worked, but offered information about how women could best challenge and be empowered to confront the conditions at their workplaces.[30]

This new women's occupational health movement had its roots in the women's health movement of the early 1970s as well as the occupational health movement that had developed during the same period. Much of the early focus of occupational health advocacy groups, however, tended to be on large, unionized, manufacturing industries, where men were predominantly employed. The women's health movement, whose primary focus, at a time when abortion was legalized and sterilization abuse issues were visible, was on issues of reproductive health, at first ignored the workplace setting, where issues of reproductive health were in fact potentially most pronounced. However, by the mid- to late 1970s, as the women's occupational health movement began to organize, a number of factors emerged to make the issues related to women at work more prominent.

For one, given the enormous increase of women in the work force and a stagnant economy, the two-income household (or the single-parent household headed by a working woman) had become an economic necessity. The cultural debate about whether women should stay at home was rapidly being eclipsed by the fact that women had no choice but to work. At the same time, new studies were demonstrating that women's work was also hazardous work. A number of jobs in which women were heavily employed turned out to have numerous occupational health problems. Office workers were confronted with a wide range of hazards and stresses, from VDT exposures to indoor air pollution to carpal tunnel syndrome; nurses and other medical workers were subject to numerous chemical and infectious hazards; electronics workers, many of them women, were exposed to a wide variety of solvents and other toxic chemicals. These hazards were poorly understood, often not regulated by OSHA or other agencies, and tended to be ignored by most unions. They only belatedly emerged as issues for unions with substantial female membership, such as the public employee unions, or for women worker advocacy groups, such as Nine to Five and the Coalition of Labor Union Women.

What most effectively mobilized the new women's occupational health movement was the reemergence of reproductive health issues in the workplace and new restrictive approaches pursued by industry. By the late 1970s, a number of companies began to institute policies that banned women from working at certain jobs within hazardous industries, often well-paying jobs primarily held by men. "It seemed," commented former OSHA head Eula Bingham, "that the more hazardous a company's operations were (i.e., the more toxic the chemicals it produced, or the higher the number of exposures suffered by its employees), the more 'protective' its management was." Bingham pointed out that such policies seemed out of place in a post-OSHA era in which the agency took into account, as with its 1978 lead standards, "the need to protect the 'functional capacities' of both women and men" by creating tighter standards and more stringent rules.[31]

For industry groups, the new OSHA standards for lead were reason enough to explore instituting restrictive practices, or what some called "fetal protection" policies. One of the more dramatic examples of this approach involved American Cyanamid's lead pigment plant at Willow Island, West Virginia, along one of the chemical corridors on the Ohio River. The company, which had not employed any women in production-related work at this plant prior to 1974, had hired by 1978 about twenty-five women among the more than five hundred workers employed at the higher-paying jobs at the facility. During 1978, these women production workers were informed at a series of meetings that the company's new "medical policy" required it to bar all fertile women between the ages of fifteen and fifty from several different job categories. Those employed at such jobs who refused sterilization would be terminated or transferred to a lower-paying job. The company claimed it had become necessary to bar women from such jobs in order to prevent "the possibility of birth defects from exposure of a fetus to hazardous substances." Fearful of losing their jobs, five women workers at the plant proceeded to undergo sterilization surgery to avoid being subject to the new company policy.

The justification for American Cyanamid's position was based less on science than a desire to avoid liability. OSHA investigators had discovered that the American Cyanamid plant was not in compliance with lead standards, but the company successfully argued before the U.S. Court of Appeals that it could not economically reduce ambient lead levels sufficiently to protect a developing fetus. The Court's action suggested that the American Cyanamid approach, similar to other company policies instituted at the time, could legally prevail in hazardous industries where women were employed. This meant that it was easier for industry to "exclude women than to clean up the workplace," as Eula Bingham put it.[32]

The events at American Cyanamid also contrasted with events at Lathrop, California, where it was discovered that the manufacture of DBCP at an Occidental Chemical plant had resulted in a reproductive hazard for men. The relatively rapid response by the regulatory agencies in suspending the use of the chemical and instituting a near total ban on its domestic production, as well as the absence of any threatened litigation on the part of Occidental, seemed to offer two crucial lessons: that reproductive hazards were a major issue for men as well as women, and that the response to an outcome of male sterility was immediate and intense. As Bingham put it, "the difference in society's response to the reproductive hazards faced by men, as opposed to women, came as a shock even to me. In the case of men, the response was to shut down the plant until the hazard was abated; when women faced similar risks, they were simply excluded."[33]

These new exclusionary policies, initiated by corporations such as General Motors, Gulf Oil, B. F. Goodrich, Union Carbide, Monsanto, Dow, DuPont, and BASF Wyandotte during the late 1970s and 1980s, became an organizing focus for a number of organizations and coalitions that began to address the women's occupational health issue. These included occupational health advocacy groups as well as the Coalition for the Reproductive Rights of Workers, organized by women in the occupational health movement. These groups argued that "hazards, not workers, should be removed" and that "employers should not be permitted to discriminate against women, or any other group of workers, in an effort to 'solve' a safety and health problem." Efforts at regulatory reform, however, were not successful, and, with the advent of the new Reagan administration, much of the focus shifted from federal agencies and Congress to the courts, where a series of cases involving fetal protection policies began to be reviewed.[34]

The most significant of these cases involved Johnson Controls, the largest manufacturer of automobile batteries in the United States. The Milwaukee-based company had adopted, in the context of concerns about lead exposures, a job exclusion policy in 1982 that applied to all women, regardless of age or plans for childbearing, except those with medical proof of sterility. Although a U.S. Court of Appeals upheld the company's policy, the U.S. Supreme Court, in a unanimous ruling in 1991, found the company had violated the Federal Civil Rights Act of 1964. A court majority, moreover, declared that the Civil Rights Act prohibited all fetal protection policies, not simply the one adopted by Johnson Controls.[35]

The Johnson Controls court case was a major victory for the women's occupational health movement insofar as it separated the issue of exclusion from the continuing concerns about job hazards, including reproductive hazards. Despite the victory, however, the question of work hazards for women loomed larger than ever. Such hazards remained prevalent within certain kinds of jobs, whether those involved with lead or other toxic substances or those creating physical and stress problems associated with workplace conditions. These hazards issues were also women's issues, not necessarily because women were more susceptible to such hazards, but primarily because sex segregation caused women to be overrepresented in such jobs. For example, women predominated at jobs that were highly repetitive or involved prolonged sitting or standing, or where a high level of stress was exacerbated by the lack of control over the conditions and nature of the work. Reproductive hazards also remained significant in jobs as varied as microchip assembly and farm work, with the hazards addressed in only limited ways and with limited research, despite visible fetal protection policies and subsequent court rulings.

By the early 1990s, the women's occupational health movement had succeeded in shifting the debates about working women from a fetal protection philosophy bound up with the assumptions about a woman's proper place to a focus on the industrial hazards themselves. "A risk to one is a risk to all" had become the movement's battle cry, although it was simultaneously pointed out that special risks were also associated with women's work. Providing a critique of the limited ways that gender issues in the workplace have been addressed, this women's movement, defined largely outside the framework of environmentalism, has also brought to the fore the question of what constitutes a safer and less hazardous environment.

The Search for a Women's Environmental Politics

The rise of the contemporary women's movement not only helped to highlight questions of sex discrimination in the workplace but also elevated gender issues for many of the social movements seeking to challenge the urban and industrial status quo. The development of consciousness raising, an activity that first spontaneously exploded on the scene at an SDS convention in December 1965, became at once a rejection of the male organizing and leadership style within these movements and a form of empowerment for women. Consciousness raising helped define women's organizational roles and what constituted a women's agenda, whether in terms of specific issues or organizational culture.[36]

The rapid rise of the women's movement in the late 1960s and early 1970s, however, had little initial connection with the rapidly expanding mainstream environmental movement of this period. There was little or no opportunity for consciousness raising or other women's movement activities in the old-line conservationist or protectionist organizations such as the Sierra Club and the Audubon Society or the new professional organizations such as the Environmental Defense Fund and the Natural Resources Defense Council. These groups had few women in leadership positions, ignored the gender implications of their form of organization, and failed to acknowledge "women's issues" as an arena for action.

One exception was Zero Population Growth (ZPG), which actively solicited professional women by linking small family size (or life without children) to support for women in their careers, although ZPG remained vulnerable to feminist criticism regarding the sterilization issue. Friends of the Earth and Environmental Action also supported abortion rights and child care programs, but these positions tended to be peripheral rather than central to their agendas. Even explicit attempts to join the two movements, including a late-1970s effort to develop a feminist/environmentalist coalition, were not able to bridge the gap between distinctive organizational agendas and priorities.

The idea for such a coalition first emerged in reaction to attempts by the timber and nuclear power industries to attract women to their anti-environmentalist positions. Helen Burke, the only woman on the Sierra Club's executive committee at the time, sought to counter these campaigns through a women's outreach program for the club. Burke was also interested in pulling together a coalition of women from both mainstream environmental groups and various feminist organizations. Those interested in this networking idea, including Environmental Defense Fund staff members Marcia Fine and Charlene Dougherty, hoped that such a network would provide support for the ERA campaign, push for a greater number of women at top-level government positions within the environmental bureaucracies, and promote a greater role for women within the environmental movement as a whole. But from the outset, this network had difficulty deciding whether to target women's issues that intersected with environmental issues, such as reproductive health and hazardous consumer products, or whether to provide environmentalist support for specific women's issues, especially the ERA. By primarily seeking to develop links with professional-oriented feminist groups such as the National Organization for Women (NOW), the effort at alliance became most defined by the question of the ERA, the exclusive area of concern and organizing for NOW in that period. Despite a short-lived effort to establish a Women for Environmental Health Coalition and a Feminist Anti-Nuclear Task Force and the appearance of a one-time EPA publication, *Women and the Environment*, the efforts to build a coalition never significantly coalesced, a casualty of the traumatic defeat of the ERA as well as the increasing professionalization and institutionalization of the mainstream environmental groups. By the early 1980s, these groups had increasingly distanced themselves from the female-led activism at the local level, while their own male leadership culture and organizational emphasis continued to prevail.[37]

During the 1970s and 1980s, the handful of women in leadership roles, such as Louise Dunlap of the Environmental Policy Center, Cynthia Wilson of Friends of the Earth, and Janet Brown of the Environmental Defense Fund, often found themselves adopting the leadership style and emphasis on management and professionalization that characterized other male-led groups within the movement. For these groups, "Washington was an all-boys club," as Dunlap put it, while local environmental activity, through independent groups or even chapters of some of the national organizations, tended to be staffed or at least informally led by women. "I tried to ignore the problem" of sexual stereotyping, Dunlap declared several years after her departure from the organization. She recalled how some of the "older, clubbier chief executive officers of the groups would even try to exclude me from White House meetings." Dunlap overcame some of those prejudices by functioning as "one of the boys" in settings such as the Group of Ten, where her participation had been assured through her personal and organizational ties with Robert Allen of the Kendall Foundation, the Group of Ten's key behind-the-scenes figure.[38]

The EDF's Janet Brown, a less visible and outspoken figure than Dunlap, nevertheless also found herself assuming a male leadership and management style. Hired for her management experience, background in fundraising, and professional associations, Brown quickly discovered that the EDF, more than most other mainstream groups, was heavily weighted toward professional staff interests and activities. During the late 1970s and early 1980s, when Brown served as the group's executive director and Group of Ten sessions began to occur, the EDF stood apart from the other groups in its heightened professionalism and lack of interest in coalitions and local activities. Brown eventually clashed with those in the organization who wanted to fully maintain the staff-directed nature of the organization and resisted Brown's efforts to relocate EDF's headquarters from New York to Washington, D.C., Brown's replacement, Fred Krupp, an attorney and executive director of a professional environmental organization based in Connecticut, succeeded in overcoming much of the staff mistrust by emphasizing fundraising activities and by heavily promoting the organization's more ideological posture regarding market-based incentives, which had been largely staff-initiated.

What struck Brown in retrospect about her experience at the Environmental Defense Fund was the absence of any gender considerations in the development of policy or approach toward management or organizational culture. Although Brown had, prior to her EDF association, developed an interest in how women relate to such questions as science, policy, and risk, she never translated those interests into the EDF setting nor was she ever encouraged to do so. "We never had the time to want to explore gender questions," Brown remarked of her EDF tenure. "It was not high enough on anyone's agenda, and it seemed much more important to raise money or expand the board or fight certain of Reagan's policies." Noting that she and Dunlap and FOE's Cynthia Wilson were all eventually fired from their positions, Brown concluded that gender considerations, like questions of equity, were simply seen as irrelevant to the framework and definition of the movement.[39]

Yet, for such issues as nuclear power and toxics, gender considerations did come into play. During the mid- to late 1970s, for example, polls indicated a gender gap with respect to nuclear power, with larger numbers of women wary of the technology and opposed to the construction of nuclear power plants. These differences in support became the focus of a major effort by the Atomic Industrial Forum, the industry trade group, which sought to link nuclear power to female lifestyle options such as the availability of gender-related consumer products. But this campaign was ultimately as ineffective as it was clumsy. Beyond the gender gap in the support for nuclear power, women played a significant role in both the community-based and direct-action wings of the antinuclear movement, a level of participation and leadership that contrasted sharply with that of the mainstream organizations.[40]

This more distinctive women's role in both antinuclear and antimilitary activities has had a long and rich history. It dates back in this century to Jane Addams and Alice Hamilton's opposition to World War I and their participation in the founding of the Women's International League for Peace and Freedom. During the early 1960s, a new organization, Women Strike for Peace (WSP), was organized to protest an escalating arms race, including the environmental hazards of nuclear testing. In its first action on November 1, 1961, the new organization undertook a nationwide mobilization of women for a one-day "strike" involving 50,000 women in more than sixty cities. Using such slogans as "Pure Milk, Not Poison" and "Let the Children Grow" and capitalizing on its middle-class constituency and especially the role of mothers within the organization, Women Strike for Peace was able to break through the anticommunist and pacifist stereotyping of the times and define itself at the center rather than at the margins of the culture. These were, according to *Newsweek*, "perfectly ordinary-looking young women, with their share of good looks, [looking] like the women you would see driving ranch wagons, or shopping at the village market, or attending PTA meetings."[41]

While Women Strike for Peace became an important part of the anti–Vietnam War movement and continued to mobilize concerning the nuclear arms race issue through the late 1960s and 1970s, its role as a women's organization was largely eclipsed by the political and cultural changes occasioned by the rise of the contemporary feminist movement during the same period. Still, despite WSP's nonfeminist appearance, the organization anticipated later challenges by feminists regarding the hierarchical, male leadership styles within the peace movement and that leadership's tendency to present its arguments in more abstract rather than personal or experiential terms.[42]

By the 1970s and early 1980s, a number of women activists influenced by feminist ideas began to shape significant parts of the anti–nuclear power movement and the revitalized anti-arms race organizing efforts. They did this in part by expressing a more personal language of opposition. Some of the key public figures in these movements, such as Helen Caldicott and Anna Gyorgy, presented what they defined as a woman's perspective on the technologies involved and their potential impacts on human and natural environments. At the same time, such actions as the 1977 occupation at Seabrook, the 1979 Diablo Canyon demonstrations, and the protests in the early 1980s against the Lawrence Livermore weapons lab in northern California were themselves infused with a protest language and organizing approach that had strong feminist undercurrents. At these actions, the process of protest—that is, the forms of action as well as the methods of decision making—became as important as the outcome of such protest. These approaches became particularly significant with respect to a series of women's protest actions during the early 1980s, when the escalation of the arms race, including the placement of U.S. Pershing missiles in Western Europe, dominated the peace and antinuclear movements.

In November 1980, 2000 women held a demonstration at the Pentagon, ringing the building with huge yellow and red puppets while shouting and dancing and chanting "Shame." During the event, several participants attempted civil disobedience by seeking to weave yarn across the building's entranceways. Though this Women's Pentagon March was small in terms of the numbers of participants as compared to some of the other anti–arms race actions in that period, it was a compelling event for those who participated as well as for a growing constituency of women organizers seeking to link peace, feminist, and environmental themes. The call to action had asserted an "ecological right" based on the Earth's—and mothers'—regenerative capacities, a right threatened by the power and activities of the Pentagon. With its evocative appeal to defend "the life of this planet, our Earth, and the life of the children who are our human future," the call also became an appeal for a new kind of movement, a new kind of environmental/feminist/antinuclear synthesis.[43]

The original idea for a Women's Pentagon March had emerged from the discussions at the Women and Life on Earth Conference held in March 1980 at the University of Massachusetts in Amherst. Attended by more than 800 people, the conference brought together veteran antinuclear activists, various women's movement participants, and several women associated with the idea that there was an organic connection between women and the environment. Lois Gibbs, immersed in the fight for relocation of Love Canal residents and yet to establish her antitoxics network, was also one of the speakers at the conference, where she defined the Love Canal events as a transformative event for women. "Women who at one time looked down on people picketing, being arrested and acting somewhat radical are now doing those very things," Gibbs said of Love Canal residents like herself.[44]

For several of the organizers of the Amherst conference, the ideas discussed and the connections established were laying the groundwork for what was being called "ecofeminism." Part theoretical construct, part literary expression, and part social movement as embodied in the Pentagon action, ecofeminism quickly emerged as an amalgam of different perspectives and approaches about the human/nature relationship and the integration of feminist, ecological, and antimilitarist ideas. The term *ecofeminism* (or *ecofeminisme*) was first introduced by the French writer Françoise d'Eaubonne in 1974 and elaborated the following year in Rosemary Radford Ruether's plea in *New Woman, New Earth* for an ecological revolution to "overthrow all the social structures of domination." With the publication of such pivotal works as Susan Griffin's *Woman and Nature* and Carolyn Merchant's *The Death of Nature*, this synthesis of feminist and ecological ideas became associated with a wide range of intellectual and literary gender-based arguments about the assault on Nature—and human environments—characteristic of the contemporary urban and industrial order. These included critiques of the uses of science and technology to subdue Nature, which consequently established a dualism between human society and the natural environment; analyses in which the historically rooted absence of fathers from nurturing and early parenting roles was linked to the way boys, as they became men, learned to want to control and exploit their environment; and spiritual arguments that had a gendered explanation about human relationships with Nature. These varied analytic approaches provided a rich and at times conflicting framework for the development of ecofeminism, situating it as both a set of ideas and an approach to action.[45]

During the early to middle 1980s as ecofeminism became a significant force among various alternative environmental groups, it also began to experience the strain of accommodating its different approaches. The most successful actions linked to ecofeminist ideas, such as the two women's marches on the Pentagon (the original November 1980 action and a related demonstration the following year) and the various peace camps modeled after the Pershing missile–related women's encampment at Greenham, England, during the early 1980s, were themselves subject to differing interpretations as to their significance. Some ecofeminists celebrated the effort to design a nonhierarchical, consensus-based process, similar to some of the antinuclear direct-action protests. This was seen as laying the basis for a "prefigurative politics," as social movement analyst Barbara Epstein has called it—that is, finding ways of constructing a feminist and ecological social order. Others valued the intent of the actions themselves: demonstrating that women could become a powerful and organized force in challenging the arms race and the slide toward environmental destruction.

Ecofeminist approaches have also tended to reflect some of the differing approaches within the feminist movement itself, especially distinctions between "radical" feminism, with its emphasis on cultural, spiritual, and personal liberation, and "socialist" feminism, with its focus on transforming gender roles in urban and industrial society. As a set of groups most connected to the anti–arms race protests of the early 1980s, ecofeminism was never able to establish an organizing approach beyond the highly intense, yet ultimately episodic anti–arms race activities of that period. While still influential in terms of its ideas, including efforts to redefine the conceptual basis for both environmentalism and feminism, ecofeminism has still failed to emerge as a fully developed social movement.

The difficulties of translating theory into action have been further compounded by the lack of relationship between ecofeminists and women activists within the antitoxics movement, one of the most dynamic forces shaping environmental politics in the 1980s and early 1990s. Despite Lois Gibbs's appearance at the Amherst conference, a significant gulf in language and organizing style separates the female-led, community-based movements from the ecofeminist theoreticians and activists most directly associated with antinuclear and feminist peace politics. Many of the women in the antitoxics groups, themselves new to any form of movement activity, have remained wary of a "feminist" label, despite a growing solidarity and celebration of their roles as women leaders within their movement.

Yet the antitoxics movement, in its language of protest, its organizational culture, and its approach to community, has fundamentally expressed a feminist approach and outlook. The reluctance to call this organizing "feminist," however, represents more than just a problem of labels. The antitoxics groups, operating primarily in the context of survival politics, have functioned largely without a developed theoretical framework, with limited strategies, and without a longer-term vision. At the same time, the divorce between organizing and organization building for the ecofeminists reflects the difficulty of creating a feminist environmentalism. And for the environmental movement as a whole, the inability to respond to the feminist question in environmentalism—what constitutes an agenda and organizing style that incorporates women's experiences equal to men's—continues to reflect the absence of approaches to gender and place capable of reshaping the movement's view of itself and its possibilities.

Chapter 7

Ethnicity as a Factor: The Quest for Environmental Justice

Tunneling to Disaster: The Gauley Bridge Episode

When Alice Hamilton launched her investigation of lead poisoning in lead smelter mines in the early part of this century, she was surprised how lightly the industrial medicine system—the company doctors, druggists, and hospitals—evaluated lead-based illnesses in their communities. During her travels, she was especially struck by the remarks of a Salt Lake City apothecary, who told Hamilton that he had never known a case of lead poisoning to occur in the neighborhood of the smelters. "I explained that was incredible," Hamilton responded. To which the apothecary replied, " 'Oh, maybe you are thinking of the Wops and Hunkies. I guess there's plenty of them. I thought you meant white men.' "[1]

By the time Alice Hamilton began to document various workplace and community hazards, ethnicity had emerged as a significant, though understated, factor in how such hazards were experienced. As one example, during the first two decades of the twentieth century, the foundry industry underwent major restructuring, including the widespread introduction of new machinery, a deskilling of the work force, an enormous increase in the hazards resulting from these changes, and a dramatic shift in the ethnic makeup of the work force as well. The number of African-American foundry workers jumped from nearly zero before 1915 to more than a third of the work force by the 1920s and as much as two-thirds in some southern cities, such as Birmingham. Many of these workers were subject to the dust-related illnesses caused by foundry work, whose conditions were similar to those of other "dirty, hot, and unpleasant" industries that increasingly employed African-American and immigrant workers in jobs that "native Americans or Americanized foreign-born white labor did not want." This ethnicity factor was seen primarily as an issue of disadvantaged groups—immigrants or people of color—working at some of the most hazardous jobs and living in some of the most polluted communities created by changes in the structure of industry and patterns of urban settlement during the Progressive Era. In this period, pollution problems rapidly developed their own discriminatory pattern, reflecting the racism and class divisions already deeply embedded in the urban and industrial culture.[2]

The most striking historical episode of environmental discrimination was the tragedy at Gauley Bridge, West Virginia. It was there, in the midst of the Depression, that hundreds of miners, both black and white, died painful deaths as a result of their work for a subsidiary of Union Carbide, the company that much later attained international notoriety for its Bhopal, India, chemical plant disaster in 1984. The sequence of events first began in the late 1920s, when plans were developed to build a dam, hydroelectric plant, and tunnel near the town of Hawk's Nest, located on the New River at the foot of Gauley Mountain. The hydroelectric facility, a major source of power for the state, would eventually illuminate the entire city of Charleston, the state capitol, thirty miles southeast of the plant. There were also plans to ship the discarded rock and sand debris consisting of a rich silica deposit for use at another Union Carbide manufacturing subsidiary.

The Union Carbide subsidiary and its contractor, Virginia-based Rinehart and Dennis, knew from the outset that construction of the Hawk's Nest tunnel, the centerpiece of this ambitious project, was likely to be hazardous. The tunnel site had an extraordinarily high percentage of silica deposits—90 percent or more, as later tests determined. The company, in fact, had selected the specific tunneling route to maximize the amount of silica for extraction and shipment elsewhere. By breathing silica dust, miners faced the risk of developing silicosis, an illness caused by tiny particles of silica becoming absorbed by cells deep within the lungs. Eventually, these cells would become damaged and scarred, breathing capacity would be reduced, and the lungs would become susceptible to infections such as pneumonia and tuberculosis. By 1930, when construction at the Hawk's Nest tunnel began, chronic silicosis was a well-known disease, one of the more widespread forms of pneumoconiosis related to breathing metallic or mineral dusts. Silicosis was considered by some to be the oldest recorded occupational disease in history, and several published articles had also noted the high susceptibility to silicosis associated with certain industrial processes, such as mining, which had expanded in scale during the first decades of the twentieth century. Major policy debates regarding silicosis, particularly compensation questions, had already emerged during that period, so the decision to proceed with the tunneling at Hawk's Nest was likely undertaken with forewarning of the risks involved.

Recruiting on a racial or ethnic basis for especially hazardous work such as silica mining was also not an unusual practice at the time. A number of industries relied on relatively docile migrant workers, who had little access to unions or community resources to enable them to challenge work decisions or hazards. In the Hawk's Nest situation, Union Carbide and its contractor sought out southern, African-American, migratory laborers, particularly from Georgia, Alabama, Florida, and the Carolinas. The companies concluded that these migrants were clearly preferable to even local African-American workers from surrounding lumbering camps and coal mines who might not tolerate the working conditions and physical abuses of the job foremen.

The work force eventually assembled at Gauley Bridge included nearly three times as many African-Americans as whites. Of the whites who worked inside the tunnel, more than two-thirds were either foremen, operators of heavy equipment, or had responsibilities for transporting the men through the tunnel. While these jobs were almost entirely held by whites, almost all unskilled work positions, including drilling and mucking, which involved the highest exposure to the silica dust, were undertaken by migrant African-Americans hired for those tasks. The job distinctions, it turned out, were not only discriminatory, but fatal. Within just a few years after construction work had commenced, hundreds of African-American miners and a smaller number of white miners began to die from exposures related to their work at the Hawk's Nest site. This catastrophic outcome, unprecedented in its scope, eventually identified the events at Gauley Bridge as "America's worst industrial disaster."[3]

What happened at Gauley Bridge also reflected the racial biases of the time. Hiring migrant African-Americans at jobs where occupational illnesses were likely anticipated, albeit as illnesses expected to develop gradually long after the dispersal of the work force, did not seem an unusual decision in the midst of the Depression. During construction, with the contractor focused on completing the work on schedule, precautionary measures such as wet drilling or respirators to suppress or protect against dust emissions were rarely undertaken, except, on occasion, for the white engineers. Certain tasks, such as clearing debris and resuming the mining directly following a blast, when the dust would have been thickest, often exposed the least protected African-American workers to the highest concentrations of dust. Later court testimony situated these divisions of tasks and precautions directly in racist terms. One purchasing agent for the contractor commented how he wouldn't give $2.50 (the cost of a single respirator) "for all the niggers on the job." "I knew they [the work] was going to kill these niggers within five years, but I didn't know they was going to kill them so quick," the same agent was quoted as saying. Though possibly apocryphal, these statements nevertheless reflected the prevailing philosophy of the companies to handle the hazards associated with the work by racially allocating the risks anticipated.[4]

What the contractor and the company had not anticipated was the speed of the onset of silicosis and the extraordinarily large numbers of Hawk's Nest tunneling workers to eventually contract the disease. According to public health historian Martin Cherniack, whose methodical investigation of the Gauley Bridge events stands as the definitive account of this disaster, the American medical profession at that time recognized only the chronic form of silicosis, in which the disease occurred gradually, and not, as at Gauley Bridge, cases in which those exposed to very high concentrations of silica dust developed an acute form of the disease and became ill within a few years or even months from initial exposure. While the unprecedented numbers of deaths at Hawk's Nest helped define this new illness for the medical profession, it also produced a series of conflicts about how to interpret what took place at Gauley Bridge and why.

Even before the tunneling project was completed in 1930, miners began to contract the disease, but accounts from later court testimony and hearings suggest a cover-up of the events. This included the disingenuous statements of the company doctors, who characterized the illness as "tunnelitis," and company payments to a nearby undertaker, who disposed of the bodies of mostly black workers in a field that was later planted over, thus eliminating the possibility of identifying the workers and evaluating the causes of their deaths. Several of the miners and miners' widows, both white and black, pursued compensation claims, and these court cases constituted the opening battle concerning the interpretation of the events. By 1933, when an out-of-court settlement was reached between the contractor, Rinehart and Dennis, and seventeen Fayette County attorneys representing Gauley Bridge plaintiffs, it had become clear that hundreds of Hawk's Nest tunnel workers had by then contracted the disease or had died. Nevertheless, that settlement, subsequently characterized as a great victory by Union Carbide officials, provided for only $130,000 to be split among 157 plaintiffs in contrast to the original $4 million in claims. Half the settlement was alloted to the plaintiffs' attorneys, and another $20,000 in attorneys' fees was secretly made available by Rhinehart and Dennis on condition that the attorneys not engage in further action and that all case records be surrendered to the defense. Though the court later stipulated that half of the additional $20,000 go to the plaintiffs, the court failed to rescind that part of the agreement that provided the case records for the defense.

Settlement monies were also allocated on the basis of racial and marriage criteria: $400 for an unmarried African-American man; $600 for a married African-American man; $800 for an unmarried white man; $1000 for a married white man; and $1600 for the families of deceased white men. Eventually, 538 damage suits were filed, 34 of which were posthumous. Of these plaintiffs, only about two-thirds actually received any form of payment, with the average settlement payment less than $400 per worker. Nor did the number of court actions fully reveal how many had been affected by the disaster itself. Though the number of deaths became a disputed issue at the time, Cherniack, through his own investigations of company records, calculated that, conservatively, 581 of the 922 African-American workers and 183 of the 291 white workers died from silicosis as a result of work in the tunnel. "The death toll of the disaster at Gauley Bridge," Cherniack concluded, "was immense when compared with any other outbreak of industrial disease in modern history."[5]

While controversies about the Gauley Bridge settlement were limited to West Virginia during the early 1930s, the issues resurfaced a couple of years later, stimulated by articles in *The New Masses*, a left-wing publication associated with the Communist Party. A new round of investigations and controversies took place, including 1936 hearings held by New York congressman Vito Marcantonio. Though these hearings generated substantial attention, including industry countercharges that the Gauley Bridge events were being exploited by "ambulance-chasing" lawyers, they failed to result in further action. The interest in Gauley Bridge soon began to fade, with the silicosis issue, particularly by the 1950s and 1960s, becoming less prominent in the debates about occupational hazards. At the same time, the question of the racial distribution of risk emphasized by *The New Masses* authors was not pursued by policy makers or occupational reform advocates, and this issue faded as well.[6]

When the events at Gauley Bridge began to be reexplored in the 1970s and 1980s, this new interest could be traced to the emergence of the occupational health movement during this same period. For these groups, the events at Gauley Bridge described an industrial disaster, linking occupational hazard issues with environmental concerns. Only in the last few years has the lesson of the Hawk's Nest disaster been explored as an example of risk discrimination. As revealing as they are tragic, the events at Gauley Bridge provide a starting point in analyzing the relationship between issues of ethnicity and environment.[7]

EXPERIENCING RISK: PESTICIDES, LEAD, AND URANIUM MINING

The ways in which ethnicity enters into the environmental experience have been primarily determined by how environmental issues and movements are themselves defined. This determination is also influenced by the longstanding assumption that the environmental movement "belongs" to upper-middle-class or elite Anglo constituencies. Reinterpreting that question along the lines of the main themes of this book—that environmental issues have emerged in the context of urban and industrial change—also broadens the notion of movement "ownership." The struggles concerning pesticides, lead pollution, and uranium mining provide useful places to begin in this respect.

During the 1960s and early 1970s, concerns about pesticide impacts among mainstream environmentalists, stimulated by the publication of *Silent Spring*, focused largely on the hazards to wildlife. Although Rachel Carson's book had also raised the issue of farmworker poisoning, it was influential primarily for its warning that pesticides such as DDT were killing large populations of birds and fish. The pesticide-related environmental activism of the Environmental Defense Fund and the membership or chapter activities of groups such as Audubon and the Sierra Club reinforced this perspective. Much of the regulatory and legislative activity that followed, including the banning of DDT, further reflected those concerns.

However, the late 1960s also witnessed the emergence of farmworker health and safety issues as part of the pesticide debates of the period, which provided an important, though less visible, historical example of environmental protest and action.

In the mid-1960s, with the termination of the Bracero Program (which had provided legal contracts for Mexican farm labor for U.S. agriculture), a major new effort was launched to organize a union for farmworkers in California. Under the leadership of Cesar Chavez, the United Farm Worker Organizing Committee—later the United Farm Workers (UFW)—sought to combine strategies of mobilization and grassroots organizing with the more traditional goal of securing a contract for union representation. The social movement quality of the organizing was directly related to Chavez's appeal to an emerging Chicano consciousness in both rural agricultural counties and urban barrios in California. The crucial historical role of Mexican labor in California agriculture combined with the significant numbers of Chicano and Mexican-born farmworkers still employed in the fields to transform the drive for farmworker unionization into *la causa*, with its enormous ramifications for ethnic identity and its themes of social justice.[8]

From the outset of its campaign, the UFW confronted enormous resistance on the part of the agricultural industry. Though the union drive focused on questions of wages and representation, the UFW, as a social movement, also sought to address the harsh conditions of work in the fields that prevailed throughout the state. A key workplace issue involved health and safety questions associated with the growers' intensive use of pesticides. For farmworkers, the explosion in pesticide use through the 1950s and 1960s created significant environmental hazards. Southern San Joaquin Valley residents, for example, told of "pesticide clouds" that rolled through the area as if a heavy fog had set in. The numbers of reported cases of pesticide poisoning also increased substantially, even though such figures failed to account for the numerous cases of poisoning unreported due to language barriers, fears about grower retaliation, and physician ignorance or unwillingness to link illness to pesticide exposure. Only a handful of California physicians and health professionals, many operating on a shoestring, sought to explore pesticide-related illness in greater depth. The small amount of research pursued primarily focused on acute reactions from handling pesticides, while information about long-term chronic effects was nearly nonexistent.[9]

Despite the limited toxicological information available, the pesticide issue offered a number of opportunities for UFW organizers. The more familiar they became with the problem of pesticides as a farmworker health hazard, the more directly they saw it connected to poor sanitary and working conditions in the fields that heightened the potential for exposure. By further insisting that the issue of pesticide use become part of the negotiating process for contracts, the union was also able to raise the issue of the workers' right to know about the toxicity of the product in question. At the same time, the UFW was fully aware of the growing interest in pesticide hazards, especially the effort to ban DDT, and hoped to link the occupational and environmental aspects of the issue to its potential as a consumer concern. The grower panic from the cranberry scare of 1965, when cranberry sales plunged after the government announced that the spray used on cranberries was carcinogenic, demonstrated to union leaders that the pesticide issue could be of enormous value in a pesticide-linked boycott campaign. Pesticides thus became simultaneously a negotiating chip and a social justice issue tied to the UFW's parallel identity as a union and social movement.[10]

By 1968, the UFW had decided to pursue the pesticide issue on several fronts at once. In Kern County, one of the leading areas of pesticide use in the state, UFW general counsel Jerry Cohen pursued the right-to-know strategy. Arriving one day at the office of the Kern County agricultural commissioner, Cohen requested information about the chemical ingredients of pesticides suspected of creating significant health hazards. The office refused to release the information, which led to the union's first major pesticide-related court action. Although the UFW lost the Kern County case, a parallel legal action in Riverside County eventually resulted in a court decision granting access to pesticide information and other public records.[11]

During this same period, the union initiated litigation to ban specific pesticides, including DDT. Though the lawsuits were hastily developed and relied on little-tested arguments, they contributed to the range of legal and political activities directed against specific pesticides. In the DDT suit, the union brought in the Environmental Defense Fund as an interested party to try to link the occupational and wildlife hazard aspects of the issue. The EDF, interested in wildlife impacts, continued to have difficulty establishing standing in cases that required plaintiffs to be directly impacted by pesticide use, as were farmworkers. The UFW suit, taken over on appeal by the UFW's legal ally, the California Rural Legal Assistance organization (CRLA), offered the EDF just such an opportunity. This UFW/CRLA/EDF alliance, according to EDF scientist and cofounder Charles Wurster, became a "marriage of convenience" rather than a strategic redefinition of the pesticide issue. Nevertheless, the suit, which ultimately became the basis of the EDF's successful petition to force the EPA to ban DDT, suggested the linkage opportunities between movements and issues.[12]

The UFW was most successful in incorporating pesticide-related health and safety language into contracts signed at the conclusion of its grape boycott campaign in 1970. This push for contract language reflected a growing union interest in the pesticide issue. During 1968 and 1969, the union newspaper *El Malcriado* published numerous articles detailing the effects of spraying or handling of pesticides. Chavez and other UFW organizers, meanwhile, continually raised the problems of pesticide hazards for farmworkers in both public and private forums. "We will not tolerate the systematic poisoning of our people," Chavez asserted in a 1969 letter to the chairman of the Southern Central Farmers Commission, nor would the union "permit human beings to sustain permanent damage to their health from economic poisons."[13]

Protection for farmworkers, the union insisted, could best be accomplished through representation and not simply through legislative or regulatory routes, since key agencies such as the California Department of Food and Agriculture were sympathetic to growers and hostile to the union campaigns for farmworker rights. When the first contracts were finally negotiated during the late 1960s, the union, thanks in part to its public campaigns, was able to secure health and safety provisions. By 1970, all contracts included a health and safety clause. Some contracts incorporated language specifically banning pesticides such as DDT, aldrin, and dieldrin, preceding the federal ban of DDT by a few years. These contracts also provided for health and safety committees with worker representation and provisions for increased sanitation and medical attention, while seeking to make available company records concerning the ingredients of the pesticides used in the fields. Ultimately, the UFW-grower contracts laid the groundwork for monitoring, evaluating, and hopefully protecting farmworkers from the complex of hazards tied to pesticide use.[14]

Though several such contracts were signed, anti-union efforts by the growers persisted through the 1970s and 1980s. Various grower tactics, such as recognition of a competing Teamsters Union, failure to renew contracts, and an increased use of immigrant labor, undermined the UFW's early successes and forced the union to protect its limited gains at representation.

This shift in focus also eventually eroded the union's ability to draw on its status as a social movement, which included its appeal on pesticides. During the late 1980s, when the union reinvoked the pesticide issue as a tactic in its new grape boycott, it linked it to pesticide contamination at the community level. But suffering from the loss of its movement status and maintaining only an uneasy relationship with community residents wary of the union's reengagement in the pesticide issue, the union had less success this time in connecting issues and movements.

This absence of linkages, such as the earlier, limited UFW/EDF "marriage of convenience," prevented the development of any new framework for defining pesticide hazards. Even the banning of DDT, while it removed a major threat to wildlife, had nevertheless failed to address the problem of occupational exposures. Pesticide replacements for DDT, though less harmful to wildlife due to their ability to break down rapidly in the environment, were almost invariably more toxic in terms of human exposure. Even where pesticide impacts became an occupational issue, as with the late-1970s revelations about the health hazards of the pesticide DBCP for manufacturing workers, the use of pesticides by farmworkers remained a less visible concern. Similarly, telone, a highly toxic DBCP substitute, was banned from use in California only after it became implicated in toxic air emissions near the city of Bakersfield rather than as a result of farmworker exposure. By the early 1990s, farmworkers continued to be excluded from pesticide policy debates, despite the importance of the pesticide issue for environmental groups. The opportunity to transform the issue, which first arose as part of a quest for social justice, remained unfulfilled, a part of environmental history that never realized its potential for redefining the movement's approach.

The question of what constitutes an environmental issue also became significant with respect to the problem of lead in the environment. Lead has long been seen as a powerful symbol of the environmental hazards of urban and industrial life. It has been the subject of long and bitter conflicts concerning industry decisions, government policies, and community and workplace standards. With increasing attention since the 1960s on lead's potential for causing harm to millions of children, lead contamination has also emerged as a social justice issue, related to the discriminatory patterns of and impacts from lead exposure and the efforts to deal with the hazards caused by such exposure.

The toxicity of lead has been known for thousands of years. It has been recognized as a cause of anemia, colic, neuropathy, sterility, coma, and convulsions, among other health problems. Despite this knowledge, the use of lead in the United States increased significantly up through the late 1970s. Annual use doubled between 1940 and 1977 to more than 1.5 million tons, due in part to a sixfold increase between 1935 and 1977 in the amount of lead additive in gasoline. Leaded gasoline, lead-acid batteries, lead pigments in paint, lead solder, lead pipe, lead in caulking and soundproofing materials for buildings, leaded glass in television sets, lead in glazes and enamels for glass containers, tableware, and cookware, lead in plastics, and lead foil wrappers to cover the corks on wine bottles have all substantially increased the sources of exposure to lead in the environment.

Since lead is a naturally occurring element and doesn't break down into something less toxic, it can remain toxic indefinitely once introduced into the environment. Nevertheless, a number of products, most notably leaded gasoline and lead paints, have been sold by industries long after information became available about their extensive impacts. As a consequence, the amount of lead in the environment capable of being absorbed into the body increased without interruption through the 1970s, creating what amounted to an epidemic of low-level lead exposures. With growing evidence that neurologic damage was associated with smaller levels of blood lead (the amount of lead detected in the blood), social movements and policy makers eventually began to focus on how the lead was entering the environment as well as who was becoming exposed. Even more than pesticides, lead came to represent a middle-class issue as a universal hazard impacting broad sections of the population as well as an inner-city issue as a specific hazard for those most likely to be exposed and for whom the consequences would be most significantly felt.[15]

Much of the attention concerning lead in the early part of this century focused on lead poisoning problems experienced by workers in the lead trades and their families. The pioneering work of Alice Hamilton on lead toxicology and occupational hazards contributed to gradual improvements in industrial hygienic standards. By the 1920s, however, new information demonstrating significant hazards associated with lead and children started to become available. New medical studies linked lead poisoning in children primarily to lead paint exposure from peeling walls, furniture, and even baby cribs. Children, one article argued, lived in a "lead world," surrounded by leaded items capable of being ingested and thus introduced into their bodies.[16]

Despite the growing literature on the subject, the paint industry ignored or suppressed information about the dangers of exposure to children and only began to replace lead with other pigments in paints after 1940. Still, lead pigments continued to show up in substantial numbers of houses painted after 1940. Some studies pointed out that as many as one-third of all dwelling units in such cities as Pittsburgh and Washington contained high concentrations of lead on walls or other surfaces.

Up until the 1960s, the problem of children eating lead paint chips was thought to be a factor in only the most severe forms of lead poisoning. Most health departments ignored lead exposure issues except when major outbreaks of lead poisoning occurred, such as the mass outbreak of the disease in the 1930s that was traced to the burning of battery casings for fuel in the home. The few studies analyzing exposure routes suggested that lead poisoning in children might have a "predilection for the poor." Childhood lead poisoning, one government report noted, was "inextricably related to dilapidated housing, where peeling lead paint and broken painted plaster were readily available." High-risk areas, or "lead belts," were seen as synonymous with inner-city slums. Such findings invariably reduced pressures on health departments to intervene. "Not unexpectedly," this government report concluded, "recurrence was the rule unless the lead paint hazard was corrected, and, with each recurrence, the prognosis became worse."[17]

During the 1960s, childhood exposure to lead paint came to be seen as an environmental problem most directly related to "the living conditions of its victims," as a 1968 statement of the Scientists' Committee for Public Information put it. For this group of scientists, which included both Barry Commoner and René Dubos (perhaps the best-known environmental scientist at the time), the scientific issues regarding lead paint poisoning were integral to its social justice dimensions: the epidemiology was clear, its victims could be predicted, its health effects could be identified, and its treatment was known to the medical profession. Yet, despite the opportunities for prevention, efforts to address lead paint exposures only emerged at the margins of the public health profession among a few "vocational" physicians and scientists linked up with inner-city activists who ultimately forced the issue onto the public agenda.[18]

The first major community-based effort to deal with lead paint poisoning emerged in Chicago in 1965. A small community organization, the Citizens Committee to End Lead Poisoning (CCELP), formed in response to neighborhood concerns about several incidences of lead poisoning in East

Garfield Park, a black community on the city's west side. The issue was dramatically highlighted at a block club meeting when a young mother announced that both her children had become feverish and convulsive as a result of lead poisoning. At a follow-up meeting attended by representatives from block clubs, social agencies, church groups, and other organizations concerned with inner-city housing questions, a pediatrician from the Medical Committee for Human Rights, whose members had previously failed to convince the Chicago city council to set up a lead screening program, discussed how and why lead exposure was prevalent in inner-city neighborhoods. A decision was made to form the CCELP to launch a preventative campaign based on screening and community canvassing to identify lead paint poisoning victims and provide for treatment or reduced exposures.[19]

The CCELP campaign, which included weekend canvassing by dozens of teenagers in conjunction with a Chicago Board of Health screening program established through CCELP pressure, had some immediate and striking successes. It provided a focus for the media and local policy makers on the lead paint issue and helped inspire several new, community-based groups and coalitions organized to deal with the issue in other parts of the country. The most prominent of the groups was the Philadelphia-based Citywide Coalition Against Childhood Lead Paint Poisoning (Citywide), involving about sixty inner-city service or advocacy organizations. Citywide not only employed screening and abatement programs like Chicago's CCELP, but focused more broadly on housing policy. This included a lawsuit brought by the organization against the federal government's Department of Housing and Urban Development (HUD) that sought to ensure that all HUD property be inspected and, if necessary, deleaded before occupancy.[20]

By 1970, dozens of inner city–based community organizations and coalitions were organizing to address lead paint issues, primarily in East Coast and Midwest cities such as Rochester, Washington, New York, and Baltimore. Though the groups continued to focus on community awareness and prevention, they also saw themselves as part of the community empowerment movements of the period, particularly those related to issues of housing and community health. Similar to the Philadelphia and Chicago groups' experiences, many of these groups were aided by New Left–inspired professional groups, such as the Medical Committee for Human Rights, that were interested in lead paint as both a social justice or equity issue and as an area in which technical information was needed and could be shared.[21]

By the early 1970s, the lead paint movement had achieved significant visibility both locally and nationally. The screening programs initiated in response to community action had led to the unanticipated discovery that many more children than anticipated had elevated blood lead levels. As evidence of a near epidemic of childhood lead poisoning continued to mount, the surgeon general, in 1970, had issued a statement emphasizing the importance of prevention and screening while calling for the elimination of certain sources of exposure. Largely in response to these events, Congress began to hold hearings on legislation introduced on the subject. Despite this national attention, lead paint problems were still perceived as inner-city rather than environmental concerns. For example, there were no representatives from the mainstream environmental groups to testify at the hearings on behalf of the legislation, which had become a major objective of the lead paint hazards movement. And although this Lead-based Paint Poisoning Prevention Act, which further limited the lead content of paints and provided for paint removal programs, was enacted in 1971, its passage ultimately reduced the level of attention concerning the issue. At the same time, new interest was developing on the question of lead in the ambient environment. A 1971 EPA report indicated that leaded gasoline could be a significant source of lead contamination, even for children. The shift in interest toward leaded gasoline was further extended when the manager for HUD's Lead-based Paint Poisoning Prevention Research program decided to shift the agency's research focus away from residential housing (despite the requirement of the 1971 Lead-based Paint Poisoning Prevention Act) and concentrate instead on leaded gasoline exposure issues.

Through the 1970s, the changing emphasis away from paint to leaded gasoline coincided with the growing recognition that lead contamination was more than just an inner-city issue. Studies during the mid-1970s demonstrated that 40 percent of blood lead sources came from the ambient environment, with leaded gasoline responsible for as much as 90 to 95 percent of the lead in the air. Lead in gasoline had already become an issue in relation to implementation of the Clean Air Act, a priority area for mainstream environmental groups such as the NRDC and the EDF. As sales of leaded gasoline began to decline, due initially to the Clean Air Act's requirement of catalytic converters on automobiles and eventually from the ban on leaded gas (stemming from the protracted campaign of the mainstream groups to force the EPA to prohibit its sale), lead levels in the air also decreased by equivalent amounts. This, in turn, resulted in significant declines of blood lead levels in children. Ultimately, the leaded gasoline campaign provided one of the most impressive success stories for the mainstream groups, a case in which environmental policy makers were forced to reduce rather than simply control or manage a pollution-based hazard.[22]

While the leaded gasoline issue took center stage, the lead paint issue seemed to fade, as federal support declined for inner-city service and advocacy groups active on the issue. At the same time, the failure within the context of environmental policy to address further the relationships among housing, community health, and the inner-city environment, where many people of color lived, only reinforced the notion that environmental issues were white, middle-class issues. By the early 1980s, at a moment of great triumph for the mainstream groups regarding the leaded gasoline fight, lead paint contamination had become a forgotten issue, with the potential for creating new kinds of environmental alliances still unrealized.

After a fifteen-year hiatus, the lead paint issue reemerged in the late 1980s. Despite the 1971 legislation, studies demonstrated that lead paint continued to represent a significant source of exposure for children, particularly where substandard housing was involved. Surveys also revealed that lead contamination (due largely to lead paint exposures) remained at high levels throughout the country, both in middle-class communities and especially in poor and minority neighborhoods, where the greatest risks continued to be found. Additional studies pointed out that while increased lead exposure (which resulted in decreased intelligence, such as lowered IQs) was "persistent across cultures, racial and ethnic groups and social and economic classes," as a *New England Journal of Medicine* editorial put it, the adverse effects of exposure remained "greatest among disadvantaged children."[23]

By the late 1980s, such studies helped stimulate a new wave of lead-related, community-based movements. These groups defined themselves more explicitly in environmental terms, raising issues of risk discrimination and the need for environmental justice. A few mainstream environmental groups, especially the Environmental Defense Fund through the work of staff scientists Ellen Silbergeld and Karen Florini, also sought to address the lead paint issue. But the primary actors in these new campaigns were alternative environmental groups for whom the ethnicity factor remained prominent. Inspired by a gathering of these groups at a 1991 conference on lead issues in Washington, D.C., the new focus on lead raised anew the question of what constituted an environmental issue, given lead's ubiquitous presence, its routes of exposure, its extensive health impacts, and the discriminatory causes of some of the risks involved. After more than three decades of community action, the issue of lead in the environment continued to be a question of social justice as well as policy.[24]

The question of how to define an environmental issue also figured prominently with respect to uranium activities in the Southwest. At the outset of the Cold War, major uranium operations were initiated throughout the Colorado Plateau and Four Corners region, including on Native American reservation lands. The first uranium ore was discovered in Native American country in Coconino County in Arizona by a Native American sheepherder, and many of the miners employed for the mining, milling, and prospecting of the uranium were from the Navajo, Hualapai, Havasupai, and Hopi tribes.

During the 1950s, uranium fever struck the Southwest. The rush to open mines and prospect for new ones paralleled the celebration at that time of the country's nuclear arms and nuclear power programs. In the process, the Atomic Energy Commission, in conjunction with major nuclear industry firms such as Kerr-McGee, actively sought to recruit Native American mine and mill workers and prospectors. These included both transient workers and those who lived in communities downwind of the sites. At the height of the boom, hundreds of mines and milling operations were operating. Coconino County alone produced more than 360,000 pounds of uranium oxide in 1956, its peak year, with more than 6 million tons of uranium ore produced and milled on all Navajo lands by 1960. Though the uranium boom collapsed in the early 1960s as a result of the glut of uranium used in the production of nuclear weapons, talk of renewed mining activity continued to resurface up through the early and middle 1970s, linked to expectations about the growth of the nuclear power industry. By 1977, New Mexico was producing almost half the nation's uranium, and estimates suggested that the expansion of nuclear power could conceivably more than double that amount during the next decade. During this period, concerns also emerged about substantial health and environmental impacts related to both uranium production activities and the nuclear testing programs in Nevada, which affected the small Mormon communities of southwestern Utah. After several decades of government misinformation and stonewalling, it was eventually recognized that the substantial hazards of uranium mining, compounded by the existence of hundreds of abandoned mines, uranium tailings, and spills, had produced an extraordinary environmental legacy of radiation-linked illnesses and deaths. Communities and landscapes would be marred for decades to come.[25]

Like the pesticide and lead exposure issues for other people of color, the uranium issue for thousands of southwestern Indians became a legacy of victimization. All through the uranium fever period, Native American miners were recruited without the AEC or the nuclear industry providing information about potential hazards. Up until the 1970s, well after most mining had occurred, Native American workers were not provided with protective gear and equipment and were often assigned to work in conditions where exposures were highest. Transient workers who were employed in a mine until it closed or until their health failed often traveled with their families, who also became exposed to the tailings, in which children from those families would play. Feelings of tribal victimization were perhaps most directly reinforced by the presence of abandoned mines and the likelihood that the scarring of the land and particularly the radioactive hazards could last for generations.[26]

Concerns about uranium's environmental impacts first surfaced in a major way for Native Americans with the 1979 radioactive tailings spill into the Rio Puerco in northern New Mexico. This massive, disastrous spill, which contaminated significant stretches of Navajo lands, occurred just weeks after the Three Mile Island accident. But, unlike the Pennsylvania accident, the Rio Puerco spill received limited attention from policy makers or mainstream environmentalists. The slow response by state and federal governments with respect to cleanup, the limited coverage by the press, and the hostile position of authorities, who denied the spill had created significant hazards and who dismissed the seriousness of the event because of the small numbers of people affected, even led tribal chairman Peter McDonald (an earlier supporter of uranium mining activities) to declare that the response to the Rio Puerco spill had racial undertones.[27]

The lack of attention to the Rio Puerco spill underlined the difficulties in getting recognition for uranium hazard issues for Native Americans in general. The romantic celebration of Native Americans among environmentalists, for example, didn't translate into specific concerns about Native American health risks or contamination of their lands. During the 1950s and 1960s, conservationist groups largely ignored uranium issues, since they were perceived as removed from the scenic resource concerns that prevailed in that period. By the 1980s, a few mainstream groups, such as the Sierra Club, expressed support for Native American compensation claims and occasionally offered resources to help in such issues, but the uranium legacy faced by Native Americans remained largely incidental to overall mainstream environmental agendas.

For several alternative environmental groups, such as Greenpeace and the National Toxics Campaign, and particularly for local groups, such as the Southwest Research and Information Center and Northern Arizona Citizens for Environmental Responsibility, concerns about uranium hazards for Native Americans tied more directly to their own agendas, including efforts to gain compensation through federal legislation. Still, these more recent positions contrasted with the failure of the antinuclear movements of the 1970s to mobilize effectively on the uranium issue, despite the frequently expressed sympathy for Native American claims about uranium hazards.

In the four decades since the boom began, the uranium legacy for Native Americans has thus evolved into another example of how Native Americans and their lands are abused by Anglo industrial forces. Uranium, according to a Navajo saying, brings only "sorrow and suffering to those who disturb it." But the act of disturbing the land brought suffering not to the mining interests but their victims. "We call ourselves 'the silent warriors of the Cold War,' " Navajo activist Michael Begay has said of the tribe's uranium experience, underlining the tribe's long and difficult process of achieving recognition and status as environmental victims. Such concerns are not just historical. During the 1980s and early 1990s, new efforts to prospect and mine for uranium on Native American lands were unveiled. These received only limited public scrutiny and mainstream environmental attention despite opposition by the Havasupai tribe and the environmental justice–oriented Tonantzin Land Institute.[28]

The uranium-based experiences of southwestern Native Americans, both as workers and as residents, raise important environmental questions, as have pesticide and lead issues. The nature of the hazards and distribution of the risks involved situate such questions within a larger context of ethnicity, livelihood, and place. Chicano farmworkers, inner-city African-Americans, and Native American uranium miners and residents of uranium country all share a common experience based on who they are as well as where they live and work. As groups with limited access to resources and power, the environmental battles they have fought have produced uncertain results. In each of these situations, mainstream environmental groups have focused on separate, though often parallel, concerns, such as pesticide impacts on wildlife, the presence of lead in the ambient environment, and concerns about energy choices and scenic resources. The issues for the African-American, Chicano, and Native American groups tended to be defined less in environmental than social justice terms. By establishing a distinction between environmental and social justice themes, these struggles have further reinforced the prevailing assumption that environmentalism continues to be a white movement.

Points of Tension: Population and Immigration

In November 1972, the Conservation Foundation convened a conference in Woodstock, Illinois, to explore the themes of race, social justice, and environmental quality. In calling for the meeting, the foundation's executive director, Sydney Howe, and several of his staff argued that environmental organizations needed to incorporate social justice themes into their agendas and to address the lack of minority involvement within their own organizations. To pursue those themes, the foundation had invited both mainstream group representatives from the Sierra Club and The Wilderness Society and a range of community activists, many of them people of color, involved in issues such as transportation, housing, and economic development. At the meeting, Peter Marcuse, a professor of urban planning at UCLA and Herbert Marcuse's son, warned participants that divorcing equity and social justice concerns from the environmental agenda threatened to create a permanent rupture between movements and would leave environmentalists vulnerable to "effective, if unprincipled attack." "Many environmentalists now believe that the movement must undertake some soul searching," the Conservation Foundation's James Noel Smith similarly commented. Smith urged environmentalists to "begin examining the motives and methods of the movement, its level of social awareness, the extent of its commitment to the agenda of social justice, and the implications of its thought and action upon all levels of society."[29]

At the time of the Woodstock conference, most mainstream environmental groups continued to remain aloof from social justice themes and people-of-color movements. Even the Conservation Foundation's brief effort to explore social justice themes was abruptly terminated when the organization's conservative board fired Sydney Howe in 1973 and replaced him with William Reilly. The future EPA director's patrician style and narrow policy focus was felt to be more appropriate by a board primarily dominated by Pew, Mellon, and Rockefeller Foundation interests. Other mainstream organizations had also expressed some tensions over social justice issues. A 1971 survey of the Sierra Club membership, for example, explored whether the organization should "actively involve itself in the conservation problems of such special groups as the urban poor and the ethnic minorities." Forty-one percent, especially older club members with higher incomes, strongly disagreed with the statement, and only fifteen percent strongly agreed. Sierra Club board member Paul Swatek, reflecting on the poll results, suggested that the club examine "the social and economic impacts of the programs which we advocate" or "be dismissed as irrelevant." But Swatek also called for a "balanced" approach to ensure that the organization not "stray far from our central conservation concern," and that the "remarkable coalition [of the] many, diverse elements that are bound together in the Sierra Club" not be threatened.[30]

The desire to keep a focus on this "central conservation concern" generated tensions between mainstream environmentalists and social justice and civil rights advocates, especially in two key areas: policy priorities and the racial implications of certain environmental themes. In the months preceding and following Earth Day, when policy makers and the media began portraying the environment as a consensus issue requiring significant resources and government attention, some civil rights leaders lashed out at this position, arguing, as the Urban League's Whitney Young put it, that "the war on pollution . . . should be waged after the war on poverty is won." In these 1970 remarks, which received wide attention at the time, Young insisted that "common sense calls for reasonable national priorities and not for inventing new causes whose main appeal seems to be in their potential for copping out and ignoring the most dangerous and most pressing of our problems." In the same vein, Norman Faramelli of the Boston Industrial Mission told a university audience that to "poor and low income families, ecology may appear to be a cop-out, a flight from social realities, and a digression from dealing with the real issues of racism and social injustice."[31]

The tensions concerning priorities (inflamed by such incidents as the burying of an automobile by San Jose State students during campus Earth Day events)[32] were significantly exacerbated when the key environmental issues of population control and immigration restrictions also took on racial connotations, a linkage that dated back to earlier protectionist and conservationist arguments. Several proponents of those earlier arguments, including William Hornaday, Madison Grant, and Henry Fairfield Osborn, were strong advocates of racially motivated exclusionist policies for environmental purposes. The sharp-tongued Hornaday, in several of his writings, linked his strong plea for wildlife protection with ad hominem racist slurs.

The New York Zoological Society director was particularly fearful that "members of the lower classes of southern Europe," such as Italians, constituted "a dangerous menace to our wild life" and could ultimately displace native-born (non-indigenous) Americans by their population "spreading, spreading, spreading." "If you are without them today," Hornaday exclaimed, "tomorrow they will be around you."[33]

The fear of an immigrant tide and its link to the erosion of "native" values, including wilderness values, also appeared in the writings of Madison Grant. A major figure in protectionist circles as president of the Boone and Crockett Club and the New York Zoological Society, as well as a leader of the Save-the-Redwoods League, Grant was also "intellectually, the most important nativist in recent American history," according to historian John Higham. In his writings, Grant sought to demonstrate that social policies in the United States were leading to a "racial abyss" and that the arrival of certain immigrant groups, including southern Europeans, Italians, and Jews, had led to the encroachment of the city on the outdoors, including wildlife. These themes were simultaneously taken up by the president of the American Museum of Natural History, Grant's good friend Henry Fairfield Osborn. In a welcoming address to the Second International Congress of Eugenics in 1921, Osborn linked nativist fears and racist arguments to the need for immigration restrictions, a major issue of debate at the time and a theme that would reappear more than fifty years later as part of the population argument related to scarce resources.[34]

Although population and immigration restriction concerns receded during the Depression years, the issues reemerged during the late 1940s with the appearance of William Vogt and Fairfield Osborn's neo-Malthusian tracts. The population versus technology debates of the 1950s framed much of the conservationist and protectionist discourse, although immigration issues and even population control tended to be less dominant than the focus on scenic resources and wildlife management issues. A number of conservationist leaders, such as Resources for the Future president Joseph Fisher, even accepted as a "virtual certainty" that there would be "more and more people in the decades ahead." The task of the movement, Fisher and others argued, was to figure out how to accommodate such growth, whether by technological innovation or by resource conservation and management, and to impose protection policies in those situations where wilderness values needed to prevail. Such an approach reflected, as Fisher put it in a speech to the North American Wildlife and Natural Resources Conference, "conservation for more and more people."[35]

Despite the hopes for resource conservation and technology, fears of a population explosion began to reassert themselves among conservationist and protectionist groups during the middle and late 1960s. Many of the fears were associated with concerns about increasing growth rates in less developed countries, a position anticipated by William Vogt in his doomsday scenario nearly two decades earlier. The population question also began to be related to the successes of technology, which rapidly increased the sources of pollution and the stresses on global resources. By 1968, when Stanford University biology professor Paul Ehrlich published his popular polemic, *The Population Bomb*, the issue of population growth was ready to become a major focus for groups interested in linking the problem of resource limits to the growing concern about "quality of life."

The Population Bomb, published by the Sierra Club through Ballantine Books, was a phenomenal success, with twenty-two printings and more than 1 million copies sold in less than two years. The book helped elevate the population issue among several mainstream groups, inspiring the creation of Zero Population Growth (ZPG), which sought to place the population issue at the center of environmental policy. At the same time, *The Population Bomb* became a polarizing text in regard to questions of race, First World–Third World relations, and eventually the volatile issue of immigration.

Ehrlich begins his book by describing an episode involving a trip to Delhi, India. Returning to his hotel in an ancient taxi "one stinking hot night," the Stanford professor passed through a crowded slum, where the "streets seemed alive with people." "People eating, people washing, people sleeping, people thrusting their hands through the taxi window, begging," Ehrlich described the scene. "People defecating and urinating. People clinging to buses. People herding animals. People, people, people. As we moved slowly through the mob, hand horn squawking, the dust, noise, heat, and cooking fires gave the scene a hellish aspect." The episode also became an epiphany for the population issue: the numbers, Ehrlich decided, will overwhelm us unless something is done. The contemporary pollution-based environmental crisis—"too many cars, too many factories, too much detergent, too much pesticide, multiplying contrails, inadequate sewage treatment plants, too little water, too much carbon dioxide" as Ehrlich characterized it—could be traced to a single, overriding causal factor: too many people.[36]

Ehrlich's argument, elaborated through a variety of doomsday scenarios, sought to fend off accusations of class or racial bias by suggesting that population control efforts needed to be directed at the white middle and upper classes in the United States and other "overdeveloped" countries, given their overuse of resources. Still, several of Ehrlich's proposals for action, including income tax changes related to family size, taxes on child-related necessities such as diapers, and the sterilization of all males with three or more children in countries such as India, had important racial and class implications. Ehrlich's analysis was also seen as complementing the position put forth by University of California professor Garrett Hardin, whose provocative metaphor of the "lifeboat" (not enough resources for everyone to share) directly offered a "have/have not" interpretation of the population control position.[37] Thus, at the very moment when key mainstream groups such as Friends of the Earth and the Sierra Club began to elevate population to a central agenda item, the population control argument, as developed by such advocates as Ehrlich and Hardin, generated charges of racism in both a domestic and international context.

The key organization at the center of these controversies was Zero Population Growth. Founded shortly after the publication of *The Population Bomb*, ZPG proved to be an immediate success, growing to more than 33,000 members in 380 chapters across the country by 1970. The group was organized primarily to provide visibility for the population issue in order to influence policy makers, the press, and environmentalists. The organization focused especially on family size by strongly promoting a maximum of two children per family as a policy base line. That base line position led ZPG to emphasize population control techniques as matters of both individual choice and social policy. The most prominent of those techniques, discussed frequently within the organization, were the various forms of sterilization or birth control, from voluntary male vasectomies to the use of the drug depo-provera.

It was the promotion of sterilization techniques that immediately and unwittingly cast ZPG and the population control movement in racist terms, reinforced by its suggestive associations with eugenics and racial breeding, which had so preoccupied William Hornaday and Madison Grant. ZPG leaders denied any racial overtones and sought to disassociate the organization from such incidents as the 1973 Montgomery, Alabama, involuntary sterilization of two young black women. But they and other population control advocates were constantly having to defend themselves on issues they often remained unwilling to confront. ZPG's emphasis on voluntary sterilization, for example, never satisfied the growing opposition to sterilization abuse that arose in the wake of the Montgomery incident. In that context, Paul Ehrlich sadly recounted how, at a 1970 Population and World Resources Conference ZPG helped organize, a people-of-color caucus walked out of the convention on the grounds that conference organizers had failed to address the racial connotations of the issues under consideration.[38]

While ZPG and other population control advocates frequently clashed with black civil rights organizations during the early 1970s, a related set of tensions emerged during the later part of the decade between the population control groups and Latino organizations concerned with immigration rights. Population control advocates began to pay more attention to immigration in the early and middle 1970s with the recognition that population growth increases in the United States were almost exclusively a result of increased immigration. While Anglo natives had a zero growth rate (or even negative growth), immigrants accounted for the margin of population growth within the country. The emerging focus on immigration also coincided with a renewed focus on U.S.-Mexico border policy less than a decade after the Bracero Program had been terminated. By 1973, the new head of the Immigration and Naturalization Service (INS), former Marine commandant Leonard Chapman, would initiate in conjunction with other border control advocates a militaristic-sounding campaign against "illegal aliens" from Mexico, claiming the country was being overrun by poor Mexicans in search of jobs and economic benefits. This campaign not only increased INS budgets but helped lay the groundwork for the emergence of a new and powerful anti-immigrant coalition that prominently included mainstream environmental population control advocates.[39]

By the late 1970s, population control was becoming synonymous with efforts to control the flow of Mexican migrants. Similar to the sterilization controversies, the focus on immigration created concerns about racist assumptions within environmentalism. Anti-Mexican sentiments surfaced throughout the movement, even associated with such iconoclastic figures as novelist and Earth First! hero Edward Abbey, who wrote disparagingly about Mexicans. Although racist attitudes were denied, and certain population control advocates sought to relate poverty and economic development issues to population and immigration problems, many mainstream environmentalists still assumed that increased population through immigration "could pose a real threat to the American quality of life as we know it today."[40]

The population control/immigration linkage was reinforced with the formation in 1978 of the Federation of American Immigration Reform

(FAIR). Key figures within both ZPG and the Sierra Club were instrumental in establishing FAIR, whose board and officers also included major corporate figures. The development of the organization escalated efforts at the judicial, legislative, and administrative levels to restrict immigration. This included legal attempts to prevent the counting of nonlegal or undocumented migrants in the 1980 census and lobbying for new immigration legislation to include employer sanctions for hiring illegal immigrants. The census litigation especially, with its far-reaching equity implications regarding entitlement programs and the loss of other potential social benefits, established FAIR as a population control organization with a social agenda, albeit one that separated rather than integrated environmental and social justice themes.[41]

Despite the continuing activities of FAIR during the 1980s and 1990s, the immigration issue eventually faded from both environmentalist and public policy agendas, particularly after a sweeping immigration reform bill was passed in 1986. Population issues, similarly, faded in this period as well, although a few of the mainstream groups, most notably Audubon during the period when Russell Peterson was its CEO, tried to keep the issue prominent through new forums and coalitions. While the Group of Ten's *An Environmental Agenda for the Future* had, largely due to Peterson's efforts, urged mainstream groups to "advance the goal of zero population growth" as integral to their activities, just three years later, in 1988, another mainstream environmental coalition document, *Blueprint for the Environment*, sought to have the federal government simply reassert its support of population and family planning assistance in the wake of Reaganite assaults. ZPG, meanwhile, in another indication of the decline of the population issue, witnessed significant losses both in membership and resources from its peak in the early 1970s.[42]

The racial implications of the population issue had also begun to recede during the 1980s, especially as the Reagan administration launched high-profile attacks against family planning and other birth control efforts both domestically and in the Third World. Third World countries such as Mexico and China initiated their own vigorous birth control campaigns, further shifting the focus away from target populations for population control toward how a decline in the birth rate might best be achieved within individual nations.

Despite this shift in focus, the positions on population control and immigration restrictions that had created a legacy of conflict during the 1970s left an undercurrent of mistrust for African-American and Latino organizations about the mainstream environmental groups. As new issues and potential arenas for both conflict and coalition (for example, toxics) emerged during the 1980s, it was clear that race and ethnicity, reinforced by the earlier history regarding population and immigration issues, were factors that could not be ignored by environmental groups.

NO LONGER JUST A WHITE MOVEMENT: NEW GROUPS AND COALITIONS

In January 1990, a group of prominent non-Anglo activists and analysts led by the Louisiana-based Gulf Coast Tenant Leadership Development Project sent a letter to each of the Group of Ten CEOs. The letter asserted that the "racism and the 'whiteness' of the environmental movement" had become its "Achilles' heel." Two months later, a second letter, initiated by the New Mexico–based Southwest Organizing Project (SWOP) and signed by more than 100 activists and representatives of community-based groups, was sent to these same organizations.[43] The two letters detailed ways in which the mainstream groups had become isolated from poor and minority communities despite evidence that the poor, and people of color in particular, were "the chief victims of pollution." The letters further argued that the Group of Ten organizations had a "clear lack of accountability" to Third World communities and a poor record of hiring and promoting minorities at both the staff and board level. The letters therefore insisted that each of the groups respond within sixty days to describe the steps they would pursue to assure that 35 to 40 percent of their staffs and significant numbers of board members would be people of color. The mainstream groups were also challenged to revise their approaches regarding such questions as rural economic development, toxics hazards in poor and minority communities, and international debt-for-nature swaps (where certain international debts would be forgiven in exchange for a country preserving a "natural" area). "The letters," recalled SWOP's Richard Moore, "were really about who had the power within the environmental movement and how it was going to be used."[44]

The January and March 1990 letters, which received substantial press attention at the time, created significant concern among mainstream groups. They raised long-standing issues about the whiteness of the groups and their organizational agendas. The racial composition of board and staff members also dramatically reflected the argument of the letter. There were no African-Americans or Asian-Americans and only 1 Hispanic among the Sierra Club's 250 professional staff. At the NRDC there were only 5 people of color among 140 professional staff, and the Audubon Society had only 3 African-Americans among its 315 staff members. Even Friends of the Earth had only 5 people of color among a 40-person staff and 1 person of color on a 27-member board. "The truth is that environmental groups have done a miserable job of reaching out to minorities," the Environmental Defense Fund's CEO, Fred Krupp, commented to the *New York Times*.[45]

In the months following the release of the two letters, several mainstream groups sought to demonstrate a new commitment to multi-ethnic hiring, board selection, and agenda setting. The NRDC actively showcased its new Los Angeles office as substantially committed to social justice and environmental racism issues, reflected in their decision to hire a staff attorney from the Western Center on Law and Poverty who had been previously engaged in anti-incinerator campaigns in poor African-American and Latino neighborhoods. A number of mainstream groups also sent out memos and issued directives seeking to locate minority candidates for staff and board positions. "I don't think anybody is as aware of the whiteness of the green movement as those of us who are trying to do something about it," the National Wildlife Federation's Jay Hair commented about this effort.[46]

The NWF, previously attacked by antitoxics groups for the presence of a Waste Management Inc. official on its board of directors, was among the more aggressive of the groups seeking to signal its multi-ethnic commitments. The organization established a new program designed to highlight environmental justice concerns, sought to recognize through award ceremonies key environmental justice figures such as University of California professor Robert Bullard, and added minorities to staff and board positions. Two years after the January 1990 letter had been sent, the organization's CEO, Jay Hair, was claiming that the NWF had the best figures among the mainstream groups, with minorities constituting 23 percent of its staff and four minority members added to its board.[47]

But the approach of such groups as the NWF and NRDC still failed to dispel the perception of the whiteness of the mainstream groups, whether reflected in staff and board composition, the agenda-setting process, or overall organizational culture. The NWF figures, for example, failed to distinguish between maintenance or secretarial workers and the professional staff. Hair refused to break down these numbers, asserting that in his organization "everybody's a professional, from the guy on the loading dock to the lawyer arguing before the Supreme Court." And although the NWF included four minorities on its twenty-nine-member board, none were community-based activists, but prominent business, political, and professional figures who blended into a board that had long emphasized access to money and power.[48]

The whiteness of the movement was also an issue that corresponded directly to the distinctions between mainstream and alternative groups. While the NRDC, for example, sought to integrate environmental justice themes into the organization's agenda, the group's heavy emphasis on professionalization served as a barrier to that process. As "law firms for the environment" and/or centers for expertise and lobbying, mainstream organizations remained divorced from community activism and grassroots organizing. Their offices were almost always located in high-rent districts, and their focus of activity was primarily the policy process in Washington, D.C., or state capitals. At the same time, the very structure of environmental policy, so central to the institutional definition of the mainstream groups, failed to account for the equity or social justice dimensions of environmental problems and the laws and regulations established to manage those problems. Despite the shift in rhetoric, the limited efforts in hiring, the various support activities for community groups occasionally undertaken, and the discussion of new agendas, mainstream groups remained caught up in the terrain and action that placed their groups apart from the new kinds of environmental politics being influenced by ethnicity, gender, and class factors.

While the January and March 1990 letters to the Group of Ten leaders challenged the mainstream groups as to their approaches, it also served to highlight the effort to establish new kinds of environmental identities. The rise of the alternative antitoxics groups most prominently influenced the development of these efforts, since the toxics problem invariably encompassed questions of work and home, where issues of ethnicity also intersected. Prior to the emergence of the antitoxics groups, other efforts that began in the 1970s, especially the Urban Environment Conference (UEC) and citizen action organizations such as the Association of Community Organizations for Reform Now (ACORN) and Citizen Action, had actively sought to develop multi-ethnic coalitions to address questions of equity and social justice.

The UEC in particular was a significant force within environmentalism through its attempt to establish links among minorities, unions, and environmental groups. Initially organized out of Michigan senator Philip Hart's office in 1971, the UEC began as a legislative and lobbying environmental counterpart to the Leadership Conference on Civil Rights. Attracting a young and enthusiastic staff interested in integrating social justice and environmental themes but operating with only a small budget and limited support from a handful of unions, the organization nevertheless expanded its reach during the middle and late 1970s. Part of the UEC's growth was linked to the availability of funds during the Carter years, particularly grant money from OSHA's New Directions Program, which enabled the UEC to become a presence in the areas of toxics and occupational health. After Reagan's election, the UEC's funding from federal sources was dramatically reduced and then eliminated, with the organization forced to curtail its technical training and support for community groups.[49]

With funds running out, the UEC decided to host one final, high-profile event: a 1983 New Orleans–based conference on toxics and minorities. The gathering was the first major attempt to identify toxics issues as issues of discrimination and social justice. Many of the conference participants, from Lois Gibbs and Dana Alston to tenant organizer Pat Bryant, themselves became pivotal figures in the emerging antitoxics movement. The New Orleans meeting also enabled community activists to exchange information and recount experiences, giving prominence to a new set of environmental themes and constituencies. But the UEC failed to survive much beyond the conference, a casualty of lack of funding and insufficient backing for its advocacy of community empowerment and the need to establish a coalition of interest groups. The New Orleans conference, while groundbreaking in its approach and themes, nevertheless failed to create, in organizational terms, the bridge between movements and constituencies so central to the UEC mission.[50]

One of the community organizing efforts highlighted at the New Orleans gathering involved a North Carolina protest against an EPA-sponsored landfill for polychlorinated biphenyl (PCB) wastes. Seeking to locate the project in rural Warren County—which had high poverty and illiteracy levels and a large African-American population—landfill advocates, including the state's governor, James Hunt, had relied on the dominant siting strategy of the period: find a site in an area that was poor and therefore job-starved, away from major urban centers, and without any recognized forms of opposition. The project's environmental necessity—the landfill, it was argued, had to be sited *somewhere*—was underlined by the use of experts to defend the siting process, without any obvious counterexpertise available to challenge that choice.

What was striking about the North Carolina PCB landfill fight was the active participation and leadership role of local African-Americans in the movement that emerged to fight the landfill, disproving the assumption of policy makers that people of color would have little interest in environmental matters. The Warren County protests also revealed that the defense of community was a prime motivating factor in the involvement of residents. The ability to overcome enormous difficulties in undertaking such a community-based mobilization (a process that culminated in a 1982 demonstration to block the PCB-loaded trucks bound for the dump site and that led to the arrest of more than 520 protesters) signaled the possibilities for this new kind of environmental protest.[51]

While the North Carolina events emphasized the importance of place in this new movement, issues of process figured significantly in the emergence of two key southern California–based groups—the Concerned Citizens of South Central and the Mothers of East Los Angeles (MOELA)—which successfully challenged waste treatment facilities planned for their neighborhoods. The predominantly African-American Concerned Citizens group was formed in 1986 because of dissatisfaction with the environmental review of a 1600-ton-per-day solid waste incinerator to be located in the group's neighborhood. Like the North Carolina PCB landfill situation, project sponsors had anticipated little community resistance to the incinerator proposal, known as LANCER, or the Los Angeles City Energy Recovery Project. As with other groups in the antitoxics movement, members of Concerned Citizens learned to rely on their own instincts and insights about possible hazards (e.g., how an incinerator might exacerbate the acute health problems already present in the community) while also becoming familiar with the technical details of the project to better challenge LANCER advocates. And while Concerned Citizens launched its effort in relative isolation from other groups in South Central or citywide environmental groups and resources, it quickly became adept at developing support within the neighborhood and connecting to the growing regional and national networks of antitoxics activists. As its organizing grew in scale and intensity, Concerned Citizens helped generate new alliances among community and environmental groups concerning the incineration issue and successfully challenged what had appeared to be an unstoppable business and political coalition of incineration advocates. Not only were the plans for LANCER shelved, but a Concerned Citizens–inspired citywide coalition ultimately forced the city to abandon its incineration strategy temporarily and pursue a higher-profile effort at recycling. In the wake of its victory over LANCER, Concerned Citizens also expanded its own community activities, focusing on community-based environmental hazards as well as housing and employment issues. Through its organizing, the group sought to join questions of process (empowerment) and place (community development).[52]

Nearby the proposed LANCER site, another community group new to environmental issues, Mothers of East Los Angeles, became engaged in a fight to stop a hazardous waste incinerator. The incinerator was to be located in an industrial zone not far from residential neighborhoods in East Los Angeles, the predominantly Mexican communities east of downtown Los Angeles. As in the LANCER situation, both the agencies and the company proposing the incinerator sought to fast-track their proposal. They hoped to avoid any significant community opposition, since the immediate site was unincorporated territory within Los Angeles County, where a conservative, white, male board of supervisors offered little resistance. But the surrounding East Los Angeles area had an important cultural and ethnic identity associated over the years with efforts to forge an urban Chicano consciousness. The proposed project also reinforced increasing concerns that East Los Angeles was becoming a refuge for such "negative land uses" as polluting industries, landfills, and incinerators, as well as a proposed state penal institution (which MOELA had first opposed) to be located not far from the incinerator site.[53]

In order to mobilize opposition against the prison (and later the incinerator), MOELA was formed in 1986 by several local activists, led by an East Los Angeles resident and grandmother of eight, Juana Gutierrez. Like Concerned Citizens, MOELA developed a strong community identity while also securing resources to challenge the unwanted facilities, including legal support from a local public-interest law firm. Successful in stalling both the prison and incinerator projects, MOELA began to deal with a range of other community-based environmental questions, including the problems associated with local businesses such as electroplating and auto body repair shops, which provided a source of income and jobs but were also significant polluters. As it expanded its reach, Mothers of East Los Angeles became one of the most prominent of the new environmental justice organizations integrating environmental concerns with issues of community identity and development.

Underlying this question of identity was the growing perception among social justice activists that environmental hazards and negative land uses were disproportionately borne by minority communities. The idea that pollution had a racial as well as a class dimension was neither a new nor particularly obscure concept among environmental analysts or movement participants. The Council for Environmental Quality, in its 1971 annual report, directly commented on how the poor and minorities were subject to significant hazards. A handful of studies during the 1970s and early 1980s that focused primarily on air pollution issues tended to reinforce this argument, while groups such as the Urban Environment Conference sought to translate the concept of the poor as victims of pollution into an organizing strategy.

Yet despite the activities of groups such as the UEC, these early perceptions about disproportionate risk failed to overcome the presumption, embedded in mainstream environmental activities and policy making, that environmental issues were simply not significant issues for communities of color. The emergence of new community-based, antitoxics groups led by people of color during the early and middle 1980s, however, sparked a renewed interest in the question of disproportionate risk and gave prominence to new studies on the subject. The most influential of these studies, a 1987 report issued by the Commission for Racial Justice of the United Church of Christ entitled *Toxic Wastes and Race in the United States*, especially crystallized the new interest in environmental justice themes. Since 1982, the commission's Special Project on Toxic Injustice had focused on the issue of community-based hazards in communities of color, functioning as a resource for such groups as the North Carolina anti-PCB protesters. Based on a locational analysis of commercial hazardous waste facilities and uncontrolled toxic waste sites, the 1987 study argued that race was the most significant variable associated with the location of the waste facilities, with large numbers of African-Americans and Hispanics, often in overrepresented numbers, living in communities where such sites were located. The commission findings were further elaborated by the case studies documented by sociology professor Robert Bullard in his book *Dumping in Dixie*, in which locational issues were also analyzed in terms of siting decisions and community response.[54]

Both the Bullard study and the Commission for Racial Justice report proved to be crucial documents in situating what commission officials called "environmental racism." While the term was used to refer to issues of disproportionate risk as well as the whiteness of mainstream environmentalism, it also served as a rallying cry for a new kind of organizing approach that identified a racial dimension in environmental issues and an environmental dimension in social justice questions. The siting of unwanted facilities such as landfills or incinerators in poor and minority communities became key to the environmental racism argument, with Bullard in particular defining such siting questions (especially in terms of African-American experiences) as civil rights issues. For some activists, civil rights and environmentalism had become, as Cora Tucker put it, "different links in the same movement."[55]

By the early 1990s, the race/environment relationship had become a major question for both mainstream and alternative environmental organizations alike. New groups, from New York City's high school–based Toxics Avengers (which sought to identify potential toxics problems in communities of color) to the Oakland, California, Pueblo (which organized around lead hazard issues), seemed to spring up overnight. These groups complemented community organizations such as South Chicago's People for Community Recovery, which had long been active in opposing particular facilities in their communities. New publications, such as Earth Island Institute's *Race, Poverty, and Environment* newsletter, also began to explore the race/environment relationship, further increasing its prominence within the overall environmental discourse.

In this context, a number of groups, most especially antitoxics organizations such as the National Toxics Campaign, underwent an extensive and difficult process of self-review regarding the race/environment relationship. Though many antitoxics groups were based in low- or middle-income communities with significant numbers of people of color, not all of them had successfully functioned as multiracial organizations. Parallel but separate organizations—one African-American or Latino, the other largely white or Anglo—would sometimes form to contest particular facilities, as occurred with respect to the opposition to the country's largest hazardous waste landfill at Emelle, Alabama. And while key antitoxics networks, such as the Citizen's Clearinghouse for Hazardous Wastes and the National Toxics Campaign, encouraged multiracial organizing and were firmly committed to a social justice perspective, they also encountered the uncertainties and complexities of race in their organizing efforts.

At the time of the January and March 1990 letters to the Group of Ten organizations, a major debate began to unfold within the National Toxics Campaign, the network with perhaps the strongest multiracial base within the antitoxics movement. This debate was heightened with the release in May 1990 of a new letter, organized through the Southwest Network for Environmental and Economic Justice and sent to several of the alternative environmental groups, such as the National Toxics Campaign, Citizen's Clearinghouse, and Greenpeace. The May letter questioned the alternative groups about staff and board composition, particularly in light of their commitments to a social justice perspective and their efforts at fundraising based on those themes. Subsequently, people-of-color members of the National Toxics Campaign board, including Richard Moore of the Southwest Organizing Project and Anthony Thigpen of the Los Angeles–based Jobs for Peace organization, challenged their group to establish even more substantial goals than the mainstream groups. This included a 50 percent target for people-of-color board and staff composition. The people-of-color caucus within the organization also proposed that the National Toxics Campaign commit resources for a training institute for organizers, designed for community-based people of color activists dealing with environmental justice themes. The National Toxics Campaign would be the source of funds and support for the project, although training institute organizers would control its budget and decision making. While still committed to the goals of multiracial organizing, the National Toxics Campaign would in effect establish its own autonomous people-of-color organizing and training operation.[56]

Similar to the National Toxics Campaign experience, organizers with the Southwest Organizing Project based in Albuquerque, New Mexico, sought to establish a regional social justice/environmental justice organization with special emphasis on empowering diverse constituencies, including urban and rural Latinos, Native Americans, Asians, and Pacific Islanders. SWOP, whose original founders had been active in the Brown Berets and other Chicano activist organizations, was established in the early 1980s as part of a campaign to address the problems of air emissions from a particle-board plant in a Latino and African-American working-class community in Albuquerque. Through its rainbowlike coalition and organizing strategies, SWOP created a regional approach that integrated the distinctive experiences of constituencies such as Latinos and Native Americans into its overall approach to environmental justice.[57]

In southeastern Louisiana, where much of the state's petrochemical industry—and significant environmental problems—could be found, another regional organizing strategy took shape. Under the leadership of the African-American–based Gulf Coast Tenant Leadership Development Project, a 200-mile walk through the state's "cancer alley" (a corridor of hazardous industries between New Orleans and the Gulf of Mexico) was organized in 1989. Although the demonstration included a range of environmental groups and constituencies, march organizers sought to directly integrate the ethnicity factor into the themes of the march by emphasizing social and environmental justice issues, including the question of disproportionate risk. Subsequent "cancer alley" marches further underlined this approach. Where the Gulf Coast group differed with other organizations was concerning the primacy of ethnicity factors and whether environmental justice groups should be multiracial or exclusively people of color.[58]

These themes reemerged forcefully at the October 1991 People of Color Environmental Leadership Summit in Washington, D.C., organized by key environmental justice advocates, including Dana Alston, Pat Bryant of the Gulf Coast organization, Benjamin Chavis of the Commission for Racial Justice, Robert Bullard, and Richard Moore of SWOP. The summit was a crucial event for this developing movement. By ratifying a statement of principles that drew together diverse themes about disproportionate risk and other forms of environmental racism, the governance of production decisions and control of toxics and other sources of pollution, and the centrality of place and cultural dimensions in environmental struggles, the summit succeeded in beginning to identify a new kind of environmental politics. It was this effort that conference organizer Dana Alston called "our vision of the environment [that] is woven into an overall fabric of social, racial, and economic justice." The organizational questions—whether the summit would establish a new, central organization or continue to serve a networking function, and whether organizing around environmental racism was best pursued by multiracial organizations or groups that exclusively involved people of color—were difficult and contentious issues still to be debated. Most importantly, the summit indicated that this new movement had passed a critical threshhold in definition, even as its unresolved questions reflected the organizational and strategic uncertainties of the alternative movement as a whole.[59]

By challenging the whiteness of the environmental movement, environmental justice advocates have successfully raised the question of constituency and the limits of the existing environmental agenda. As a civil rights issue, environmental racism has pointed to the existence of disproportionate risks and how they relate to the policy arena, although the analysis doesn't directly address the question of how to reduce or eliminate such risks. A civil rights perspective on the environment has also left partially unanswered the question of vision, or how a new, social justice–oriented environmentalism can address daily life experiences. Although the ethnicity factor remains central to this redefinition of environmentalism, such issues as gender roles, the importance of place, and the organization of work and production are critical as well. Still, the rise of the environmental justice groups and their challenge concerning environmental racism suggests that reconstituting environmentalism in order that it become more than just a white movement remains a central organizing task.

Chapter 8

A Question of Class: The Workplace Experience

A Steelworker's Discovery

When Larry Davis was a boy, living in the unincorporated town of Wheeler, Indiana, population 400, he loved to go into the woods by an old farm near the edge of town. With his buddies, he'd play in the creek, wade through the marsh, and camp out under the stars in the summer. When he was in high school, he began to notice tanker trucks that would pass over the old road on the way to the farm, but didn't think much of it. It wasn't until later, after he had been laid off from Bethlehem Steel's Burns Harbor plant in northern Indiana, that he had the opportunity to look into the matter of the old farm in Wheeler.

Once laid off, Davis had begun to follow the growing scandal about the Midco site in the city of Gary. The Midco dump, located in former dune and swale areas near Lake Michigan and the Grand Calumet River, was Indiana's version of the Love Canal. Problems at the site had first come to the community's attention in 1981, when hazardous wastes, already leaking from several thousand fifty-five-gallon drums, had been spread through town by the floods from a major storm. During the next several days, kids who'd been playing in the puddles left by the storm received chemical burns, as did workers exposed to the flood waters. Local residents could also smell chemical odors drifting through their neighborhoods. As a result, residents formed a new antitoxics group, and their protests ultimately forced the EPA and state agencies to target the site for cleanup through the Superfund process. But what Davis also learned in that spring of 1982 was that the "cleanup" of Midco simply meant shipping the wastes to other sites around the state, including the little-known landfill in Wheeler near where Davis had played as a boy.[1]

With time on his hands as a result of his layoff, Davis started to look into the Wheeler situation. At one point, he traveled to Chicago to the EPA's Region V headquarters to try to find additional information about the site and the wastes being deposited less than a mile from his parents' home. Sifting through files about to be discarded and rummaging through various other documents, Davis reconstructed the maneuvering and minimal safeguarding that characterized not only the Wheeler landfill but other waste sites in the state as well. Becoming involved in the newly formed antitoxics network and subsequently going back to school to learn more about the technical aspects of the issues, Davis soon found himself becoming a self-taught, self-motivated environmental expert. His passage to activist expert became complete when he got his old job back at the steel mill and decided to translate his new environmental awareness directly to the circumstances of his work.

The Bethlehem Steel plant at Burns Harbor contained the gamut of environmental hazards found in the steel industry. As an integrated steel plant, there were significant emissions and workplace hazards associated with coke ovens; there were waste disposal problems tied to the finishing end of the production process, as well as the use of blast furnaces to extract iron from the ore for steel making; and there were numerous problems related to hazardous materials, such as solvents, used in production. Certain materials and processes used at the Burns Harbor plant had created significant workplace hazards from various chemical exposures, generated toxic air contaminants and liquid wastes that impacted the surrounding community, and produced a range of environmental concerns, such as the heavy use of the solvent methylchloroform, or III TCA, a significant ozone depleter. As Davis would later discover, the toxics use problems related to solvents and other hazardous materials had been basically ignored by the company and not addressed as an issue by the union.[2]

When Davis returned to his old job loading barges, he decided to approach the president of United Steelworkers Local 6787 about getting involved in environmental issues at the plant. Davis was most interested in participating in two key committees: the Energy and Environment Committee and the Safety and Health Committee. The first was largely dormant after its previous head had been laid off, and the second tended to be relatively inactive. The lack of activity, even for the steelworkers, did not seem surprising. The international union, the United Steelworkers (USW), had been, during the late 1960s and 1970s, one of the few unions advocating health and safety reform while also being supportive of certain environmental laws, such as the Clean Air Act. The USW, an important backer of OSHA legislation, had even sought to incorporate health and safety language in contract negotiations, such as its 1974 agreement with U.S. Steel mandating engineering controls to limit carcinogenic emissions from the coke ovens at U.S. Steel's Clairton, Pennsylvania, plant. Still, like other unions, the USW focused more on bread-and-butter than occupational hazard issues. And while USW members participated in local environmental coalitions, such as the Pittsburgh-based GASP (Group Against Smog and Pollution) organization, the union's efforts concerning environmental and workplace hazard issues contrasted with the steel industry's sharply focused campaign to block environmental and occupational health and safety regulations and legislation. By the early to middle 1980s, during the Reagan administration, when OSHA programs were being dismantled, when unions in general were under attack, and when many of the big steel companies were restructuring partly by abandoning some of their older plants, even unions such as the USW were becoming reluctant to press hard on issues concerning the environmental and occupational hazards of production.[3]

The situation that Davis found at Burns Harbor revealed some of the difficulties of addressing environmental hazard issues within the plant. For one, the Energy and Environment Committee was never recognized by the company. Therefore, any activities by committee members had to be undertaken on their own time. After Davis took over the committee and sought to revive it, he tried to meet with the company's environmental representatives, but to no avail. Davis then decided to pursue his own research on company activities, exploring the environmental hazards at the plant. He discovered a history of no inspections, few enforced regulations, and successful maneuvering by the company, in conjunction with the steel industry, to be exempted from many of those same regulations. Yet there had been countless incidents of materials spilled, improper handling of hazardous materials and wastes, and situations where materials and substances used in the plant and likely to be quite hazardous had yet to be regulated, given OSHA's unwillingness to intervene in hazardous product use. Some of the most toxic materials, Davis complained, were handled and used "like popcorn."[4] In his frustration over the company's lack of response, Davis thought about filing a protest with state agencies and federal regulators, but that suggestion was dismissed by Burns Harbor veterans. "Hell, the company likes to go that way," one old-timer told Davis, pointing to Bethlehem Steel's readiness to litigate, with just one coke oven suit generating hundreds of thousands of pages of depositions.

Davis didn't find the situation much better with respect to the Safety and Health Committee. Safety issues tended to be treated in a more straightforward manner, with some attention given to areas where accidents were most likely to occur. But the health issue, Davis felt, remained unexplored. "No one really knows what's happening in the plant," Davis later said of these issues. "If the company were willing to do a decent health study, we might be able to begin to identify the kinds of problems that need to be addressed. Instead, it becomes a little like detective work identifying workplace hazards, which is a lot of what I've ended up doing."

Davis eventually became the chair of the Safety and Health Committee. He began urging the union to integrate environmental issues more directly into the safety and health framework and include language about environmental issues in the local union's contract with the company. By 1990, after much debate, environmental issues were made a primary concern for the USW as well as for the international union itself. At the USW's Twenty-fifth Constitutional Convention held in Toronto, Canada, environmental themes were broadly encouraged, with the union's Health and Safety Department converted to the Health, Safety, and Environment Department. "If you have a modern facility that produces less pollution, you certainly have a more secure job and you have a cleaner community to live in," Davis told the convention, and his statement was included as part of a report by the union on the convention's embrace of environmental issues.[5]

Though he now asserts that he is "proud of my union," Davis still worries about the strong tendency toward denial that prevails among workers and hampers the development of a powerful workplace environmentalism. Davis continues to encounter ignorance about hazardous materials, an unwillingness to address the problem of environmental risk, and a widespread apathy about the inevitability of workplace conditions. The problem of workplace passivity is of particular concern for Davis, given his strong identification with the activist culture of the antitoxics movement. As a new environmentalist, Davis has become most passionately engaged in community-based environmental issues, joining with other community activists to challenge industry interests rather than just focusing on the Burns Harbor plant, where he faces numerous obstacles even in identifying the problems that exist.

The evolution of Larry Davis's activism demonstrates how toxics issues so often become more compelling as community issues than as workplace issues, despite the fact that production hazards precede and frame the hazards experienced at the community level. At Love Canal, the production at a Hooker Chemical plant of chemicals such as vinyl chloride that later contaminated the surrounding houses and elementary school entailed serious workplace hazards. These occupational exposure issues were never successfully addressed by the company, state and federal agencies, the union at the plant, the Love Canal Homeowners Association, or the workers, who remained reluctant to link their own situation with a community at risk.[6] It has been this disjuncture between the hazards of production as experienced by a work force and those same hazards within the general environment, experienced as a community issue or as a question of harm to the natural environment, that most directly reveals the split identity between movements.

The way to overcome that split, Larry Davis has decided, is not to forgo his community identity but to extend it to his job and to the workplace. The work force at the steel plant at Burns Harbor, Davis argues, is like a community resting atop a hazardous waste site. If the people at the plant fail to address the problems or dismiss the environmental issue as something "out there" not connected to their own situation, the issue will nevertheless force its way back into their lives, whether in terms of the hazards experienced or from the pressures of community sentiment. Similarly, if the people in the community fail to include workplace hazards and job questions as part of their perspective, they will find it even more difficult to contend with the potent industry argument of jobs *versus* the environment. To situate environmental issues as production choices and thereby create the link between jobs *and* the environment has become the crucial task for Davis and other workplace and community activists. It is a task that also requires finding ways in which community and workplace movements can be joined as well.

SETTING THE STAGE: FROM ASBESTOSIS TO BLACK LUNG

The environmental hazards of the workplace have been an area of investigation and concern since the rise of the industrial city. The research pioneered by Crystal Eastman, Alice Hamilton, and Florence Kelley during the Progressive Era provided an early glimpse of the extent and breadth of environmental problems within the workplace. Whether detailing the hazards of the "sweat" system or hard-rock mining, radium watch-dial painting or foundry work, these early occupational health advocates were simultaneously interested in social reform, seeing industrial hazards as a public health issue. Similarly, trade unions such as the International Union of Bakery and Confectionary Workers and advocate groups such as the Women's Trade Union League sought to directly link unsanitary workplace conditions with community hazards. For these advocates, occupational health came to be defined as a social issue at the intersection of workplace and community environments.[7]

By World War I, the professionalization of the public health field, with its de-emphasis of environmental factors and the development of programs such as workers' compensation, began to undercut this social definition of occupational hazards. The industrial health field became dominated by a more restrictive practice of industrial medicine, a class-based legal system of review and compensation for workplace injuries or illness, and a growing tendency to separate issues of worker health from community or public health. Industries were able to redefine what constituted occupational hazards through the company doctor, who became most responsible for characterizing occupational illness, and by influencing the research and analysis of occupational hazards through industry-dominated groups such as the Industrial Hygiene Foundation.

By the 1920s and 1930s, the Progressive Era linkage of social reform, medicine, and the investigation of occupational hazards had given way to this less visible, more "professional" system, which kept issues—and movements—apart from both sides of the factory gates. This was a period that also witnessed intense and extraordinary episodes of occupational disease, such as the Gauley Bridge disaster. Another such episode, particularly related to the efforts to control the definition of what constituted a hazard, involved workers in the asbestos industry and the role of that industry's leading firm, the Johns-Manville Corporation.

Today, asbestos conjures up strong environmental associations, generated in part by the publicity surrounding the court cases involving Johns-Manville during the 1970s and 1980s. As with lead, the dangers of asbestos had been recognized as far back as the first century B.C., when Roman slaves became victims of diseases associated with the mining and weaving of asbestos. Known for its thermal insulation properties, asbestos increased significantly in use during the early twentiethth century following improvements in mining technology and the discovery of new uses for short fiber asbestos discards from mining. These new uses, which included insulation paper for homes, brake lining for cars, and insulation material for the construction industry, helped transform asbestos into one of the most widely used industrial and commercial products after World War I. By the middle 1930s, with production levels having doubled in just ten years, additional markets aggressively pursued, and a sophisticated public relations apparatus in place to "woo the public" by touting the industry's "social responsibilities," the asbestos industry was ready to take its place as a major twentieth-century success story.[8]

The success of the industry, however, like the parallel contemporary success stories of the leaded paint and leaded gasoline industries, was vulnerable to the claim that industry products were hazardous. This was particularly the case for those who inhaled asbestos fibers. By 1924, with the publication of a study in the *British Medical Journal* linking asbestos dust exposure with asbestosis, a chronic and sometimes fatal lung disease, information was becoming available about the hazards of asbestos use. In 1935, at a time when production increases were causing a major jump in occupational illness among asbestos workers, a U.S. Public Health Service report based on an earlier industry-financed study pointed to the high correlation between asbestos work and incidences of asbestosis. This astounding study, however, was significantly altered in its final published form by Johns-Manville officials, who eliminated the numbers from the study and downplayed the seriousness of the hazards involved.[9]

As later court documents revealed, the tampering with the Public Health Service report was just one example of a systematic effort by asbestos industry officials, lasting more than three decades, to control the research about asbestos hazards in order to limit industry liabilities. This effort included placing overwhelming constraints on independent researchers, directing and modifying industry-funded studies, and developing secret agreements with researchers to allow asbestos companies to withhold studies from publication. At the same time, these same companies, through the prevailing company doctor system, continually understated the problem to their work forces. One striking 1949 memo from Kenneth Smith, Johns-Manville's corporate medical director, to company superiors even suggested that X-ray evidence of asbestosis was withheld from the company's workers so exposed workers would continue to work up to the point of their becoming disabled from asbestosis. "They have not been told of this diagnosis," Smith wrote, "for it is felt that as long as the man feels well, is happy at home and at work, nothing should be said." Along these same lines, Charles Roehmer, a Paterson, New Jersey, industrial commissioner, recounted a 1940s conversation with Johns-Manville CEO Lewis Brown and his brother, Vandiver Brown, the firm's corporate counsel. "One of the Browns made this crack (that Unarco managers were a bunch of fools for notifying employees who had asbestosis)," Roehmer recalled the conversation in a later deposition, "and I said, 'Mr. Brown, do you mean to tell me you would let them work until they dropped dead?' He said, 'Yes. We save a lot of money that way.' "[10]

The control of information about asbestos hazards was all the more consequential, given that production levels increased by more than a factor of eight between 1935 and 1950. At the same time, large increases in lung cancer brought on by asbestosis were also recorded among asbestos workers. Yet there was virtually no research in this area, since many of the key figures who dominated the field of occupational medicine from the 1920s through the 1950s were also associated with the industry-sponsored suppression of information about asbestos hazards. Meanwhile, industry interests were able to significantly limit their liability by having states amend their workers' compensation statutes to eliminate the right of workers to bring suit because of asbestosis, making it instead a compensable disease, with average settlements as small as a few thousand dollars each.

It was not until the early 1960s that this corporate lock on information was finally pried open, when independent researchers led by Dr. Irving Selikoff at New York's Mount Sinai Medical Institute documented through union welfare and retirement records just how widespread this occupational hazard had become. Selikoff himself helped to inspire a new occupational health movement of doctors and other health professionals seeking to challenge the dominant industrial medicine system and its approach toward industrial hazards. The emergence of this new movement of health professionals paralleled changes within the environmental movement, where questions of the manipulation of information and corporate malfeasance were also becoming issues. Strikes by asbestos workers focused on health and workplace hazards, along with a dramatic increase in individual and class-action suits, eventually forced asbestos companies to release information on occupational hazards already compiled but never made public. It was this sequence of events, including extraordinary disclosures of corporate deceit and cover-up, that helped link the issues of occupational exposures, environmental hazards, and the public's right to know and launched a new workplace-related movement.

Similar issues emerged concerning the hazards of coal mining, especially the sharp increase in serious chronic respiratory problems, or black lung disease. During the turbulent 1960s, when the dimensions of asbestos hazards were first being revealed, the black lung movement—the most important grassroots movement of workers confronting an occupational hazard at the time—also exploded on the scene. Led by retired miners and widows of miners, this movement would in the space of a few years reshape the United Mine Workers and help change occupational health policy. Movement groups would also play a central role in forcing a reevaluation of the relationship of black lung to the hazards of mining and the overall industrial workplace environment as well.

By the middle to late nineteenth century, the black lung of Emile Zola's coal miner in *Germinal* and the "black spittle disease" described by Friedrich Engels in *The Condition of the Working Class in England* had become "powerful symbols of the dark and dangerous human effects of a rising industrial capitalism," as black lung chronicler Barbara Ellen Smith noted. Hazardous conditions in the mines paralleled the hazards of community life in the mining towns, with their poor sanitation, overcrowded housing, contaminated water supplies, and high incidences of typhoid fever and gastroenteritis. But it was the respiratory problems related to the work environment, most especially pneumoconiosis (literally, "dust-containing lung" disease) and silicosis (the inhalation of free silica dust), that remained disturbing issues through the late nineteenth century and into the twentieth century. By 1910, a consensus had emerged that occupational disease for mine workers—"the insidious undermining of health through the breathing of dust from machine drills, vitiated air, [and] poisonous gases from explosives and from the rocks," as one mining publication put it—had become even more significant than the higher-profile issue of mine accidents.[11]

Though the mining industry and the government was most focused on the silicosis problem up through the 1940s, coal worker's pneumoconiosis remained an important, although largely unaddressed, issue. For coal miners, respiratory problems were a fact of life related to the working conditions in the mines. But the industrial medicine system refused to make that link, more often attributing miner health problems to non-work-related causes. What altered this situation most dramatically for both the miners and industry was the restructuring resulting from the 1950 industry-wide contract between the United Mine Workers and the coal operators, an agreement that also had significant environmental health ramifications.

The 1950 agreement involved a crucial trade-off for the union between higher wages and benefits for their members on the one hand and technological innovation for the industry on the other. The unimpeded introduction of machinery into the mines intensified the exposure to certain hazards while also instituting more specialization, an increased division of labor, and a reduced labor force. For example, where mining had previously been done by "muscle," it was now undertaken by machines. Workers such as runners and drill men (who operated the machinery that cut into the coal face) found themselves constantly surrounded by unprecedented levels of coal dust. Consequently, serious chronic respiratory problems jumped dramatically during the 1950s and 1960s, and mining communities throughout the Appalachian region most affected by work force reductions and the increase in respiratory hazards were also severely impacted by the changes in the mines. These communities began to lose cohesion, their land marred by the new mining practices and their demographics altered by the increasing numbers of black lung retirees and widows.

Part of the 1950 agreement called for an industry-financed but essentially union-controlled Welfare and Retirement Fund to replace the existing system of company doctors. The fund immediately attracted progressive physicians and public health advocates who saw its development as an opportunity to renew the quest for a publicly controlled health system. Many of these advocates were skeptical of the industrial medicine system's approach toward occupational hazards in the mines and were ready to contest the more exclusive focus on silica in order to raise concerns about the impacts of coal dust. But while the fund's staff campaigned extensively for recognition of coal miner's pneumoconiosis within the medical field, the information gathered was never shared with the miners themselves. Nor did the fund's leadership press for modifying production processes to reduce the level of dust in the mines. The issue of coal miner's pneumoconiosis remained a medical issue debated among medical professionals, not a social or political issue related to the compensation system or mining production technology. At the same time, the union leadership, which controlled the fund, was increasingly seen as corrupt in its activities and distant from its own constituency.

By the late 1960s, protests against environmental hazards in the mines, the role of the union, and the impacts on mining communities emerged from several different sources. These included a growing rank-and-file protest against the union leadership led by Jock Yablonski, a West Virginia union district leader who was sympathetic to the claims of black lung victims; a disabled miners' and widows' movement that challenged the Welfare and Retirement Fund's eligibility requirements for black lung–related pensions and medical coverage; and community-based activities, primarily funded through the War on Poverty's Office of Economic Opportunity, which focused initially on such issues as strip mining and land reclamation. In the spring of 1967, the Association of Disabled Miners and Widows, calling itself the "organization of living dead men trying to help each other," was the first of the black lung protest groups to form. The black lung issue was further galvanized when Welfare and Retirement Fund medical official Dr. Lorin Kerr gave an impassioned speech at the 1968 UMW convention. Kerr spoke of the "constant stream of wheezing, breathless coal miners coming to an area office in Morgantown seeking relief from their struggle to breathe." "Never in my earlier professional career," Kerr told the assembled miners, "had I observed or heard of a single industry with so many men who seemed to be disabled by their jobs."[12]

Through 1968, the black lung issue continued to flare up throughout the Appalachian region. It also began to receive national attention, including a February 1968 article by Ralph Nader in the *New Republic* that offered a blistering attack on the industry, the union, and the minimal federal occupational safety and health bureaucracy. Then, suddenly, at 5:15 A.M. on November 20 of that year, an explosion at the Mountaineer No. 9 coal mine of the Consolidated Coal Company in Farmington, West Virginia, dramatically shifted the national debate about environmental hazards in the mines and other industries as well. Unsuccessful efforts to save seventy-eight miners trapped in the mines extended through the Thanksgiving holidays and received heightened media attention. Though the black lung movement was only tangentially affected by the Farmington disaster, it intensified national concern about occupational hazards and provided the black lung protesters with a direct opportunity to influence policy debates within the West Virginia Legislature and the U.S. Congress.[13]

In January 1969, the movement took another major step with the formation of the West Virginia Black Lung Association. A few weeks later, between 3500 and 5000 miners and miners' widows attended a statewide rally in Charleston. One of the group's most visible supporters was Ralph Nader, who linked the black lung movement to the issue of how the disease was defined. No longer should miners "tolerate puppet-like physicians, employed by the coal operators, disgracing their profession by saying black lung is not all that bad and miners can learn to live with it," Nader declared in a telegram read at the rally. "From now on, the coal miners—whose bodies are being abused and whose lungs are being destroyed—are going to be the ones who say how bad black lung is."[14]

In February 1969, as the West Virginia Legislature debated the issue of compensation for black lung, a wildcat strike by coal miners erupted. At first, strikers spoke of creating a broader occupational health movement that would incorporate workers from other industries and threatened to launch a general strike in the foundries, chemical plants, and other industries that surrounded Charleston. Although the general strike idea never caught on, miners' actions, including additional wildcat strikes, framed the debates in Charleston and later in Washington, D.C., where Congress was considering a Coal Mine Health and Safety Act. With insurgency in the mines reaching a fever pitch through 1969, both the West Virginia Legislature and U.S. Congress passed their respective legislation. The federal legislation was signed into law less than twelve hours after an unauthorized strike of miners spread through the West Virginia mines. A year later, Congress passed the Occupational Safety and Health Act, crucial workplace legislation helped by the miners' actions and the media attention generated by the Farmington tragedy.

Though unintended, passage of the OSH Act served as a culminating point for the black lung struggles associated with the establishment of a more expansive occupational health regulatory system and its recognition of workplace hazards. Such changes paralleled, at the same time, the recognition of pollution within the general environment and the establishment of an eco-bureaucracy. Despite these achievements, however, the black lung movement soon came to be buffeted by two contrasting realities that tempered its successes and influenced its future direction. These involved the rank-and-file insurgencies that exploded after the union leadership–directed murder of dissident leader Jock Yablonski (one day after the Coal Mine Health and Safety Act was signed into law) and that tore apart the union through much of the 1970s; and the difficult and time-consuming work of filing claims and monitoring the compensation system that the black lung movement had helped establish. Like other movements of the 1960s, the black lung movement's legacy had less to do with these kinds of organizational influences than with its critical role in helping redefine the work/environment relationship. Participants in a movement born and bred in industrial conflict but eventually absorbed by the complex, institutional demands of a compensation system, black lung activists remain best recognized for their inspiring new movements dedicated to changing the ways industrial hazards are defined and addressed.

WORKPLACE POLITICS

Fifty years after the eclipse of the reform movements of the Progressive Era, with their overlapping concerns about issues of work, community, and environment, new industrial-related movements reappeared during the 1970s to try to reconstruct that neglected link between workplace and community environments. Passage and implementation of the 1970 Occupational Safety and Health Act most influenced and shaped these movements, while also serving as a workplace counterpoint to the construction of an environmental policy system.

The idea of developing occupational health and safety legislation first emerged during the mid-1960s, when officials within the Johnson administration sought to develop quality-of-life policy objectives in advance of the 1968 election. These included an emphasis on air and water pollution issues as well as consumer product safety issues made more visible thanks to Ralph Nader's auto safety campaigns. Johnson's chief policy advisor, Joseph Califano, had organized an interdepartmental effort to identify potential "accident prevention" legislation, including occupational safety issues. Despite an initially lukewarm response from the unions, Johnson decided to raise the question of occupational hazards directly in a May 1966 speech before the International Labor Press Association. The president warned of the problems related to post–World War II production technologies and asked the labor movement to go beyond bread-and-butter issues to "join with us in the effort to improve the total environment." In 1967 and 1968, top officials in the Department of Labor, including Secretary Willard Wirtz and Assistant Secretary Esther Peterson, who were influenced by the emerging black lung and uranium mining hazard issues, drafted legislation to allow the federal government to regulate workplace hazards. Although the legislation didn't pass Congress while Johnson was still president, occupational safety and health legislation was reintroduced in 1969 by the Nixon administration as part of its "blue collar" political strategy.[15]

Through 1969 and 1970, a range of different groups became involved in the effort to secure passage. These included Ralph Nader and his activist researchers, who sought to systematically document the failure of existing workplace hazard policies; a handful of union activists led by Anthony Mazzochi of the Oil, Chemical, and Atomic Workers Union and including Jack Sheehan of the Steelworkers and Frank Wallick of the Auto Workers; and a few of the more activist environmental groups, such as Environmental Action and Friends of the Earth. Industry groups mobilized in opposition to the OSH Act, although their efforts later became more pronounced over implementation and rule making. When the OSH Act, with provisions authorizing enforcement of standards, unannounced inspections, and research into potential health and safety problems, was signed into law in December 1970, a new era of concern regarding workplace hazards seemed ready to emerge.[16]

The passage of the OSH Act, and the establishment of the Occupational Safety and Health Administration (OSHA) to administer the act, helped stimulate the budding workplace environment movements that had emerged during the middle to late 1960s. For example, two organizations established as part of the Nader network—the public interest research groups (PIRGs) and the Health Research Group (affiliated with the Public Citizen organization)—began to focus on occupational health issues in a more programmatic way during the early 1970s after the OSH Act was signed into law. They did this by publishing reports, including their major background study *Bitter Wages*; by publicizing serious workplace hazards, such as a cancer epidemic at a Rohm and Haas plant in Bridesburg, Pennsylvania; and by lobbying at the state and local level for specific occupational safety and health measures. Nader himself played a key role in highlighting occupational hazard issues and seeking to link labor, consumer, and environmental groups.[17]

The Bridesburg situation also pointed to the influential role of public interest advocacy, when unions had a limited or nonexistent role in confronting occupational hazards. Since the early 1940s, Rohm and Haas, a large chemical manufacturing company, had been involved in the research and production of bis-chloromethyl ether (BCME) and chloromethyl ether (CME), both widely used for different chemical processes, including water purification. Despite Rohm and Haas's access to data suggesting by the late 1960s that an extraordinarily high proportion of workers directly exposed to BCME were contracting cancer, the company failed to provide researchers with direct exposure data to measure the prevalence of carcinogenicity among BCME workers. When a New York University scientist made his findings about his BCME inhalation studies on rats available to Rohm and Haas executives and other chemical industry officials at a highly charged meeting in July 1971, the officials reacted with disbelief and anger. One industry official argued that such animal testing methods were "irrelevant" and "invalid" since the chemical hadn't been directly "injected in man."[18]

By 1973, the Bridesburg workers still remained largely without recourse—or knowledge—about their own situation, especially since the facility where the BCME had been produced was not unionized and there was little opportunity to collectively explore concerns or possible actions. That year, however, researchers from Temple University School of Medicine published an article in the *New England Journal of Medicine* presenting evidence, based on their analysis of the Bridesburg plant, that BCME was an occupational hazard capable of causing lung cancer. Unable to obtain information directly from the company, the researchers had instead relied on detailed information about the work force from the observations of a plant worker then dying of lung cancer.[19]

When the *New England Journal of Medicine* article was published, it immediately caught the attention of the Public Citizen–affiliated Health Research Group (HRG), which had begun to focus on occupational hazard issues. The HRG, along with the Oil, Chemical, and Atomic Workers Union (OCAW) had recently filed a petition with OSHA requesting an emergency temporary standard to eliminate worker exposures for ten specific carcinogens, including BCME. The inclusion of BCME was based on the rat inhalation studies, but not on any specific knowledge about the tragic events unfolding at Bridesburg. Although the Bridesburg plant was never mentioned in the *New England Journal* article, the HRG researchers quickly determined which plant was involved. Since no union existed at Bridesburg, the HRG became the key advocate obtaining more information about the facility, including damaging information from the company itself about what it knew and when it knew it. By 1976, BCME exposure in Building Six at Bridesburg had become another symbol, along with the cases of asbestos and black lung disease, of the extent to which information about occupational hazards was suppressed. By publicizing such information, HRG activities helped extend support for key legislative initiatives, such as the Toxic Substances Control Act, and new occupational health standards and policies.[20]

The role of the Health Research Group paralleled the development of a number of other public health, science, and medical professional organizations that sought to undertake research about workplace hazards and make their own expertise available to worker groups. These included scientist and physician networks such as the Scientists' Committee for Occupational

Health and the Medical Committee on Human Rights, as well as community-based organizations of activists and professionals such as the various Committees on Occupational Safety and Health (COSH groups) that sprang up in the early to middle 1970s. These groups shared a common goal in seeking to empower workers addressing occupational hazard issues. These efforts were pursued much in the manner of the Workers' Health Bureau of the 1920s with its attempt to link public health professionals and physicians, social reformers, and unions directly concerning specific occupational health issues.

At the time of the passage of the OSH Act, most unions paid little attention to occupational hazard issues, with health-related concerns primarily tied to the development and funding of the various insurance, welfare, and pension funds that became prominent in the post–World War II era. Despite indications that health and safety had become a major area of interest and concern among broad sections of the work force by the late 1960s, both industry groups and unions tended to "either overlook or downgrade its importance," as the UAW's Frank Wallick put it. Wallick, the editor of the UAW's Washington newsletter and one of the few union officials then exploring both environmental and occupational hazard issues, suggested that part of the disinterest stemmed from "a feeling of fatalism about working conditions" among union officials, who preferred to focus on economic issues as "winnable" demands in contract terms.[21]

This lack of interest within the union hierarchy was further exacerbated by the chain-of-command style of decision making among larger unions such as the UAW. During the late 1960s and early 1970s, for example, Wallick's newsletter contained descriptions of air-monitoring devices union locals could purchase to document emissions within their plants. For this initiative, Wallick was reprimanded by the international union's Safety Department, which insisted that management would never allow the equipment to be used. Similarly, the union sought to centralize all complaint procedures under the new OSH Act through the union's regional service representative, who would decide whether to request a formal OSHA inspection. "Without this chain of command," UAW Safety Department official Lloyd Utter remarked at the time, "we would run the risk of having self-appointed experts in hundreds of plants operating singly, on the one hand, or of being inundated with requests here at headquarters in Detroit, on the other."[22]

There were important exceptions to this approach, most notably the Mine, Mill, and Smelter Workers Union in the 1940s and 1950s and the Oil,

Chemical, and Atomic Workers Union in the late 1960s and early 1970s. The Mine, Mill union, originally the Western Federation of Miners, was a leading advocate of health and safety in the lead and zinc mines of the Midwest and Southwest, especially in the tri-state area of Missouri, Kansas, and Oklahoma, where deaths from silicosis and tuberculosis had reached epidemic proportions by the 1940s. But the union's successful efforts in mobilizing on occupational hazard issues were undermined when the union became subject to unprecedented political attacks on its radical leadership. These included congressional investigations, raids by other union locals, and expulsion from the CIO. When Mine, Mill was taken over by the United Steelworkers in 1967, its disappearance into the larger union also coincided with the fading of the silicosis issue that the Mine, Mill union had once made so prominent on its agenda.[23]

Following the disappearance of the Mine, Mill union, the Oil, Chemical, and Atomic Workers became the most prominent trade union engaged in occupational health issues. At the OCAW's 1967 convention, the union adopted health and safety as a major organizing focus, an approach necessitated by the safety and hazard issues prevalent within chemical plants, oil and gas production and processing facilities, and nuclear power plants where the OCAW sought representation. By 1970, the union began to hold meetings in local districts concerning hazards in the industrial environment as a means of helping workers better identify the hazards they experienced. As part of this effort, Ray Davidson, editor of the union's newspaper, was given a six-month leave to interview workers about their specific job hazards. Davidson's book based on these interviews, *Peril on the Job: A Study of Hazards in the Chemical Industry*, underlined the crucial role of workers in identifying such hazards.[24]

Despite the OCAW's important commitment to occupational health and safety issues, its support didn't always translate into funding and staff resources. Much of the OCAW activity was generated at the local level in conjunction with the efforts stimulated by Tony Mazzochi, the OCAW's legislative director, who became the most influential figure within the new occupational health movement emerging during the early years of OSHA. Mazzochi's approach at once emphasized the importance of worker-generated activity and the potential links that could be established between workers and public interest and professional groups with an interest in the occupational health issue. Toward that end, Mazzochi and his assistant, Steve Wodka, established "plant situation teams" to teach local workers "how to control their own work environments," as Mazzochi put it, by incorporating their newly acquired information about occupational hazards into bargaining and grievance procedures. Key to the Mazzochi/Wodka approach was the worker support function of the public interest and professional groups, such as the Scientists' Committee for Occupational Health and the various COSH groups, and their willingness and ability "to overcome the aura of mystery surrounding their profession" and both learn from and provide information to groups of workers at the local or district level.[25]

The emergence of COSH groups in cities throughout the country during the 1970s, many of them organized by one-time New Left activists responding to quality-of-life concerns in the workplace, indicated a growing potential for a multilayered occupational health movement. The first COSH group, the Chicago Area Committee for Occupational Safety and Health (CACOSH), was organized in January 1972 from a conference on occupational hazard issues. The conference had been put together by members of the Chicago chapter of the Medical Committee on Human Rights (MCHR), a group of New Left–oriented young medical professionals. Cosponsors for the event included local unions and regional union leaders as well as the University of Illinois School of Medicine. Two other key COSH groups that emerged about the same time were MassCOSH (the Massachusetts Coalition for Occupational Safety and Health) and Philaposh (the Philadelphia Area Project on Occupational Safety and Health). MassCOSH was an outgrowth of the Job Safety and Health Project of Urban Planning Aid, one of the prominent late-1960s War on Poverty groups influenced by the push for empowerment (then defined as "participatory democracy"), which became a central feature of the COSH approach. Philaposh had been put together by Rich Engler, an MCHR organizer who worked closely with the OCAW's Mazzochi during that union's strike against Shell Oil, an action in which occupational hazard issues figured prominently. Engler and other MCHR activists launched Philaposh through training sessions on occupational health and safety for members of twenty different local unions. Philaposh became a model for other COSH groups in pursuing the kind of linkage strategy advocated by the OCAW's Mazzochi.

Though the COSH groups provided an important support function, not all unions looked favorably on them. In Pittsburgh, the development of PACOSH (the Pittsburgh Area Committee for Occupational Safety and Health) during 1972 and 1973 came to be opposed by the United Steelworkers international leadership, who saw the group as too independent and potentially disruptive of union activities. The international union discouraged participation in PACOSH by local unions and union members and thus effectively undermined the organization before it got off the ground. Other COSH groups deliberately sought to encourage rank-and-file movements to use occupational safety and health issues to challenge their union's leadership. These COSH groups thus became subject to the charge of "dual unionism" and were strongly opposed by the union bureaucracies. Despite these examples, most COSH groups by the middle to late 1970s had established working relations with local unions. Union officials served as COSH directors, and a few unions provided limited funding for the organizations. By the end of the decade, with COSH groups and COSH-related groups organized in several dozen cities and states, and OSHA's New Directions Grant Program offering important financial support, the success of the COSH form of organization demonstrated how linkage strategies continued to be a central dynamic of the occupational health movement.[26]

The success or failure of linkage was also critical to the question of whether labor and environmental groups could more effectively coalesce. During the late 1960s and early 1970s, nearly all mainstream environmental groups had been absent from the debates about passage of the OSH Act and its subsequent implementation.[27] The absence of ties was further compounded by the mistrust between groups, expressed in terms of their differing class base and organizational objectives. Mainstream environmentalists were especially opposed to the labor movement's strong backing of economic growth and development strategies such as the construction of large, central-station power plants that negatively impacted the natural environment. Union officials, on the other hand, remained wary of the upper-middle-class composition of the mainstream groups and their relative disinterest in the issues of jobs and economic security. A few unions, such as the UAW, had offered modest support for the 1970 Earth Day events and were active in lobbying for the Clean Air Act, and a handful of mainstream environmental groups expressed interest in discussing areas of mutual concern with the unions. Nevertheless, divisions between the groups remained substantial, even as environmental and occupational health policies began to be constructed along parallel, but separate, tracks.

During the early and middle 1970s, tentative efforts were made to explore possible ties between the groups. Ralph Nader proved to be especially effective in seeking to coalesce interests, partly through the activities of his own network groups. The OCAW linkage strategy also extended to efforts to secure environmental support for the union's actions, most notably the Shell Oil strike, which galvanized both union and environmental activists committed to a workplace/environmental connection. The Shell strikers' demands included the availability of data about occupational hazards, access to information about illness and death, and company-funded release time for a union occupational safety and health representative at the plant. Through Tony Mazzochi's efforts, an outreach campaign was launched to public interest and community-based groups focused on occupational hazard issues, along with an appeal to mainstream environmental groups to back the Shell workers. Several groups, including The Wilderness Society and the National Parks and Conservation Association, signed a joint statement with a relatively vague offer of support. More significantly, the Sierra Club decided to directly back the strikers on the occupational hazard issue. Despite the limited intent of the Sierra Club's support statement, it nevertheless generated a fair amount of criticism within the organization from those who worried that the "Club's extension of its activities beyond traditional wilderness and resource conservation" served to diminish "Club credibility."[28]

What most concerned both union officials and mainstream environmental leaders was industry's growing and effective use of the threat of job loss in its attacks on environmental and occupational hazard regulations. During the late 1960s and early 1970s, the job loss argument had less effect in the context of an expanding and overheated economy fueled by Vietnam War–related expenditures. These years, argued University of California at Berkeley occupational health analyst James Robinson, "were a unique period in postwar American history during which business boomed and job opportunities abounded." A "labor market bonanza" allowed workers to switch or quit their jobs in unprecedented numbers and underscored the growing quality-of-life concerns as opposed to job security issues within the work force. With the protracted recession that began in late 1973 and extended into 1975 on the heels of the oil-related energy crisis, labor markets contracted and the question of job loss became paramount. These changes enabled industry groups to mobilize more effectively against the new environmental and occupational safety and health regulations put in place prior to the recession. The slogan "No work, no food—Eat an environmentalist," first heard in steel-producing communities during the 1973–75 recession and later in timber-producing regions, symbolized the potential fissures between environmental and worker constituencies.[29]

The initial response to the industry campaign by labor and environmental groups, particularly the large industrial unions such as the UAW and the Machinists, was to try to expand the tentative efforts at coalition building.

The UAW, Steelworkers, and a handful of other unions, for example, supported the Urban Environment Conference as a way to bridge worker, environmental, and minority constituencies. Several unions also supported environmental initiatives such as the Clean Air and Clean Water acts, their interest stimulated as much by the job-creating aspects of these technology-based laws as by their cleanup objectives. By the mid-1970s, a number of unions and environmental groups, along with the EPA and the Council on Environmental Quality, were touting the job benefits of the manufacture, installation, and operation of pollution control equipment and facilities.[30]

The job loss/job creation argument also extended to the energy arena. The nuclear power industry, oil- and gas-refining producers, coal operators, and electrical utilities used the occasion of the energy crisis to link job loss with opposition to power plants or other facilities. In a classic battle of the period, supporters of a proposed port facility for liquified natural gas for the Los Angeles area—whose backers included both industry interests and the local county federation of labor led by the construction trades—spoke ominously of the collapse of the regional economy and widespread unemployment if the facility were not built and its additional energy generation not made available. But local opponents successfully raised safety and environmental considerations to block the facility, and the various economic and job loss predictions (estimated as high as 750,000 jobs) proved to be totally groundless. Nevertheless, energy-related scare campaigns were commonplace during the 1970s and forced energy activists and other environmentalists to confront the job loss question more directly.[31]

Echoing the job-creation potential of the pollution control field, alternative energy advocates, led by economist Hazel Henderson and including Barry Commoner, argued that the more labor-intensive solar energy approach would create new kinds of jobs. These arguments were also central to the development of Environmentalists for Full Employment (EFFE), an important organization seeking to establish environmental/labor coalitions in the late 1970s.[32]

The emergence of EFFE was important in two respects. First, it provided an opportunity for trade unions to work closely with an environmental organization committed to supporting certain key strategic objectives of the labor movement, such as full employment and labor law reform. At the same time, EFFE worked closely with several alternative environmental groups, especially antinuclear activists seeking ways to broaden the environmental agenda by approaching the nuclear issue as representative of larger social questions. The existence of EFFE further enabled mainstream environmental groups to avoid the complex task of building and maintaining relations with worker constituencies and unions by simply passing that responsibility to EFFE itself.

EFFE's bridge role was initially developed at a national gathering at the UAW's Black Lake center in Michigan in the spring of 1976. Hosted by EFFE, the UAW, and the Urban Environment Conference, along with more than 100 other local, state, and national groups, this "Working for Environmental and Economic Justice and Jobs" conference sought to explore the basis for coalescing different movements, with such participants as the UAW's Leonard Woodcock, David Brower (then with Friends of the Earth), and Russell Peterson (then at the Council on Environmental Quality) speaking "eloquently of the conferees' common cause," as one of the participants described it. Once the EFFE was organized, key union leaders such as Douglas Fraser of the UAW and William Winpinsinger of the International Association of Machinists looked favorably upon the activities of the group and granted substantial access to EFFE staff officials Richard Grossman and Gail Danecker to help with resources and common campaigns. EFFE was also successful, through its Labor Committee for Safe Energy and Full Employment, in bridging antinuclear and alternative energy groups with those unions outside the top AFL-CIO leadership and the construction trades most receptive to the alternative energy argument. EFFE publications, such as its *Jobs and Energy* monograph and *Fear at Work*, a more systematic analysis of industry "job blackmail" strategies, became pivotal texts underlining the problems and opportunities for a labor/environmentalist coalition. Yet within just a few years of EFFE's 1976 ground-breaking conference at Black Lake, the efforts toward a labor/environmental alliance had been placed on the defensive, the tentative coalitions subject to powerful pressures toward divergence.[33]

For one, it had become clear by the late 1970s that the approach of mainstream environmentalists had not broadened sufficiently to encompass workplace and social justice issues. Though some groups and environmental leaders saw the need for a coalition with labor on certain issues, these were perceived as external ties, not directly associated with a change of direction or mission or a broadening of the constituency for the mainstream groups. With only a few exceptions, such as Environmental Action, occupational hazard issues were not seen as part of the environmental agenda, nor was there an appreciation during the 1970s of the volatile issue of community-based hazards such as those experienced at Love Canal and

Stringfellow. Though defensive about charges of elitism and class composition, mainstream environmentalists failed to extend their environmental policy interests to include the OSHA-based occupational regulatory system. They were unwilling to review broader questions of industrial organization as an environmental issue. Even the early 1980s' attempt to put forth a collective "industrial policy" (broadly resembling the Group of Ten's *Environmental Agenda for the Future*) focused on arguments about international competitiveness and trade policy rather than examining core structural issues such as the role of military and toxics-related production systems.[34]

While environmentalists kept to their narrower agenda, the position of the labor movement, even among more progressive unions, remained bound by union acceptance of the structure of industry decision making. At the Black Lake conference, AFL-CIO head George Meany's special assistant, Tom Donahue, indicated that the need for jobs took precedence over any evaluation of the hazardous nature of the industries creating those jobs, including energy facilities such as nuclear power plants. Those few union officials, such as Jack Sheehan of the Steelworkers and Tony Mazzochi of the OCAW, most closely identified with occupational health issues even found themselves accused of being surrogate environmentalists, not sufficiently interested in the job losses resulting from plants presumed to be shut down for environmental reasons. With Tony Mazzochi defeated in his bid for the OCAW presidency, with the Carter administration in full-scale retreat regarding its alternative energy and environmental programs, and with OSHA under siege from its industry critics, the possibilities for further labor/environmental coalitions were quickly fading less than ten years after passage of the OSH Act. With the 1980 election of Ronald Reagan, that divergence of interests intensified.

REAGAN REDUX

Despite the uncertainties for a labor/environmental alliance, occupational health through the late 1970s continued to be a compelling issue. It involved significant public interest group activity, remained a priority area of concern for growing numbers of the work force, received key coverage in the press, and was an expanding area for new policy initiatives, particularly under Eula Bingham's leadership at OSHA during the Carter years. During the late 1970s, Bingham introduced a number of new programs, including a cotton dust standard primarily impacting textile manufacturing and a generic cancer policy aimed at establishing criteria for designating a particular substance used in the workplace as a carcinogen. Most importantly, OSHA's new lead standard represented a crucial policy breakthrough by elevating worker health considerations based on the available science as the central locus of standard setting. Efforts also increased at OSHA's research counterpart, the National Institute for Occupational Safety and Health, providing new evidence of the extent of workplace hazards in the post–World War II era. And the increasing interest in toxics and hazardous wastes, which potentially linked occupational and community health concerns, suggested that the new policies and interest in occupational hazards could become a central part of the new and expansive focus on the environment. "The 1970s were an exciting period for those of us involved in the occupational health issue," recalled Andrea Hricko of the Health Research Group. "The press was attentive, we were able to generate a great deal of public concern, and the unions were becoming more sympathetic to the idea that the issue deserved priority. We felt we were at the threshold of establishing a powerful new movement, of developing a way to address democracy in the workplace, when certain occupational health policies began to be eroded in the last years of the Carter administration and especially after the Reagan administration came into power seeking to undermine much of what we had begun to construct."[35]

With the new Reagan administration, the frontal attack on environmental policies led by James Watt and Anne Gorsuch paralleled an even more aggressive onslaught against occupational safety and health policies that were far more vulnerable than their environmental counterparts. Led by Thorne Auchter, Reagan's appointee to head OSHA, and set in motion by Reagan's pledge during the 1980 campaign to abolish the agency and pursue widespread "regulatory reform," the Reagan attacks signaled an effort to undo even the previous limited gains concerning occupational hazards.

The immediate focus for Reagan's approach to occupational issues was OSHA. His appointee, Thorne Auchter, had himself come into conflict with OSHA when his Florida construction company was cited for forty-eight different safety violations, including six serious ones. Auchter immediately sought to reorient the agency toward a cooperative relationship with industry. He accomplished this by reducing or eliminating fines and otherwise cutting back on enforcement; by doing away with worker "empowerment" programs, such as Eula Bingham's New Directions Grant Program, which had significantly aided the COSH groups; by altering educational efforts, such as the agency's controversial recall of 100,000 booklets describing the hazards of brown lung caused by breathing cotton particles in textile factories; and by reversing other major policy initiatives established under Bingham, such as the generic cancer policy.[36]

The debates about the cotton dust standard and brown lung were indicative of the shifts taking place in occupational issues during the late 1970s and early 1980s. Brown lung, or bysinnosis, similar to black lung (pneumoconiosis) and white lung (asbestosis) disease among coal miners and shipyard workers exposed to asbestos, referred to a well-known occupational health hazard. Bysinnosis is a chronic lung disease long prevalent in cotton mills, but its causes have nevertheless remained contested terrain, with textile manufacturers denying that the dust generated in the mills related to lung disease among their workers. The textile industry's company doctors reinforced the owners' contention that lung problems experienced by mill workers had some other medical explanation. With little union representation in states such as North Carolina and South Carolina, where cotton production was concentrated, textile workers were forced to rely on a bureaucratic and largely unresponsive workers' compensation system that defined problems in individual terms as lost wages rather than as a need for structural changes in a production process.

Despite these difficulties in challenging the prevailing approach concerning bysinnosis, new union drives helped textile workers begin to mobilize on the brown lung issue. Inspired by the earlier success of the black lung movements, a Brown Lung Association was established in 1975 to demand a more responsive workers' compensation system and to pressure OSHA to intervene directly to reduce dust levels. In response, OSHA set a cotton dust standard in 1978 that was immediately attacked by the textile industry on the basis of its high cost.[37]

By the time the Reagan administration came to power, OSHA found itself subject to a far more hostile environment concerning the regulation of workplace hazards such as bysinnosis. Thorne Auchter, in one of his first major policy decisions at OSHA, sought to withdraw the cotton dust standard for further cost-benefit review. This decision immediately led to demonstrations at the offices of the Department of Labor in Washington, D.C., and in thirty-six cities across the country organized by the Brown Lung Association, COSH groups, and supportive unions. The brown lung groups also joined together with the black lung and white lung associations to create a "Breath of Life Organizing Committee" to fight for safer work environments, adequate compensation for disabled workers, and specifically to force the Reagan administration to maintain the cotton dust standard. In the midst of this battle, the U.S. Supreme Court upheld the cotton dust standard in June 1981 and forced the Reagan administration to lift its attempt at withdrawal. Auchter, however, continued to chip away at the implementation of the standard by exempting certain categories of workers and by delaying the installation of engineering controls, actions that paralleled the fate of other OSHA rules and standards.[38]

During the Reagan years, the occupational health issue also began to fade from public attention. The 1980s, retrospectively characterized by many as the "greed decade," was a time when the Reagan administration launched an unprecedented attack against unions, which began to experience substantial declines in membership. At the same time, vulnerable constituencies, such as workers employed in hazardous industries, became further isolated from the centers of decision-making power and influence in the society. Attempts to create linkages between the environmental and workplace movements had also stalled during the decade. Both groups had been largely unable to counter the powerful job loss arguments of industry and failed to overcome the social and cultural divisions that separated them. COSH groups, for example, continued to be limited by their uneven relationships with unions and the lack of any recognition (including their own) that they could and should be considered environmental groups as well. Mainstream environmental organizations, on the other hand, continued to ignore occupational hazard issues, maintaining their focus on the natural environment and community hazards, but not on the production systems themselves.

The efforts to link worker and alternative environmental groups also remained problematic. During the late 1970s and early 1980s, at the peak of the antinuclear protests, antinuclear groups basically avoided addressing the problem of job loss associated with plant shutdowns. Similarly during the late 1980s, the old-growth forest campaigns in northern California and the Pacific Northwest, including the "Redwood Summer" demonstrations sponsored by Earth First! in the summer of 1990, came to be seen as cultural conflicts invariably pitting workers against environmental activists. Even in the antitoxics area, community groups failed to overcome the jobs-versus-environment argument, as they insisted that community and protection-of-place issues take priority over job issues. Unions often intensified these divisions by their own insistence that job protection take precedence over other issues and by their failure to contest industry decision making as it impacted not just the workplace but the community and the natural environment as well. By the end of the 1980s, the divisions between groups were more entrenched than ever, even as the issues they confronted, built into the very fabric of the urban and industrial order, indicated that the question of production would remain central to the workplace/environment debate.

NEW STRATEGIES: BREAKING THROUGH THE EXISTING DISCOURSE

The union campaign, as with so many other previous job actions, began defensively, seeking to protect jobs and union status. The huge West German chemical company, BASF, had decided in 1984 to restructure its Geismar, Louisiana, facility. In the process, it hoped to undercut the local Oil, Chemical, and Atomic Workers union by including as part of contract negotiations a one-year wage freeze, reductions in health care coverage, and the ability to hire nonunion contractors. BASF had already successfully purged OCAW unions at several of its other plants in the United States. When negotiations between the company and the union broke down, BASF, claiming it feared potential sabotage, escorted 370 of the workers outside the plant, locked the gates, and vowed to not let the workers—or the union—return.[39]

The BASF lockout, which lasted more than five-and-a-half years, transformed a bitter and protracted battle between management and labor into a new kind of environmental campaign directed against the second-largest chemical producer in the world. The plant involved was located in Louisiana's infamous "cancer alley." By locking out the OCAW workers, BASF unintentionally established the basis for a labor/environmental alliance seeking to address BASF's role at every stage of its production activities.

At the time of the lockout, the OCAW had already experienced major declines in membership, due in part to the restructuring of the chemical industry. The union's leadership was desperate to counter the BASF lockout at Geismar, possibly in the form of a corporate campaign. First developed by organizers in the textile industry, corporate campaigns relied on consumer, community, and public interest support networks to undertake boycotts and disrupt shareholder meetings at the national level to help defeat a company at the local plant level. The OCAW turned to corporate campaign veteran Richard Leonard, who, on Tony Mazzochi's advice, hired a young organizer named Richard Miller to help put together the campaign against BASF.

Although both Miller and Leonard had little experience dealing with environmental issues, the BASF situation seemed ripe for an environmentally oriented campaign. Involved in the production of a wide variety of chemicals, BASF released significant amounts of toxic chemicals directly into the environment at its Louisiana facility and at other sites across the country. BASF was also part of the huge complex of petrochemical facilities located in an area where 25 percent of the country's commodity chemicals were produced and where hazardous wastes were routinely and extensively injected into the ground. Yet when Miller and Leonard arrived at Geismar the year following the lockout, there were few signs of any environmental or public interest presence or even awareness about toxics issues. The OCAW organizers had little idea of how to proceed, but were urged to pursue an analysis of the chemical industry and its role in the state by William Fontenot, an environmental specialist in the state attorney general's office. Fontenot also helped lay the groundwork for the initial ties between the OCAW and the handful of community activists and mainstream environmentalists interested in an alliance with the workers locked out at Geismar.

To succeed, the union had to be "born again," as Richard Miller put it.[40] As the BASF campaign took shape, it began to capture the imagination and tap into the resources of local residents and union members, as well as a number of alternative environmental groups throughout the country. The OCAW organizers helped establish, with varying degrees of success, several new groups within the state, including the predominantly African-American Ascension Parish Residents Against Toxic Pollution, the COSH-derived Louisiana Workers Against Toxic Chemical Hazards (LA WATCH), the Geismar-based Clean Air and Water Group, and the Louisiana Coalition for Tax Justice. Campaign organizers also became adept at monitoring BASF, even obtaining access to internal corporate and environmental documents that allowed the union to impeach the company's statements on a number of different occasions. The campaign was especially effective in its environmental efforts, such as challenging BASF's deep-well injection of hazardous wastes at Geismar and other locations where the chemical giant engaged in such practices.

The BASF campaign was both broad in scope and imaginative in character. It took on an international dimension when OCAW organizers successfully established ties with international groups such as the German Greens regarding the pollution stemming from operations of the company's

German plants as well as other, nonenvironmental issues and themes, such as BASF's Nazi origins, when the company first prospered. But the most striking aspect of the campaign was how local links were forged between community activists and BASF workers. These were not simply coalitions of necessity but represented new forms of environmental/workplace alliance, aided significantly by the more assertive role of the union. Some of the organizations established, such as the Louisiana Coalition for Tax Justice (which challenged tax breaks for toxics generators) and the multiracial Clean Air and Water Group (which challenged the lack of infrastructure in Geismar), provided new models for organizing and action that in turn helped to redefine the constituent base and nature of environmental politics in the state.

While the BASF campaign suggested new kinds of linkages, these were created in the context of a lockout that made OCAW especially responsive to new ways to counter the anti-union strategies that had become so effective during the Reagan and Bush administrations. The lengthy and ultimately successful outcome of the BASF campaign—a new contract was finally signed in December 1989 that allowed reinstatement of 110 of the original locked-out workers—didn't necessarily signify that this new form of labor/environmental alliance could now more easily overcome the continuing divisions between groups and constituencies. The BASF campaign was exceptional in its ability to link a union more responsive to occupational hazard issues and community groups more responsive to the focus on production activities within the petrochemical industry. The "victory" itself still reflected the defensive character of the specific lockout battle and didn't simply translate into a new strategy for challenging production decisions as they impacted the hazards of the workplace and the general environment.

Still, the BASF campaign offered hope in an otherwise bleak period for those advocating a new workplace/environmental politics. The campaign's most significant lesson was demonstrating that new kinds of alliances and restructured movements could be created, an idea that received increasing attention from a range of community and workplace groups during the late 1980s and early 1990s. Meetings on this theme, such as an annual Labor and Environment Conference organized by the Citizens' Environmental Coalition of Albany, New York, began to proliferate, and a few groups, such as the Labor-Community Strategy Center in Los Angeles, explicitly defined their task as reorienting both movements toward a new basis for action.[41]

These tentative new efforts at alliance, similar to the experience of groups such as EFFE and the COSH organizations during the 1970s, remained fraught with uncertainties and the need to overcome the long-standing mistrust among constituencies. The experience of the Silicon Valley Toxics Coalition (SVTC) was indicative of the opportunities—and problems—of the more recent efforts aimed at restructuring the workplace/environment relationship. The SVTC was initially established in 1982 as an offshoot of the Santa Clara, California, COSH group (SCCOSH) in response to the discovery of groundwater contamination caused by the Fairchild company, a major semiconductor company operating in this area of high-tech firms. While the Fairchild episode led to the development of a community-based movement, the SVTC, through the efforts of its founder, Ted Smith (whose wife, Amanda Hawes, was director of SCCOSH), sought to maintain strong ties with the local labor movement regarding the fast-developing toxics issue. The SVTC's first board president, Peter Cervantes Gautschi, was also the head of the Santa Clara Central Labor Council.

Despite these personal and organizational ties and the SVTC's goal of incorporating workplace issues as part of its community-based organizing approach, job loss kept reemerging as a difficult and sometimes intractable issue. The SVTC, which promoted the slogan "Ban Toxics, Not Workers," was most successful when it worked directly with unions at specific plants, such as its joint efforts with a progressive local of the Machinists Union at a Westinghouse plant in Sunnyvale. Located near one of the few affordable housing tracts in Silicon Valley, this plant had been responsible for extensive groundwater contamination, and the area was designated a Superfund site. Based on the theme "Clean Up the Plume, Clean Up the Plant," union and community organizers together raised common demands about the cleanup process, hoping to avoid the trap of developing competing objectives.[42]

Without ties to sympathetic unions, however, the SVTC, like other groups, was forced to confront the clash among constituencies. This occurred most notably when peace groups targeted a United Technologies plant in the Santa Clara Valley that was producing one stage of the Trident missile. The local peace organizations had decided, as part of a nationally based anti-Trident campaign, to try to close down the plant through demonstrations. The SVTC was interested, since the UTC plant had a long history of worker safety and health and community hazard issues. As plans for the action proceeded, SVTC director Smith received a call one day from the president of the United Technology's union local. "I understand you're the son-of-a-bitch who wants to take away food from my daughter's table," Smith recalled the union official shouting at him on the phone. "We'll break your heads if you try to come to our plant to demonstrate." Faced with a potentially embarrassing clash, Smith went to the Central Labor Council to seek to mediate. A meeting of all the parties, including local peace groups, the union local, and the SVTC, was convened and a compromise was ultimately arranged. The demonstration site was to be shifted from the plant gates to company headquarters, thus averting a possible clash between workers and demonstrators. Though a disaster was avoided in this instance, the difficulties in developing a workplace/environment alliance remained paramount.

The SVTC's experiences also demonstrate the need for groups to directly and specifically address the issue of job loss in the context of polluting industries or plants whose very activities are challenged, as in the case of United Technologies' production of Trident missile components. Even the SVTC's success at the Westinghouse plant in Sunnyvale has been primarily an organizing success, with little to show in terms of reducing the toxics generated or restructuring the plant's operations. The job loss problem is most keenly felt when environmental regulations or community-based demands fail to include provisions for placement, retraining, or the creation of equivalent types of jobs lost when a facility is shut down. In this context, several key labor and environmental activists, led by Tony Mazzochi of the OCAW and organizers with the National Toxics Campaign, formulated during the late 1980s the concept of a "Superfund for Workers" to deal specifically with the question of job loss.

Superfund for Workers defines job loss as a pivotal issue for both labor and environmental groups. It seeks to demonstrate the inadequacy of environmentalist positions on this question, such as the argument that job losses from environmental regulations can be offset by new jobs generated by the pollution control requirements for those same regulations. This approach, the Superfund advocates counter, does not acknowledge that such new jobs primarily involve low-paying secretarial work or shipping and receiving clerks, or highly specialized electrical or electronics engineering jobs not likely to be filled by production workers whose own jobs have been eliminated through a mandated cleanup. A number of pollution control jobs are also "dirty," such as those tied to waste management operations, which are themselves targets of community opposition. Other environmentalist positions—that plants shut down because of factors other than environmental regulation, or that polluting or toxic industries can be converted to nonpolluting forms of production—still fail to address the immediate consequences of job loss, particularly the elimination of higher-paying union jobs. The Superfund for Workers idea was put forth specifically to address those consequences and in the process to inspire a new kind of social claim concerning the jobs/environment issue.

As spelled out in articles and position papers, the Superfund for Workers proposal calls for up to four years of assistance to allow for a change of career for workers who lose their jobs through environmental or occupational regulations or due to a decline in hazardous military-based production. Such assistance includes income and benefits set at average union levels in a particular region, plus tuition for universities or for other programs to help prepare workers for new careers. In this way, Superfund for Workers can be defined as a job transformation strategy for workers in hazardous industries, but only indirectly as an industrial restructuring initiative.[43]

As a policy alternative, Superfund for Workers seems problematic. It is costly and requires substantial political support to aid production workers in hazardous industries whose position had already significantly eroded during the 1980s. Union membership, constituting the key group affected by such a program and likely to be its strongest advocate, also experienced sharp declines during the Reagan and Bush years, shrinking to as little as 16 percent of the work force by the end of the decade. Nevertheless, the Superfund for Workers idea, its advocates feel, has a compelling logic of its own. It is rooted in the fact that large numbers of industries, particularly in the military sector, are invariably going to be forced to shut down or clean up in the years to come. The costs for such cleanup are themselves astronomical. The Superfund analogy is intentionally evoked to suggest that a program of such magnitude is capable of being established, particularly if political pressure equivalent to the emergence of the community-based antitoxics movement can evolve on behalf of displaced worker constituencies.

This environment-related job loss/job retraining issue already emerged during the debates about the 1990 Amendments to the Clean Air Act. Legislative opponents of the Clean Air Act, led by Senator Robert Byrd of West Virginia, introduced an amendment to the legislation, the Relief for Terminated Workers Act, that proposed benefits for a three-year period to coal miners displaced as a result of the acid rain provisions in the Clean Air Act Amendments. The amendment offered only limited retraining or educational benefits, as this approach, while central to the career shift concept of Superfund for Workers, remained secondary to Byrd's primary goal of exposing the high costs of job loss as a way to attack the overall legislation.[44]

Although Byrd's amendment failed by one vote, another part of the legislation that did win passage, the Clean Air Employment Transition Assistance Program, or Title XI of the Amendments, established benefits for displaced coal workers equivalent to unemployment insurance payments or the amount set by the poverty level, whichever was higher. The retraining programs were essentially designed to locate low-wage jobs for their participants, with an average placement wage of $5.69 an hour. Such a program, Superfund for Workers advocates complained, "explicitly directs workers into poverty."[45]

For the Superfund for Workers advocates, justifying the costs of developing an adequate benefits and retraining program remain secondary to the proposal's central function as a strategy for mobilization. To succeed as an organizing strategy, Superfund for Workers requires major political support. For those like Tony Mazzochi, who successfully reestablished a major role for himself in the OCAW by the late 1980s after his bitter 1979 defeat for the union presidency, Superfund for Workers represents a surrogate call for a new labor party tied to the emergence of a working-class populism in the 1990s and beyond.[46]

For alternative environmental groups, such as the National Toxics Campaign and the Silicon Valley Toxics Coalition, the Superfund for Workers idea has offered a vehicle for political action, primarily as an argument to be used in confronting especially hazardous industries. Beyond the question of community and occupational hazards or job loss, the existence of such industries raises the broader, more comprehensive concern about industrial restructuring. Job loss, these groups argue, is already caused by an existing restructuring process, whether an environmental cleanup or plant shutdown. Antitoxics groups are thus constantly obliged to address not simply the problem of displaced workers but what happens to a hazardous plant or facility or even an industry itself. Most community-based groups engaged in specific battles challenging particular facilities or waste operations have not directly focused on the overall hazards of a production system or the systemic problems of waste generation. And while antitoxics groups have been able to counter the charge that they are not sufficiently expert to address such problems, they still need a way to confront the charges of parochialism or "NIMBYism" ("Not in My Backyard") that portray them as groups with criticisms but no solutions.

The key to the search for solutions has been the toxics use reduction concept popularized by several of the antitoxics groups. Initially conceived at a retreat in New York State in the summer of 1987 attended by organizers from the National Toxics Campaign, the PIRGs, Greenpeace, and Clean Water Action, a new campaign was launched to advocate changes at the front end of the production system. The campaign was designed to distinguish toxics use reduction from the mainstream/environmental policy system approach of managing environmental problems once they exit into the general environment as wastes, water discharges, air emissions, or toxic products. This new campaign, with its focus on the generation and use of toxics, was to be called "Toxics Use Reduction Now," or "Our TURN Now."[47]

Though the "Our TURN Now" slogan never caught on, toxics use reduction nevertheless emerged as a major alternative environmental policy approach and an effective organizing strategy for the antitoxics groups and others confronting the question of what kind of change they wanted. Despite its potential for radical industrial restructuring, toxics use reduction advocacy has not been marginalized the way other alternative conceptual approaches have been attacked and trivialized in previous years. For one, the substantial opposition to industry activities and waste facilities combined with the high costs of pollution control and financial liabilities established through such legislation as Superfund have created far greater receptivity among industry groups and government regulators for some form of waste reduction or waste minimization strategy. The EPA's introduction in the late 1980s of the ambiguous and poorly defined (and implemented) concept of pollution prevention reflects the impact of the toxics use reduction idea in influencing environmental policy in this area. Some alternative groups, particularly the PIRGs, Greenpeace, and the National Toxics Campaign, have inspired new legislation, particularly at the state level, that has incorporated toxics use reduction approaches. A number of mainstream environmental groups have also sought to integrate the reduction concept into their own environmental policy agendas, even as key groups such as the EDF continue to work closely with the EPA to emphasize a voluntary, noninterventionist strategy to accomplish such goals.[48]

Conceptually, toxics use reduction also provides an opportunity to redefine relations among constituent groups. As an approach that emphasizes reduction of toxics or environmental hazards at the work site, in relation to products and as wastes, it suggests a common focus for workplace groups concerned with occupational issues, consumer movements focused on product hazards, and environmental groups dealing with community hazards or with problems in the natural environment. As a framework for environmental policy, toxics use reduction, with its emphasis on addressing issues at the design stage and altering certain processes and materials used in production, directly incorporates Barry Commoner's call for the social governance of production decisions, a concept that had largely disappeared from the environmental discourse with the demise of the New Left.

Despite the success of the toxics use reduction idea in influencing environmental debates, it remains more tentative as an actual organizing strategy, often removed from the realities of the production system that divide workplace and environmental groups. Similar to the Superfund for Workers idea and the lessons of the BASF campaign, however, the elevation of toxics use reduction among environmental groups suggests that a new kind of politics is capable of evolving in the 1990s. With their implicit call for industrial restructuring, worker empowerment, and a redefinition of the work/environment relationship, these new strategies can form the basis for new kinds of social movements. The barriers to the development of these new movements still remain substantial, though the question of the environment, by the late 1980s and early 1990s, has clearly been extended to the production realm, no longer limited to the focus on individual behavior that characterized the second Earth Day in 1990. This new focus on production and the renewed efforts at addressing questions of the environment in daily life suggest that a synthesis based on the issues of gender, ethnicity, and class is available for an environmentalism still searching for a more encompassing and inclusive identity.

To understand the importance of the growing participation of women, people of color, and industrial workers in environmental struggles, it is necessary to understand how such groups experience those struggles. They see their communities used as dumping grounds for toxic wastes because they lack the power to prevent it; they find only the most hazardous jobs available, if they can find jobs at all; and they discover that the circumstances of their work are set by managers and engineers who do not have to suffer the hazards of the job floor.

Women entering the environmental movement through grassroots and community organizations challenge practices and processes that mainstream environmental groups have taken too much for granted. As a crucial starting point, they challenge the idea that expert knowledge in defining environmental problems and in negotiating for their resolution should take priority over the experiences of people in their everyday lives. They suggest that people can learn to map their own environments from their own experience, contesting the errors and overgeneralizations of experts from government and business. In this way, people learn to trust their own observations, knowing that the experts who may claim to know much about a problem abstractly, know little from direct experience.

Women engaged in environmental struggles are often fighting for the circumstances of their own lives and those of their families. They cannot make the easy compromises that trade one position off against another, nor can they lose a battle, lose interest, and move on to another issue or location. As a result, they are often critical of the current willingness of mainstream environmentalists to compromise with their opponents. As Penny Newman defines it, "The compromise between a healthy baby and a dead baby is a sick baby." For these women, a struggle for a livable environment is a personal, not a professional, commitment.

Much of what characterizes the actions and experiences of women in environmental struggles also appears among communities of color. The particular vulnerability of communities of color arises out of the historical experience of social and spatial segregation such communities face. These communities are often at the margin—in barrios, on reservations, in inner-city ghettos, in abandoned regions of the South—out of sight, out of mind. A solid waste incinerator sited in an inner-city African-American neighborhood removes garbage and its disposal from suburban households and wealthier urban areas; a hazardous waste disposal site in the rural South removes industrial wastes from the northeastern industrial cities where they are produced and where their products are consumed. These facilities are presented to the surrounding communities as possibilities for economic development, not environmental contamination. But at best they represent economic development *and* pollution, and the environmental contamination often exceeds and outlasts any economic development benefits.

Communities of color have brought to the environmental discourse the language of civil rights and environmental justice. From their own histories of struggle for access, participation, and the right to self-determination, they offer a model based on the claim that discrimination along racial or ethnic lines is unfair—that people of color have the same right to life, and to quality of life, as do other Americans. But ending environmental discrimination along ethnic lines is not yet sufficient to reworking the urban industrial order that remains the source of such discrimination.

Such problems are often most directly experienced in the workplace. Concentrated effects of hazardous technologies are encountered at jobs where exposures to hazardous substances are at levels which may be tens, hundreds, or thousands of times greater than for people in communities that surround these plants. For workers, the choice is too often hazardous work or no work at all, since they have achieved little influence over changes in production technology or product design.

The separation between the environmental and labor communities, reinforced by the different class positions and personal histories of the institutional agents of these movements, has impeded communication that would lead to a rediscovery of the intimate relationship between workplace hazards and environmental problems. The policies of the mainstream environmental groups, too often concerned with environmental management without regard to its employment effects (a model that seeks to contain pollution within the plant rather than to rework the production system to reduce workplace and environmental hazards), have left these groups open to the charge that they put the environment ahead of jobs. Labor activists have reason for their common perception that environmentalists overlook the job circumstances of working people. At the same time, labor unions, often set into old forms of dispute concerning wages and work rules, have not taken up the opportunity for developing a constructive coalition with environmental groups. Larry Davis's insight that "if you have a modern facility that produces less pollution, you certainly have a more secure job and you have a cleaner community to live in," has not yet become a substantial component of either union or mainstream environmental agendas or strategies.

The common claims of movements based on experiences affected by the realities of gender, ethnicity, and class offer the potential of a political base strong enough to confront the sources of economic, industrial, and political power and their role in environmental destruction. Gender, ethnicity, and class are not distinct from one another in people's lives: women work in hazardous jobs and care for men and children who may return home from contaminated workplaces and schools; people of color, both men and women, discover that they are living in poisoned neighborhoods where their babies eat lead chips, and they can find jobs only in fields or factories where the hazards of uncertain employment intersect with the hazards of the work itself. Industrial practices contaminate workers, and industrial wastes contaminate neighborhoods, water, and air. Differences in power based on distinctions of gender, ethnicity, and class mean that some people disproportionately suffer the effects of economic and environmental decisions, and other people disproportionately reap the benefits of those decisions. As those who are most at risk join the environmental movement and expand its definition, their participation also broadens the possibilities for social and environmental change.

Conclusion

Environmentalism Redefined

Opportunities for Change

When Vice-president-elect Albert Gore convened the session addressing environmental and technology issues on the afternoon of the last day of the December 1992 Economic Summit in Little Rock, Arkansas, there was already speculation among the press and policy makers that there would emerge, as the *Los Angeles Times* put it the day after the summit, a "green tilt" to the new administration. Even more than the election of Jimmy Carter sixteen years earlier, there was enormous anticipation that change—a Clinton buzzword during the campaign—was in the air.

Speakers at the session had proposed a number of environmental policy approaches, some of which were likely to be contradictory: pollution taxes; a nonadversarial relationship among industry, environmentalists, and government; pollution prevention strategies; rationalizing regulatory and permit processes. There was even talk of a "new paradigm of environmental governance," as Gus Speth (NRDC cofounder and Jimmy Carter's chair of the Council on Environmental Quality) put it. Such talk inevitably reflected raised expectations, a feeling that so much more was possible than at any previous period during environmentalism's short but complex history as both movement and policy system manager and watchdog.

One month later in his inauguration speech, Clinton spoke of "forcing the spring" by his election. This evocation of change, which had been so prominent at both the economic summit and in pre-inaugural events, was beginning to have a life of its own. For one, it was suggesting in environmental terms that the historic divisions between jobs and environment or economic development and environmental governance, as the Clinton campaign rhetoric had insisted, would now, at last, be overcome.

The mood of change, however, does not necessarily reflect the realities of change. The divisions over policy, between movements and policy makers, and among movements themselves, are potentially as sharp and contentious as ever. What becomes different is the level of expectation, the promise of change to come. And such a promise has arrived at a moment in time when the possibilities of environmental change have become increasingly linked to the redefinition of the environmental movement and its capacity to transform American society itself.

We live today in a period of great upheaval, when environmental issues increasingly reflect crucial social and economic choices, and when new opportunities for change are emerging both within the movement and throughout society. Such opportunities for change have turned out to be unexpected, broad in scope, potentially far-reaching, and radical in their implications. They involve questions of technology and production, decision making and empowerment, social organization and cultural values. They reflect changes at the global level—most significantly, the end of the Cold War. They also reflect local struggles and victims, as demonstrated by new constituencies and new social claims that may influence the direction of environmentalism. Of all these changes, it is the passing of the Cold War that has the greatest consequences.

More than thirty years ago, the Students for a Democratic Society argued forcefully in *The Port Huron Statement* that the Cold War had become the defining event of the post–World War II order. Cold War realities, the SDS manifesto declared, affected the "whole living conditions of each American citizen," with "defense mechanisms" in the nuclear age altering the very "character of the system that creates them for protection."[1]

The Cold War, pervasive at the time this statement was written, dominated public life for more than four decades. Government budgets became heavily weighted toward military expenditures. The new technologies of such Cold War–related industries as aerospace, electronics, and petrochemicals have all left huge legacies of pollution. And for much of the post–World War II period, the rhetoric of national mission became intertwined with Cold War–linked themes concerning national security, access to resources, and strategic competition.

The Cold War was a pivotal factor in influencing the state of the environment. It stimulated the nuclear industry, whose severe environmental impacts have included the contamination of local communities where nuclear facilities (from mines to generators) have been located. It shaped energy and industrial policies that both exhausted resources and intensified the development of an economy significantly based on the use of toxic or hazardous materials and processes.

The Cold War also influenced the state of environmental politics. During the early years of the Cold War, key conservationist figures, such as Fairfield Osborn of the Conservation Foundation and Joseph Fisher of Resources for the Future, incorporated Cold War themes such as national security into the conservationist argument. If materials shortages threatened to "impair the long-run economic growth and security of the United States and other free nations," as the 1952 report of the President's Materials Policy Commission explicitly warned, then, according to Joseph Fisher, the "availability of a wide variety of resource materials [was] essential to our defense and security." A strong conservationist approach was therefore connected to Cold War agendas and presented as integral to sustaining what Fairfield Osborn called the values that make "American life what it is."[2]

Response to the Cold War also differentiated between the types of environmental groups that emerged in the post–World War II era. The opposition to nuclear testing that developed in the late 1950s and early 1960s began a direct line of dissent through the New Left to the antinuclear and antitoxics movements of the 1970s and 1980s. These contrasted with the national security–oriented approaches of the cold-warrior conservationists and with later arguments by mainstream environmentalists that nuclear issues especially were not appropriate subjects for an environmental movement.

Yet the Cold War remained, for environmental groups, a reality from which it was never quite possible to escape. Through the 1970s to middle 1980s, while mainstream environmentalism grew and matured, Cold War politics still influenced the direction of policy. The imperatives of the Cold War, reinforced by the massive military buildup that was begun in the last years of the Carter administration and was escalated dramatically during the Reagan years, remained dominant in setting the priorities for government expenditures and in structuring economic and industrial policies. For mainstream environmentalists, a continuous Cold War created uncomfortable choices: either deny Cold War realities or accommodate to the culture of the Cold War in policy debates.

In the area of nuclear weapons policy, for example, most mainstream groups (with the significant exceptions of the NRDC, Friends of the Earth, and Environmental Action) avoided involvement with the issue. Distancing themselves from the tactics and perspectives of the anti–nuclear power protesters, mainstream groups remained uneasy about the efforts of organizations such as Mobilization for Survival to link explicitly energy and Cold War–based military and arms race issues. Even the NRDC, which promoted policies for managing an arms race that had spun dangerously out of control, tried to avoid association with the direct-action wing of the movement.

In contrast, among the alternative groups, the escalating arms race of the 1970s and 1980s generated new and important organizing efforts. Community-based groups emerged to contest particular facilities, such as the weapons production center at Hanford, Washington, and the Rocky Flats facility near Denver, Colorado. Other groups focused on the impacts of proposed new weapons systems or high-level radioactive waste dumps or opposed specific programs such as the mobile, land-based MX missile in southern Utah. The emergence of many of these grassroots groups, which became integral to the rise of the antitoxics movements of the 1980s, helped transform the question of the Cold War from an issue of ideological conflict to one concerning the environmental hazards of military production. By the late 1980s, what had begun as a series of local protests against the pollution generated by hazardous facilities had evolved into a full-blown movement exposing the country's most extraordinary toxics scandal. Military facilities—weapons production centers, military bases, and other units of the military's production complex—had become the largest and most unregulated source of pollution and toxic hazards and a major polluter overseas. Documented problems included the nearly 15,000 toxic waste sites, with about 100 of those areas listed on Superfund's National Priorities List of most contaminated sites; extensive contamination at nearly all of the military's major foreign installations; contract arrangements based on a system of military specifications that encouraged the use of hazardous materials; and the almost blanket exemption of military facilities from environmental laws and regulations. The Cold War had left a toxic legacy, as the military created "a chemical plague over the land through its misuse and disposal of dangerous chemicals."[3]

The strategic limits of existing environmental policy in addressing this toxic legacy became more visible and intolerable with the sudden passing of the Cold War in the late 1980s and early 1990s. How to clean up polluting industries—many spawned and nurtured by a Cold War system—no longer seemed a question of how best to manage such pollution, but, more appropriately, how to restructure the country's toxics-centered, military-based economy. The long-standing, presumedly utopian desire of antinuclear movements for economic conversion—turning swords into plowshares, or toxic military facilities into socially useful, clean-technology operations—had suddenly become, in the wake of the Cold War's demise, an issue of some urgency and practicality. Questions about the future role—or restructuring—of Cold War–related military production have been transformed in the new post–Cold War era into a broader set of issues linked to the concepts of pollution prevention and toxics use reduction raised by the alternative environmental groups. The end of the Cold War exposed a production system toxic at its core and an environmental politics that continually limited the possibilities for change. The conversion of production to *what*? Controlled by *whom*? These have become the central questions for post–Cold War environmentalism. How to advocate and organize concerning such issues and the related question of what kind of post–Cold War environmental politics might emerge have therefore become defining points for the movement.

As the end of the Cold War has created enormous opportunities for restructuring the urban industrial order, it has also provided an opportunity for new visions about environmental change and social transformation. Not since the period of social rebellion during the 1960s, with its search for a new quality of life, have circumstances appeared to create the basis for such new visions. During the 1960s, many of the generational challenges that arose were associated with a desperate search for community in the midst of Cold War realities. Participants in the youth movements, including but not limited to students, were essentially children of the Cold War. Today, the end of the Cold War presents opportunities for a post–Cold War generation to establish a new framework for change.

Contemporary environmental groups, especially the mainstream organizations, have largely ignored young people, including students. During the rise of mainstream environmentalism in the 1970s and 1980s, young people and students served primarily as ground troops for the particular activities of an evolving environmental movement, such as electoral initiative campaigns and mega-events such as Earth Day. Even organizations such as the PIRGs or the National Wildlife Federation's "Cool It" program (which focused on issues such as recycling and global warming) were organized directly as youth and student-based efforts, and relied on older professional staff or the larger mainstream organizations to set their agendas or establish their directions.

The relative lack of environmental activity among young people began to change in the late 1980s, linked to the upsurge in environmental activism in general. Several new student environmental groups especially took root, in some cases inspired by the rise of the alternative community-based and direct-action groups and the growing interest in class, ethnicity, and gender-related questions in environmentalism. In 1988, one of those groups, based at the University of North Carolina at Chapel Hill, began to explore the possibility of establishing a network with other student environmental groups. Taking the name Student Environmental Action Coalition (SEAC), the students placed a notice to this effect in *Greenpeace* magazine. To their astonishment, they received responses from some 200 campus-based environmental groups, many less than a year old themselves. An informal association of student groups, based on phone, fax, mail, and computer networks, was then established. A conference, hosted by the University of North Carolina group, was organized for October 1989. Entitled "Threshold," the meeting represented the first national student environmental conference ever held.[4]

With more than 1700 students in attendance and spirited talk of new organizing and action, Threshold far surpassed everyone's expectations. It appeared to transform SEAC overnight into a major new player within the environmental movement. Conference speakers, including Barry Commoner, David Brower, and John O'Connor of the National Toxics Campaign, found a receptive audience hungry for information and strategic direction.

After Threshold, SEAC became a loose amalgam of campaigns, network functions, campus organizing activities, and occasional lobbying work. One of the network's most successful undertakings was the campus environmental audit, a program modeled after a study by UCLA graduate students of the environmental problems and issues associated with that university. More than 200 student activist groups undertook audits of their own campuses as part of Earth Day 1990 activities, with SEAC integrating the audit concept into a newly established university accountability campaign, paralleling SEAC's own corporate accountability theme. With its network mushrooming, foundation interest in the new organization blossoming, and active courtship by both mainstream and alternative environmental groups increasing, SEAC prepared to emerge as a viable national organization. SEAC leaders hoped that process would culminate at "Catalyst," the group's second annual student environmental conference, to be held in October 1990 at the University of Illinois campus in Champaign-Urbana.[5]

The Catalyst gathering turned out to be an even headier experience for the students than the previous year's Threshold event. This time, 7600 students from all fifty states registered, with additional representatives from eleven foreign countries. The presence of so many potential student activists was inspiring. The numbers alone seemed to suggest that this new student environmental force might change the shape of environmental politics in the years to come. The mass media also began to take note, with several publications commenting warily on the unabashed radicalism of SEAC and its affinity with the New Left. *Forbes* magazine, for example, labeled the group's constituents as "the children of the children of the 1960s" and warned readers that the anticorporate, activist-oriented SEAC had "a clear line of descent from SDS."[6]

Within the year, however, SEAC unexpectedly found itself at an impasse. Overwhelmed by the attention, distracted by its instant celebrity status, unprepared to consolidate and extend initial student interest into a coherent campus organizing strategy, and uncertain about its own status as a national player as it continued to function as a loose collection of campus contacts and activities, the organization appeared unable to contend with its enormous growing pains.

These growth pains were compounded by the continuing division among environmental groups. Although mainstream groups welcomed student participation, they worried about the potential radicalism of student activists and the tendency of the activists to accuse the Group of Ten of having sold out concerning issues and tactics. And the alternative groups, while identifying with the activism of students, remained wary that these were students likely to be trained as future professionals who in turn would likely be ignorant of community or workplace experiences. Unlike SDS, whose formative role within the New Left helped define a larger social movement, the new student activism thus finds itself emerging at a point in time when the environmental movement has already become divided and an institutionalization process has narrowed certain definitions of the movement's role. Student groups such as SEAC and youth groups in general therefore tend to see themselves as limited in their capacity to set new directions or establish a new spirit or framework for movement activity.

Despite these constraints, a student and youth-based environmentalism has the capacity to play an important role in creating new opportunities for environmentalism. One crucial arena for action is the university itself, as student environmental groups seek to transform this powerful institution, which is a major environmental player through its infrastructure activities (waste generation and disposal, resource use, land use policies) and, most importantly, through the influence of its research on environmental policy. A student-based environmentalism also has the potential to help revive campus activism in general. By the early 1990s, SEAC and other campus environmental groups were becoming "the biggest and often the only source of campus activism attempting to stimulate a new movement," as Swarthmore College activist Heather Abel put it.[7] The 1992 presidential election, with its large youth vote (some of it environmentally related) for Clinton and Gore, also suggested an opening toward activism, which had been suspended during the 1980s but seemed ready to emerge once again.

Beyond the possibilities for activism, young people have come to represent the first generation to be aware of the dimensions of environmental crisis and its relationship to everyday experience. Young people are not simply becoming early eco-activists (children taking up environmental values as moral absolutes has become increasingly common and noted), but they are also increasingly aware that the economic and social choices they face about their future work and lives have become more and more limited. At a moment when the end of the Cold War creates new possibilities for economic and environmental restructuring, young people seem destined to become the first generation likely to experience downward mobility combined with more degraded environments. The environmental crisis becomes the metaphor for the larger social crisis, while a generational politics, including a student and youth-based environmentalism, becomes one crucial route for linking issues and movements that must be linked if that crisis is to be addressed. A youth politics, like the post–Cold War future to which it is now intricately tied, thus offers another kind of opening for change, part of a renewed activism capable of undertaking its own "long march through the institutions" in its quest for environmental redefinition and transformation.[8]

Interest Group or Social Movement?

What is an environmental group? Who is an environmentalist? How might different kinds of environmental groups influence the state of the environment—and the state of society? These charged questions lie at the heart of the problem of how to define environmentalism and its future direction. Does environmentalism represent new kinds of social movements, democratic and populist insurgencies seeking a fundamental restructuring of the urban and industrial order? Or is it a set of interest groups influencing policy to better manage or protect the environment and help rationalize that same urban and industrial order? Is it a left-wing movement in disguise, as George Will and other conservative commentators have complained? Or is environmentalism apolitical, concerned with science rather than ideology, as Sierra Club executive director Carl Pope has argued? Is it a movement primarily about Nature or about industry, about production or about consumption, about wilderness or about pollution, about natural environments or about human environments, or do such distinctions themselves indicate differing interpretations of what environmentalism has been and ought to be?[9]

In describing the complexity, distinctive roots, and different contemporary forms of environmentalism, this book provides an analysis of environmentalism as both interest group and social movement. The origins of environmentalism derive in part from the development of a set of social claims about Nature and urban and industrial life. During the Progressive Era, one set of groups came to be associated with the rise of resource agencies and environmental bureaucracies, adopting a perspective related to the uses of science and expertise for a more efficient and rational management of the natural environment and the urban and industrial order. Similarly, concerns about wilderness and protection of the environment came to be linked to the activities of specific constituencies who saw themselves as "consumption users of wildlife" (as Thomas Kimball of the National Wildlife Federation described NWF members) or consumers of other environmental amenities. At the same time, there were groups that saw themselves fighting for the public interest by challenging powerful special interests, such as the timber companies and mining interests, in their attempt to exploit the environment for private gain.[10]

These Progressive Era conservationist and protectionist groups were similar to and contrasted with urban and industrial groups of the same period that focused on the environmental conditions of daily life. Such groups as settlement house workers, regional planning advocates, and occupational health activists shared a faith in scientific analysis and the power of rational judgment pervasive in that era. But these groups also sought to empower powerless constituencies and organize for the minimal conditions of well-being and community identity in a harsh urban and industrial age.

With the significant social and technological changes of World War II and the postwar era, environmentalism became increasingly identified with changes in the patterns of consumption associated with an expanding urban and industrial order. Scenic resources came to be valued as recreational resources, although the mobilizations in defense of wilderness at Dinosaur National Monument and the Grand Canyon tentatively (and inadequately in terms of outcome) sought to define a public interest value in protecting Nature for its own sake. The changes in consumption also helped to shape the rebellious politics of the New Left and the counterculture, with their concerns about quality of life and the hazards of technology. Consumption thus became at once an economic activity, with a defined set of interests associated with it, and a social condition giving rise to the new social movements of the post–World War II period.

By Earth Day 1970, some confusion about the new environmentalism and what it represented had set in. This reflected its eclectic origins, which included 1960s New Left and counterculture activism, recreational/leisure time–oriented protectionist politics of the post–World War II period, and the resource management emphasis of several of the old-line conservationist groups. Even the trend toward professionalization that so quickly prevailed among environmental groups during the 1970s nevertheless incorporated both adversarial and system management perspectives on how to achieve environmental change. Environmentalists were activists *and* lobbyists, system opponents *and* system managers. But by the late 1970s, most of the professional groups had become thoroughly linked to the environmental policy system, a system designed to manage and control rather than reduce or restructure the sources of pollution and other environmental ills. Though at times contesting how policies were implemented, the ties that developed between professional or mainstream groups (mainstream in part because of those very ties) and the environmental policy system itself made mainstream environmentalism especially vulnerable to the charge of interest group politics.

Through the 1980s and into the 1990s, mainstream groups have been continually subject to criticism about their interest group status. Mainstream environmentalists are defined as elitists in class terms, upper-middle-class whites who have a lack of concern for job loss. At the same time, the national focus of the mainstream groups, their emphasis on professionalization, and their unwillingness to organize specific constituencies or in specific communities have provided an opening for blatantly transparent pro-industry campaigns, such as the Sagebrush Rebellion of the late 1970s and the more recent "wise use" movement to erode public support for certain environmentalist goals. The contemporary "wise use" movement is particularly instructive in its effort to appropriate long-standing conservationist language about efficient "use" of resources and then seeking to charge the mainstream groups with hostility to the "use" concept itself.

Today, the interest group identity for mainstream environmentalism seems more entrenched than ever. With their chief executive officers, targeted mailings, and continuing emphasis on lobbying, litigation, and the use of expertise, mainstream groups have replicated certain key aspects of interest group organization. Even the shift in focus of groups such as the EDF away from environmental "command and control" regulatory policies to market incentives has only reinforced the interest group connection. Through market arrangements such as pollution credits, the environment becomes a commodity to be traded and sold, with environmental groups seen as defining their objectives increasingly in terms of their economic value.

As they have grown in size and reach, mainstream groups have successfully secured for themselves a seat at the negotiating table and are now powerful organizations contending with other powerful interest groups. For example, since 1990, a "three-way" negotiating process concerning water policy in California has taken place. The three parties to the negotiations include urban interests (signifying large urban water agencies), agriculture interests (associated with the large landowners and irrigation districts), and environmental interests (consisting largely of mainstream environmental groups interested in protecting the Sacramento Bay Delta environment in northern California). These three-way negotiators have sought to reach a consensus position about water policy for the state. For the environmental interests, the key objective in this regard has been the creation of an Environmental Water Authority with the power to purchase water to help protect the Delta environment. The key to the three-way process has thus evolved into the effort to establish a political process for the purchase and sale of water among contending interest groups. In the process, environmental negotiators, defined as a quasi–economic interest group, begin to lose their public interest status as a social force fighting on behalf of groups not represented at the table.[11]

This interest group association most directly separates the mainstream groups from alternative environmental groups, grassroots networks, and community-based groups, as well as direct-action and more ideologically oriented "green" organizations. The use of the term *alternative* for environmental groups suggests different scenarios for these organizations. On the one hand, it implies that they are outside the mainstream, different from the way that environmentalism has come to be defined, especially in terms of its interest group status. It is in this context that many activists within alternative groups, such as antitoxics organizers, don't like to call themselves environmentalists. "Calling our movement an environmental movement," Lois Gibbs declares, "would inhibit our organizing and undercut our claim that we are about protecting people, not birds and bees."[12] Similarly, Richard Moore of the Southwest Organizing Project and Tony Mazzochi of the Oil, Atomic, and Chemical Workers Union, among others, have argued that people of color and workers are turned off by the environmentalist label, suggesting the term conjures up associations with middle- and upper-class Anglo yuppie types seen as consumers of Nature or policy technicians.

At the same time, alternative approaches are noteworthy for their critical view of the existing urban and industrial order, similar to ways that the new social movements of the 1960s defined their activities. Such a view is critical in the sense that environmental problems are seen as rooted in the structures of production and consumption in society. Such a viewpoint also extends to the critique of mainstream interest-group forms of environmental organization and activity. In this context, the term *alternative* suggests a social change movement, as well as efforts to construct a prefigurative movement seeking new forms of community and technology. As opposed to mainstream environmentalism's interest group identity, alternative groups can be seen as inheriting the mantle of the new social movements, or at least the tradition of social claims about equity, empowerment, and daily life concerns.

As part of this critical tradition, antinuclear groups challenged the nuclear industry during the 1970s, questioning the technologies employed and advocating replacing such environmentally harmful technologies with a clean production system. The direct-action antinuclear groups became especially wedded to a vision of a decentralist, democratic, solar energy–based society, connecting their vision to earlier New Left and counterculture ideas about transforming a post-scarcity society. Similarly, antitoxics groups, which first mobilized against local landfills and incinerators, expanded their claims to talk about empowerment and environmental democracy as well as the new sense of community that their struggles had helped define. Concepts such as toxics use reduction and environmental justice, which became central to the antitoxics campaigns, also spoke to the need for restructuring industry decision making to account for industrial impacts on daily life.

Unlike the mainstream groups, the social claims of these alternative groups directly address questions of gender, ethnicity, and class. Many, though not all, of the alternative groups have established a feminist perspective that is both more interactive and democratic. By pursuing a message of environmental justice and equity, some alternative groups, especially those in the antitoxics area, have necessarily become focused on issues of discrimination and racism embedded in the toxics economy. And with the tentative efforts to link workplace and community environmental concerns, some alternative groups have begun to shift the definition of environmentalism away from an exclusive focus on consumption to the sphere of work and production. By elevating issues of work and production, the dynamics of ethnicity and gender, and questions of community and empowerment, a reconstituted environmentalism has the capacity to establish a common ground between and among constituencies and issues, bridging a new politics of social and environmental change.

Who, then, will speak for environmentalism in the 1990s and beyond? Will it be the mainstream groups with their big budgets, large staff, and interest group identity? Can alternative groups, many of whom reject the label "environmentalist," lay claim to a tradition that has yet to be considered environmental? Will the opportunities for transformation, heightened by a mood of change associated with the end of the Reagan and Bush era and the raised expectations of the 1992 elections, translate into more vigorous and broader-based social movements that can influence and frame "a new paradigm of environmental governance"? Can mainstream and alternative groups find a common language, a shared history, a common conceptual and organizational home?

The figures of Alice Hamilton and Rachel Carson provide a clue. These compassionate, methodical, bitterly criticized women, accused of being sentimentalists, biased researchers, and pseudoscientists, opened up new ways of understanding what it meant to be concerned about human and natural environments. They were figures who transcended the narrow, limiting discourse of their eras, forcing their contemporaries to realize that much more was at stake than one industrial poison or one dying bird. Their language was transformative, their environmentalism expressed in both daily life and ecological dimensions.

To understand these women as connected rather than disparate parts of a tradition helps answer the question posed by Dana Alston at the People of Color Environmental Leadership Summit discussed in the introduction to this book. How should environmentalism be defined? Alston asked. Should we keep to the narrow definitions that have provided environmental legitimacy for some groups and ideas to the exclusion of others? To learn the lessons of Rachel Carson and Alice Hamilton and how they are linked in their concern for the world we live in helps begin that process of redefining and reconstituting environmentalism in the broader terms that Alston urged. It involves a redefinition that leads toward an environmentalism that is democratic and inclusive, an environmentalism of equity and social justice, an environmentalism of linked natural and human environments, an environmentalism of transformation. The complex and continuing history of this movement points the way toward these new possibilities for change.

Afterword: A Note on Method

In researching and drafting my other books and writings, I've always enjoyed collaborating with others. Several of my books are coauthored, and I've established rich and lasting relationships with many of the people with whom I've worked. Although I'm the sole author of *Forcing the Spring*, I also became engaged with a number of people with whom I shared a dialogue about the book or parts of it; I worked closely with people who contributed to the research for specific chapters or parts of chapters; and I had the good fortune of receiving comments from those who read early drafts of chapters or the manuscript as a whole. These are people who collaborated in particular ways, enriching a process of writing through support, feedback, and interaction.

Several people contributed to the research for the book. Laura Pulido, at the time a doctoral student at the UCLA Graduate School of Architecture and Urban Planning, undertook major research (in conjunction with her own dissertation) on the pesticides, lead, and uranium mining section of Chapter 7. She conducted several of the interviews cited in the chapter, and her insights and analysis about the complexities and importance of race, class, and ethnicity with respect to environmentalism were valuable to the crafting of that section and the chapter as a whole. Carol Kuester worked with me on Chapter 6, concerning gender, helping with specific research tasks and in discussing broad chapter themes. Sabrina Gates (aided by Pamela Pickens) proved to be an overall research trouble-shooter, locating documents and sources necessary for the text. Leslea Meyerhoff undertook an interesting investigation of the corporate response to Earth

Day 1970, and DeShawn Clayton explored the background material with respect to the Gauley Bridge disaster.

Several people provided important, sometimes invaluable documents and materials, as well as insights about their own activities. These include Lois Gibbs, Penny Newman, Russell Peterson, Sydney Howe, Helen Burke, Richard Moore, Richard Miller, Andrea Hricko, Marianne Brown, Michael Picker, George Coling, Dana Alston, David Diaz, Ted Smith, Larry Davis, and Jack Sheehan. I conducted interviews with more than 125 participants within or connected to the environmental movement, broadly defined. I spoke at length with several participants on more than one occasion and with others who requested that I not identify them by name. I also drew on material from interviews conducted for some of my other writings and identified those cases in the notes section.

A number of colleagues, friends, associates, and movement participants shared their thoughts and criticisms about the manuscript as a work in progress. Lois Gibbs, Margaret FitzSimmons, Maureen Smith, and Barbara Berney read the entire manuscript or substantial parts of it, while Emily Abel, Ellen Dubois, Andrea Hricko, Marianne Brown, Laura Pulido, Louis Blumberg, Ellen Friedman, Larry Davis, Michael Picker, and Penny Newman read one or more of the chapters and provided important feedback. Margaret FitzSimmons, my colleague at the UCLA Urban Planning Program, especially made crucial suggestions about shaping the overall manuscript and entered into an invaluable dialogue with me about the nature of the movement and its roots. Margaret's wonderful analytic mind, her strong sense of language, and her passion about the subject demonstrate what collaboration, in its richest meaning, is all about.

I also had the opportunity of working with Barbara Dean as my editor. Barbara's insights and discussion about this manuscript, given its complexity of subject, breadth of detail, and need for readability and analytic clarity, represented to me as a writer what is most valuable about an editor's role.

With these valuable collaborations, I was able to draw together *Forcing the Spring* from a wide range of experiences and research work. It is a book that has fully absorbed me and become a part of me, not simply because I'm the sole author or that I spent the length of time that I did on the research and writing, but because it connects so directly with my own experiences. *Forcing the Spring* is a book about history, social movements, theoretical arguments, and *experiences*, including the ability to draw on and learn from my own experiences. These include my participation in the New Left and

SDS specifically; my seven years as a dissident director of the Metropolitan Water District of Southern California (the largest and most powerful water agency in the West); my work with and on behalf of a wide variety of environmental groups (both mainstream and alternative); my work at UCLA, where I supervised a number of projects that touched directly on some of the themes of the book; my community work, which has included confrontations regarding waste disposal problems at the neighborhood level and advising government agencies on environmental policies; and my writing on topics, including profiles of individuals, that eventually led me to this project. Putting together *Forcing the Spring* has taken five-and-a-half years, and the process has helped me more fully appreciate the value of trying to be a participant/observer, researcher/activist, analyst/storyteller.

As I was researching and writing *Forcing the Spring*, I decided to dedicate this book to my children, Casey and Andy. I wanted to do this partly because I decided that the environmentalism I was in the process of analyzing and describing is directly about daily life experiences, including, prominently, the relationships and experiences of parents and children. And, as I learned from this book, environmentalism is also about a past that needs reclaiming and about a future that needs vision, and about a place for my kids that makes life worth living.

Notes

Introduction

1. Michael Fischer's remarks are from his speech, "We Need Your Help," given at the session "Our Vision of the Future: A Redefinition of Environmentalism," People of Color Environmental Leadership Summit, Washington, D.C., October 26, 1991.
2. John Adams's talk was given at the session "Our Vision of the Future," People of Color Environmental Leadership Summit, Washington, D.C., October 26, 1991.
3. Dana Alston's speech was given at the session "Our Vision of the Future," People of Color Environmental Leadership Summit, October 26, 1991. See also Louis Head and Valerie Taliman, "Reshaping the Environmental Movement," *Voces Unidas* (Southwest Organizing Project), vol. 1, no. 4, p. 9; author's 1992 interviews with Dana Alston and Michael Fischer and 1991 interview with John Adams.

Chapter 1

1. Marshall's comments are from "Wilderness as Minority Right," an August 27, 1928, article for the *Service Bulletin* of the U.S. Forest Service, cited in James M. Glover, *A Wilderness Original: The Life of Bob Marshall* (Seattle: The Mountaineers, 1986), p. 96.
2. The "region which contains no permanent inhabitants" quote is from "The Problem of the Wilderness," *Scientific Monthly*, vol. 30, February 1930, p. 148. The "vast lonely expanse" quote is from *Arctic Village* (New York: Harrison Smith and Robert Haas, 1933), p. 198. See also George Marshall, "Bob Marshall and the Alaska Arctic Wilderness," *The Living Wilderness*, vol. 34, no. 111, Autumn 1970, pp. 29–32.
3. The "people can not live" quote is from "Recreational Limitations to Silviculture in the Adirondacks," *Journal of Forestry*, vol. 23, no. 2, February 1925,

p. 176. In this article, Marshall argued for continued protection of virgin forest in the Adirondacks, both for its proximity to the New York metropolitan area as a recreational resource and as a source of inspiration in its undeveloped state. "The finest formal forest, the most magnificently artificially grown woods," Marshall wrote for his audience of professional foresters, "can not compare with the grandeur of primeval woodland. In these days of over-civilization, it is not mere sentimentalism which makes the virgin forest such a genuine delight" (p. 173).

4. In a Forest Service publication, the agency promoted Forest Service lands as places where there were "few rules and few crowds" and where you could "bring your family, have a picnic, gather wood for your campfire (but be careful with it), hike in the mountains—or take a genuine wilderness trip into some of the ruggedest country on the continent." See Russell Lord, ed., *Forest Outings* (Washington, D.C.: U.S. Forest Service, 1940), cited in James M. Glover, *A Wilderness Original*, p. 95.
5. Cited in *The Nation*, December 20, 1933, p. 696. The quotes from *The People's Forests* (New York: Harrison Smith and Robert Haas, 1933) can be found on pages 123 and 211.
6. The "in which there shall be no roads" quote is from a draft of a text for the Forest Service that Marshall prepared and is reproduced as "Protection at Last for Wilderness" in *The Living Wilderness*, vol. 5, no. 5, July 1940, p. 3.
7. The Bauer/Marshall correspondence is described in James M. Glover, *A Wilderness Original*, pp. 218–19.
8. Murie sought to directly associate Marshall with this elite posture in his essay "Wilderness Is for Those Who Appreciate," *The Living Wilderness*, vol. 5, no. 5, July 1940, p. 5. On the redbaiting of Robert Marshall, see Benjamin Stolberg, "Muddled Millions: Capitalist Angels of Left-Wing Propaganda," *Saturday Evening Post*, February 15, 1941, p. 9; "High Federal Aides Are Linked to Reds at House Hearing," *New York Times*, August 18, 1938, p. 1; "WPA Union Called Communist Plan," *New York Times*, April 18, 1939, p. 3:2.
9. For an outline of Robert Marshall's will, see Stephen Fox, "We Want No Straddlers," *Wilderness*, Winter 1984, p. 10, and James M. Glover, *A Wilderness Original*, p. 272. See also the letter to the editor written by his close friend Gardner Jackson after Marshall's death in *The Nation*, December 2, 1939, p. 635.
10. The "best preserve the timber" quote is cited in Leonard Arrington, *Great Basin Kingdom: Economic History of the Latter-Day Saints* (Lincoln: University of Nebraska Press, 1958), p. 54. For background on the Mormon stewardship concept, see also Leonard Arrington, *Building the City of God: Community and Cooperation Among the Mormons* (Salt Lake City: Feramorz Fox and Dean L. May, Deseret Book Company, 1976).
11. For background on John Wesley Powell's approach, see John Wesley Powell,

Down the Colorado: Diary of the First Trip Through the Grand Canyon (New York: Dutton, 1969) and John Wesley Powell, *The Exploration of the Colorado River and Its Canyons* (New York: Dover, 1961). See also Wallace Stegner, *Beyond the Hundredth Meridian: John Wesley Powell and the Second Opening of the West* (Boston: Houghton Mifflin, 1954) and Donald Worster, *Rivers of Empire: Water, Aridity, and the Growth of the American West* (New York: Pantheon, 1986).

12. Pinchot's comments are cited in his autobiography, *Breaking New Ground* (New York: Harcourt, Brace, 1947), p. 23.
13. Timber-cutting issues in the West had been anticipated by the rise and fall of lumber production in the Great Lakes region from the 1850s through the 1880s. Whole stands of timber, such as the white pines of Michigan, Wisconsin, and Minnesota, for example, were virtually destroyed, with loggers leaving behind what William Cronon characterized as "a literal wasteland." See William Cronon, *Nature's Metropolis: Chicago and the Great West* (New York: W. W. Norton, 1991), p. 202.
14. Cited in John Muir, *Our National Parks* (Boston: Houghton Mifflin, 1917), p. 363.
15. The Boone and Crockett Club was founded in 1886 at a dinner party given by Theodore Roosevelt for his sporting friends, including his wildlife mentor, George Bird Grinnell, the editor of *Forest and Stream* magazine. Members were required to be "American hunting riflemen" and were drawn from political, economic, military, and professional elites. They included such figures as Henry Cabot Lodge, J. Pierpont Morgan, and Elihu Root. Pinchot joined in 1897, sponsored by Roosevelt. See James B. Trefethen, *Crusade for Wildlife—Highlights in Conservation Progress* (Harrisburg, Pa.: Stackpole, 1961) and James A. Tober, *Who Owns the Wildlife: The Political Economy of Conservation in Nineteenth Century America* (Westport, Conn.: Greenwood Press, 1981), p. 189.
16. The Roosevelt speech is cited in Gifford Pinchot, *Breaking New Ground* (New York: Harcourt, Brace, 1947), p. 190.
17. "The main thing is that the land, as well as what grows on it, must be used for the purpose for which it is most valuable," Pinchot wrote of the multiple-use concept. "On it may be built stores, hotels, residences, power plants, mills, and many other things. All these are advantages to National Forests, because they help to get the fullest use out of the land and its resources. Railroads, wagon roads, trails, canals, flumes, reservoirs, and telephone and power lines may be constructed . . . as long as they do not unnecessarily do damage to the forests." Cited in Grant McConnell, *Private Power and American Democracy* (New York: Knopf, 1966), p. 35.
18. Through Muir's influence, the Sierra Club, in a statement to the Governors' Conference on Conservation, argued that "the moral and physical welfare of a nation is not dependent alone upon bread and water. Comprehending these primary necessities is the deeper need for recreation and that which satisfies

also the esthetic sense." Cited in Stephen Fox, *John Muir and His Legacy: The American Conservation Movement* (Boston: Little, Brown, 1981), p. 130. See also *Proceedings of a Conference of Governors: In the White House*, May 13–15, 1908 (Washington, D.C.: Government Printing Office, 1909).

19. The term "industrial dependents" is used in W. J. McGee, "Conservation of Natural Resources," *Proceedings, Mississippi Valley Historical Association*, 1909–1910, p. 364. One key reform organization of the period, the American Association of Labor Legislation, argued that its demand for health insurance, a major Progressive plank, had, as one of its aims, the "conservation of human resources," analagous to the "conservation of natural resources." See Paul Starr, *The Social Transformation of American Medicine* (New York: Basic Books, 1982), p. 245.
20. The "economic justice and democracy" quote is from J. Leonard Bates, "Fulfilling American Democracy: The Conservation Movement, 1907 to 1921," *Mississippi Valley Historical Review*, vol. 44, no. 1, June 1957, pp. 29–57. See also Clayton R. Koppes, "Efficiency, Equity, Esthetics: Shifting Themes in American Conservation," in *The Ends of the Earth: Perspectives on Modern Environmental History*, edited by Donald Worster (New York: Cambridge University Press, 1988), pp. 230–51.
21. In a December 1919 article in the *Journal of Forestry* entitled "The Lines Are Drawn" (vol. 17, no. 8, p. 900), Pinchot wrote, "The fight has now begun. . . . Forest devastation will not be stopped through persuasion, a method which has been thoroughly tried out for the past twenty years and has failed utterly. Since otherwise they will not do so, private owners of forest land must now be compelled to manage their properties in harmony with the public good."
22. William Greeley's testimony before the House Committee on Agriculture in 1921 is cited in George T. Morgan, Jr., *William B. Greeley: A Practical Forester, 1879–1955* (St. Paul, Minn.: Forest Historical Society, 1961), p. 48. Grant McConnell, in *Private Power and American Democracy*, his study of the influence of private interests on government agencies such as the Forest Service and Bureau of Reclamation, argued that the Forest Service's lack of accountability and its "simple insistence upon the virtues of administrators as wardens of the public interest led deviously but certainly to ties with the special interests, opposition to which had been the point of Progressive beginnings" (p. 50).
23. For background on the changing role of the Bureau of Reclamation, see Karen Smith, *The Magnificent Experiment: Building the Salt River Reclamation Project, 1890–1917* (Tucson: University of Arizona Press, 1986); Donald Worster, *Rivers of Empire*; William Warne, *The Bureau of Reclamation* (New York: Westview Press; 1973); Robert Gottlieb, *A Life of Its Own: The Politics and Power of Water* (San Diego: Harcourt Brace Jovanovich, 1988).
24. The multiple-use argument around pollution also extended to the issue of air emissions. A 1952 pamphlet, "A Rational Approach to Air Pollution Legisla-

tion," issued by the Manufacturing Chemists Association, asserted that it was "logical to regard the atmosphere as one of many natural resources which are being and should be used to technical and economic advantage." Cited in John C. Esposito (project director), *Vanishing Air: Ralph Nader's Study Group Report on Air Pollution* (New York: Grossman, 1970), pp. 260–61. On the water discharge issue, see, for example, the panel discussion on the "Effects of Pollutants in Water Supplies" at the May 19, 1960, Annual Conference of the American Water Works Association, especially the paper presented by Ralph E. Fuhrman, the executive secretary of the Water Pollution Control Federation, "Sewage and Industrial Wastes," in *American Water Works Association Journal*, vol. 52, no. 11, pp. 1349–53; also Robert Gottlieb, *A Life of Its Own*, pp. 162–67.

25. One indication of this attitude was the support provided by certain wildlife advocates for the military in their anti-Indian buffalo extermination campaign. "The extermination of the buffalo would do more to solve the Indian problem than the Army had done in thirty years," testified General Phil Sheridan, a Boone and Crockett Club member and commander of the U.S. Army in the Southwest (cited in James B. Trefethen, *Crusade for Wildlife*, p. 91). See also Alston Chase, *Playing God in Yellowstone: The Destruction of America's First National Park* (Boston: Atlantic Monthly Press, 1986); and Helen Hunt Jackson, *A Century of Dishonor: A Sketch of the United States Government's Dealings with Some of the Indian Tribes* (Boston: Roberts Brothers, 1886).
26. The "legitimate demands on the forests" quote is from John Muir, *Our National Parks*, p. 379.
27. The full quote—"These temple destroyers, devotees of a ravaging commercialism, seem to have a perfect contempt for Nature, and instead of lifting their eyes to the God of the mountains, lift them to the Almighty Dollar. Dam Hetch-Hetchy! As well dam for water-tanks the people's cathedrals and churches, for no holier temple has ever been consecrated by the heart of man"—appears in Muir's famous tract *The Yosemite* (New York: Century, 1912), pp. 261–62.
28. For background on Owens Valley, see William Kahrl, *Water and Power* (Berkeley: University of California Press, 1982); Robert Gottlieb and Irene Wolt, *Thinking Big* (New York: Putnam, 1977); and the J. B. Lippincott papers (Los Angeles Public Library).
29. The "know the land it is lived in" quote is from Mary Austin, *Land of Little Rain* (Garden City, N.Y.: American Museum of Natural History, Anchor Books, 1961), p. 103.
30. See Stephen Fox, *John Muir and His Legacy*, pp. 117–18.
31. The "pursues his game" quote appears in Robert Roosevelt's 1862 text *The Game Birds of the Coasts and Lakes of the Northern States of America*, as cited in James A. Tober, *Who Owns the Wildlife*, p. 46.

32. Cited in William T. Hornaday, *Wild Life Conservation in Theory and Practice* (New Haven: Yale University Press, 1914), p. 191. The Audubon Society (then the Audubon Association) was also engaged during the early 1900s in the campaign against "the alien gunner" and helped pass legislation in New York prohibiting aliens from carrying firearms in "public places." Legislation was also developed in other states such as Pennsylvania that actually prohibited immigrants from killing wild game and made possession of firearms a crime. See Environmental Law Institute, *The Evolution of Natural Wildlife Law* (Washington, D.C.: Government Printing Office, 1977), p. 47. See also Frank Graham, Jr., with Carl W. Buchheister, *The Audubon Ark: A History of the National Audubon Society* (New York: Knopf, 1990), p. 87.
33. The Audubon organization emerged in the late 1880s through a promotional campaign in George Bird Grinnell's *Forest and Stream* magazine. The publication offered a free membership card if the participant pledged not to kill any wild birds for food or destroy their nests or eggs and not to wear any bird feathers as dress ornaments. While the campaign enrolled 37,000 people by August 1887, financial pressures on the magazine caused it to abandon the campaign two years later. In 1895, the organization was reconstituted as the National Association of Audubon Societies for the Protection of Wild Birds and Animals and developed a stronger protectionist stance toward birds and wildlife until the Pearson era modified that approach. During the Pearson years, Audubon expanded its ties with commercial interests, such as the Winchester Repeating Firearms Company, which had become convinced that "as the game decreases, our business grows less," and that game "management" as opposed to wildlife preservation was the best vehicle to protect its economic interests. This marriage of interests was reflected in part by Audubon's approach to protection of the bald eagle, bag limits for hunters, and the potential extinction of a variety of bird species. See Carl W. Buchheister and Frank Graham, Jr., "From the Swamps and Back: A Concise and Candid History of the Audubon Movement," *Audubon*, January 1973, pp. 6–45. See also Carl W. Buchheister with Frank Graham, Jr., *The Audubon Ark*, pp. 60–73, and Irving Brant, *Adventures in Conservation with Franklin D. Roosevelt* (Flagstaff, Ariz.: Northland Publishing, 1988).
34. For a discussion of the role of the railroads, see Alfred Runte, "Pragmatic Alliance: Western Railroads and the National Parks," *Environmental Journal*, April 1974, pp. 14–21.
35. See Robert Shankland, *Steve Mather of the National Parks* (New York: Knopf, 1951), p. 97. The Mather quote comes from Stephen T. Mather, "The National Parks on a Business Basis," *American Review of Reviews*, April 1915, pp. 429–30. See also Alfred Runte, *National Parks: The American Experience* (Lincoln: University of Nebraska Press, 1979).
36. Cited by Richard Wade in his preface to Peter J. Schmitt, *Back to Nature: The*

Arcadian Myth in Urban America (New York: Oxford University Press, 1969). The "Making a Business of Scenery" article, written by Robert Sterling Yard, is from the June 1916 issue of *The Nation's Business*, vol. 4, no. 6, pp. 10–11.

37. See Robert Shankland, *Steve Mather of the National Parks*, p. 148.
38. See Alfred Runte, *National Parks: The American Experience*, pp. 170–71. See also Alston Chase, *Playing God in Yellowstone*, and Horace M. Albright (as told to Robert Cahn), *The Birth of the National Park Service: The Founding Years, 1913–33* (Salt Lake City: Howe Brothers, 1985).
39. Cited in Alfred Runte, *National Parks*, p. 159. The Park Service also became quite adept at using publicity to attract park-goers, an approach that also came to disturb preservationist critics. See Donald N. Baldwin, *The Quiet Revolution: Grass Roots of Today's Wilderness Preservation Movement* (Boulder: Pruett Publishing, 1972).
40. The Leopold quote is from "The Wilderness and Its Place in Forest Recreational Policy," *Journal of Forestry*, vol. 19, no. 7, November 1921, p. 719.
41. See Aldo Leopold, "Forestry and Game Conservation," *Journal of Forestry*, vol. 16, no. 4, April 1918, pp. 404–11. The "tragedy of prescribed lives" quote is cited in Curt Meine, *Aldo Leopold: His Life and Work* (Madison: University of Wisconsin Press, 1988), pp. 154, 182. "The hunting instinct" quote is from Aldo Leopold, "The National Forests: The Last Free Hunting Grounds of the Nation," *Journal of Forestry*, vol. 17, no. 2, February 1919, p. 150.
42. The unpublished manuscript, "The Social Consequences of Conservation," is cited in Curt Meine, *Aldo Leopold: His Life and Work*, p. 296.
43. The "rudimentary grades of outdoor recreation" quote is from Aldo Leopold, *A Sand County Almanac, and Sketches Here and There* (New York: Oxford University Press, 1949), p. 178. The Darling letter is cited in Curt Meine, *Aldo Leopold: His Life and Work*, p. 363.
44. The "His experience of urban problems" quote is from Curt Meine, *Aldo Leopold: His Life and Work*, p. 322.
45. The "To change ideas" quote is from the *Journal of Wildlife Management* (July 1940) and the "science of land-health" quote is from *Living Wilderness* (July 1941) and are cited in Curt Meine, *Aldo Leopold*, p. 408. The "low altitude desert tracts" quote is from Leopold's article "Origin and Ideals of Wilderness Areas," *Living Wilderness*, July 1940, p. 8. See also *Round River: From the Journals of Aldo Leopold*, edited by Luna B. Leopold (New York: Oxford University Press, 1953).
46. The "Free enterprise" and "Day of Judgement" quotes are from William Vogt, *Road to Survival* (London: Victor Gollancz Ltd., 1947), pp. 133, 78.
47. The "technologists who may outdo themselves" quote is from Fairfield Osborn, *Our Plundered Planet* (Boston: Little, Brown, 1948), pp. 194, 200–201.
48. See "Where Is Man? A Mid-Century Appraisal," *Time*, vol. 53, no. 15, April 11, 1949, pp. 27–30.

49. Nolan's talk on "The Inexhaustible Resource of Technology" is reproduced in *Perspectives on Conservation: Essays on America's Natural Resources*, edited by Henry Jarrett (Baltimore: Johns Hopkins Press, 1958), pp. 49–66. Similarly, two other Resources for the Future economists, in a 1963 publication, argued that "the cumulation of knowledge and technological progress is automatic and self-reproductive in modern economies, and obeys a law of increasing returns. Every cost-reducing innovation opens up possibilities of application in so many new directions that the stock of knowledge, far from being depleted, may even expand geometrically." See Harold J. Barnett and Chandler Morse, *Scarcity and Growth: The Economics of Natural Resource Availability* (Baltimore: Johns Hopkins Press, 1963), p. 235.
50. Cited in Samuel H. Ordway, Jr., *Resources and the American Dream: Including a Theory of the Limit of Growth* (New York: The Ronald Press, 1953), p. 25. See also Fairfield Osborn, "60,000 More Every 24 Hours!" (Princeton, N.J.: The Newcomen Society, Princeton University Press, 1951).
51. The "principle of Growth" quote is from President's Materials Policy Commission, *Resources for Freedom: A Report to the President* (Washington, D.C.: Government Printing Office, 1952), p. 3. The "faster automobiles" quote is from Samuel H. Ordway, Jr., *Resources and the American Dream*, p. 40. See also Fairfield Osborn, *The Limits of the Earth* (Boston: Little, Brown, 1953), and Samuel H. Ordway, Jr., *Prosperity Beyond Tomorrow* (New York: The Ronald Press, 1955).
52. The Conservation Foundation's statement of purpose emphasizes the promotion of "greater knowledge about the earth's resources" by ascertaining "the most effective methods of making them available and useful to people," especially through examination of "population trends and their effect upon environment." See "Purposes," *Conservation Foundation Annual Report* (New York: Conservation Foundation, 1959), p. 1.
53. In later years, Albright would write that he had resisted efforts of the Eisenhower administration to disassociate RFF from New Deal figures and strategies, though the proceedings of the conference suggest otherwise. See Horace M. Albright, *The Birth of the National Park Service*, and Donald C. Swain, *Wilderness Defender: Horace M. Albright and Conservation* (Chicago: University of Chicago Press, 1970), pp. 294–99.
54. Eisenhower's welcoming remarks are reproduced in *Resources for the Future, The Nation Looks at Its Resources: Report of the Mid-Century Conference on Resources for the Future, December 2–4, 1953* (Washington, D.C.: Resources for the Future, November 1954), p. 9.
55. See Resources for the Future, *Annual Report* (for the year ending September 30, 1954) (Washington, D.C.: Resources for the Future, December 1954), p. 2.
56. See Reuben G. Gustavson, "Resource Problems and Research Possibilities,"

in Resources for the Future, *Annual Report* (for the year ending September 30, 1955) (Washington, D.C.: Resources for the Future, December 1955), p. 4.

57. See Resources for the Future, *Annual Report* (for the year ending September 30, 1957) (Washington, D.C.: Resources for the Future, December 1957), p. 3.
58. See Ralph Turvey, "Side Effects of Resource Use," in *Environmental Quality in a Growing Economy: Essays from the Sixth Resources for the Future Forum*, edited by Henry Jarrett (Baltimore: Johns Hopkins Press, 1966), p. 49.
59. See M. Mason Gaffney, "Welfare Economics and the Environment," in *Environmental Quality in a Growing Economy*, p. 88.
60. The "turning point of historic significance" quote is from Grant McConnell, "The Environmental Movement: Ambiguities and Meanings," *Natural Resources Journal*, July 1971, p. 433.
61. This membership-related service orientation of the Sierra Club had even turned the organization into a developer and landlord through its investment in lodges and other facilities for club skiers in such places as Yosemite, Kings, and San Antonio canyons. See Michael P. Cohen, *The History of the Sierra Club: 1892–1970* (San Francisco: Sierra Club Books, 1988), p. 80.
62. The "just aginners" quote is from a letter written September 15, 1955, by J. W. Penfold, the western representative of the Izaak Walton League, to C. Edward Graves, the western representative of the National Parks Association; the letter is located in the David Brower file, Bancroft Library, University of California at Berkeley. In that same file is a July 1955, twenty-page analysis on Glen Canyon, which, as Brower commented in an October 31, 1955, letter to Raymond Morley, "raised all kinds of embarrassing questions about the dam." The material, however, was never released publicly, reflecting the decision among conservationist groups that "conservation interest was limited to conservation matters, and therefore opposition would stop when Echo Park was dropped." The Brower letter is also located in the David Brower file, Bancroft Library, University of California at Berkeley. On the Dinosaur fight, see also Stephen Fox, "We Want No Straddlers," *Wilderness*, Winter 1984, pp. 5–19.
63. National Park Service head Conrad Wirth recalled Brower commenting that the club's new anti–Mission 66 position was reflective of "a different Sierra Club." Cited in Conrad L. Wirth, *Parks, Politics, and the People* (Norman: University of Oklahoma Press, 1980), p. 359.
64. An editorial about passage of the act, printed in The Wilderness Society publication *The Living Wilderness* (no. 86, Spring-Summer 1964, p. 2), commented that the "Wilderness Act has been called a benchmark in our civilization. Indeed it is. For only in a civilized culture, in a climax period of man's intellectual, social, economic, and forward grace, could a wilderness preservation concept capture the national mind and be made a law of the land." David Brower, writing in the *Sierra Club Bulletin* (vol. 49, no. 6, September 1964, pp. 2–3) about the same time, was more sanguine. "This was no easy compromise

to accept, nor are conservationists happy about it," Brower stated. "Nevertheless, the Wilderness Bill is a major recognition of the importance of wilderness to the American people." Brower's "be the end of a series" quote is cited in Michael P. Cohen, *The History of the Sierra Club*, p. 330.

65. The "Whereas the long campaign for the Wilderness Act" quote is from Michael McCloskey, "Wilderness Movement at the Crossroads, 1945–1970," *Pacific Historical Review*, vol. 41, no. 3, August 1972, p. 354.
66. The "neither pro-reclamation, nor anti-reclamation" quote can be found in a draft letter to the editor from David Brower to the *Deseret News*, dated January 5, 1956, and located in the David Brower file, Bancroft Library, University of California at Berkeley. The full-page ad that appeared in both the *New York Times* and *Washington Post* concluded with the admonition "Now only you can save Grand Canyon from being flooded for profit," with attached coupons for readers to send to elected officials. Within twenty-four hours after the ad appeared, the IRS delivered a letter to the club's San Francisco office warning that any future donations might be ruled nondeductible. That action in turn led to a series of court and legislative battles, at the same time that the club increased both its visibility and membership, which further exacerbated its internal conflicts. See "Sierra Club in a Three-Way Tax Fight," *Sierra Club Bulletin*, vol. 54, no. 5, May 1969, p. 8.
67. Even during the Dinosaur fight, conservationists had argued that coal- and nuclear-fired plants were preferable power sources to hydroelectric energy facilities that threatened scenic resources. See the letter from David Brower to the *Santa Barbara News-Press*, April 4, 1955, in the David Brower file, Bancroft Library, University of California at Berkeley.
68. On the Diablo Canyon conflict, see Jason MacNeill Andrews, "Did the Sierra Club Bargain with the Devil at Diablo?" *Big Sur Gazette*, July-August 1979, pp. 9–11; Robert A. Jones, "Fratricide in the Sierra Club," *The Nation*, May 5, 1969, pp. 567–70; and Michael P. Cohen, *The History of the Sierra Club, 1892–1970*. John McPhee's compelling account of the Diablo fight, which first appeared in *The New Yorker* and was later published as *Encounters with the Archdruid* (New York: Farrar, Straus and Giroux, 1971), still represents the classic description of this conflict.

Chapter 2

1. The "I chose medicine" quote is from *Exploring the Dangerous Trades: The Autobiography of Alice Hamilton, M.D.* (Boston: Little, Brown, 1943), p. 38.
2. The "ideal place from which to observe" quote is from Barbara Sicherman, *Alice Hamilton: A Life in Letters* (Cambridge, Mass.: Harvard University Press, 1984), p. 4. The typhoid epidemic issue is also discussed in Wilma Ruth Slaight, "Alice Hamilton: First Lady of Industrial Medicine" (Ph.D. diss., Case Western University, 1974), pp. 24–27.

3. The "here was a subject" quote is from Alice Hamilton, *Exploring the Dangerous Trades*, p. 115.
4. The phossy jaw episode is discussed in Alice Hamilton, *Exploring the Dangerous Trades*, pp. 116–18. See also R. Alton Lee, "The Eradication of Phossy Jaw: A Unique Development of Federal Police Power," *The Historian*, vol. 29, no. 1, November 1966, pp. 1–21; and Alice Hamilton, "Industrial Diseases: With Special Reference to the Trades in Which Women Are Employed," *Charities and the Commons*, vol. 20, September 5, 1908, pp. 655–59.
5. The "foremen deny" quote is from a letter from Alice Hamilton to Jesse Hamilton dated February 26, 1910, as cited in Barbara Sicherman, *Alice Hamilton: A Life in Letters*, p. 157.
6. The "wash hands or scrub nails" quote is from Alice Hamilton, *Exploring the Dangerous Trades*, p. 122. The concept of industrial lead poisoning being "inevitable" is discussed in Alice Hamilton, *The White Lead Industry in the United States, With an Appendix on the Lead-Oxide Industry. Bulletin of the Bureau of Labor, No. 95* (Washington, D.C.: Government Printing Office, 1912), p. 190.
7. The "It seemed natural and right" quote is from Alice Hamilton, *Exploring the Dangerous Trades*, p. 269. The duty "to the producer" quote is from Alice Hamilton, *Industrial Poisons in the United States* (New York: Macmillan, 1925), p. 541.
8. The "laboratory material" and "The quicker the solvent" quotes are from Alice Hamilton, *Exploring the Dangerous Trades*, p. 294. See also Alice Hamilton and Gertrude Seymour, "The New Public Health," *The Survey*, vol. 38, no. 3, April 21, 1917, pp. 59–62, and Alice Hamilton, "The Scope of the Problem of Industrial Hygiene," *Public Health Reports*, vol. 37, no. 42, October 20, 1922, pp. 2604–08.
9. Alice Hamilton, "What Price Safety? Tetra-ethyl Lead Reveals a Flaw in Our Defenses," *The Survey Midmonthly*, vol. 54, no. 6, June 15, 1925, p. 333.
10. Harvard established three informal conditions for Hamilton's appointment: no use of the Harvard Club, no football tickets, and no participation in the commencement procession. Women were admitted to the medical school only after 1945. See Wilma Ruth Slaight, "Alice Hamilton: First Lady of Industrial Medicine," p. 135, and Barbara Sicherman, *Alice Hamilton: A Life in Letters*, pp. 209–18.
11. The "race poisons" discussion is from Alice Hamilton, *Industrial Poisons in the United States*, p. 110.
12. The Mencken quote is from an essay of his in the *Boston Herald* and is cited in Joseph M. Petulla, *American Environmental History: The Exploitation and Conservation of Natural Resources* (San Francisco: Boyd and Fraser, 1977), p. 189.
13. Jane Addams, *Twenty Years at Hull House: With Autobiographical Notes* (New York: New American Library, 1960), p. 8. A nearly identical quote can be found in Addams's 1893 essay "The Subjective Necessity for Social Settlement." Two

years later, settlement resident Agnes Sinclair Holbrook wrote similarly, in *Hull House Maps and Papers* (New York: Thomas Y. Crowell, 1895), of the social conditions in the nineteenth ward of Chicago: "Little idea can be given of the filthy and rotten tenements, the dingy courts and tumble-down sheds, the foul stables and dilapidated outhouses, the broken sewer pipes, the piles of garbage fairly alive with diseased odors, and of the numbers of children filling every nook, working and playing in every window-sill, pouring in and out of every door, and seeming literally to pave every scrap of yard" (p. 5).

14. Cited in John Duffy, *The Sanitarians: A History of American Public Health* (Urbana: University of Illinois Press, 1990), p. 176.
15. Lewis Mumford, *The City in History: Its Origins, Its Transformations, and Its Prospects* (New York: Harcourt, Brace, 1961), pp. 459–60. See also Craig E. Colten, "Historical Questions in Hazardous Waste Management," *The Public Historian*, vol. 10, no. 1, Winter 1988, pp. 7–20.
16. Cited in Nelson Blake, *Water for the Cities: A History of the Urban Supply Problem in the United States* (Syracuse: Syracuse University Press, 1956), p. 258.
17. The "peculiar filth" quote is from Joel A. Tarr, "Industrial Wastes and Public Health: Some Historical Notes, Part 1, 1876–1932," *American Journal of Public Health*, vol. 75, no. 9, September 1975, p. 1059. See also Stuart Galishoff, "Triumph and Failure: The American Response to the Urban Water Supply Problem, 1860–1923," in *Pollution and Reform in American Cities, 1870–1930*, edited by Martin Melosi (Austin: University of Texas Press, 1980).
18. See, for example, Alice Hamilton's *Industrial Poisons in the United States*, pp. 124–201, and Craig E. Colten, "Historical Questions in Hazardous Waste Management."
19. The commission's report is cited in Irving Fisher, "A Department of Dollars vs. a Department of Health," *McClure's Magazine*, vol. 35, no. 3, July 1910, p. 329.
20. See John Duffy, "Social Impact of Disease in the Late 19th Century," in *Sickness and Health in America: Readings in the History of Medicine and Public Health*, edited by Judith Walzer Leavitt and Ronald L. Numbers (Madison: University of Wisconsin Press, 1978), pp. 395–402. See also John Duffy, *The Sanitarians*, p. 196.
21. The "promote health" quote is from John Duffy, *The Sanitarians*, p. 206. The early twentieth-century public health official is Charles Chapin, the health commissioner of Providence, Rhode Island, and his statement is quoted in James H. Cassedy, *Charles V. Chapin and the Public Health Movement* (Cambridge, Mass.: Harvard University Press, 1962), p. 96. In the same vein, public health historian Paul Starr comments that "the more narrow focus of bacteriology also provided a rationale for public health officials to disengage themselves from commitments to moral and social reform," in Paul V. Starr, *The Social Transformation of American Medicine* (New York: Basic Books, 1982), p. 189. See also Barbara Gutmann Rosenkrantz, *Public Health and the State:*

Changing Views in Massachusetts: 1842–1936 (Cambridge, Mass.: Harvard University Press, 1972), pp. 97–127.

22. Cited in William G. Christy, "History of the Air Pollution Control Association," *Journal of the Air Pollution Control Association*, vol. 10, no. 2, April 1960, pp. 126–37.
23. On the historical background to the solid waste issue, see Martin Melosi, *Garbage in the Cities, 1880–1980* (College Station: Texas A&M University Press, 1981); David Gordon Wilson, "History of Solid Waste Management," in *The Handbook of Solid Waste Management*, edited by David Gordon Wilson (New York: Van Nostrand Reinhold, 1977); George E. Waring, Jr., "The Utilization of City Garbage," *Cosmopolitan*, vol. 24, no. 4, February 24, 1898, pp. 405–12; George E. Waring, Jr., "Out of Sight, Out of Mind: Methods of Sewage Disposal," *Century*, vol. 47, no. 25, April 1894, pp. 939–48; Louis Blumberg and Robert Gottlieb, *War on Waste: Can America Win Its Battle With Garbage?* (Washington, D.C.: Island Press, 1989).
24. Marshall O. Leighton's comments are cited in Joel A. Tarr, "Industrial Wastes and Public Health," p. 1060.
25. Cited by Henry Steele Commager in his Introduction to Jane Addams's autobiography, *Twenty Years at Hull House*, p. x.
26. The charter provisions are cited in Margaret Tims, *Jane Addams of Hull House, 1860–1935: A Centenary Study* (New York: Macmillan, 1961), p. 49. In a pamphlet describing the mission and purposes of Hull House that was published in 1893 and reproduced in *Hull House Maps and Papers*, the settlement is said to require "an enthusiasm for the possibilities of its locality, and an ability to bring into it and develop from it those lines of thought and action which make for the 'higher life.' "
27. The letter, written in April 1892, is from Florence Kelley to Friedrich Engels and is cited in *Notes of Sixty Years: The Autobiography of Florence Kelley*, edited and with an introduction by Kathryn Kish Sklar (Chicago: Charles H. Kerr Publishing Company, 1986), p. 12. See also Virginia Kempfish, "The Hull-House Circle: Women's Friendships and Achievements," in *Gender, Ideology, and Action: Historical Perspectives on Women's Public Lives*, Janet Sharistanian, editor (Westport, Conn.: Greenwood Press, 1986), pp. 185–222.
28. The "greatest menace" quote is from Jane Addams, *Twenty Years at Hull House*, p. 202. Addams's role in the garbage issue is also discussed in Margaret Tims, *Jane Addams of Hull House, 1860–1935*, and Allen F. Davis, *Spearheads for Reform: The Social Settlements and the Progressive Movement, 1890–1914* (New Brunswick, N.J.: Rutgers University Press, 1984).
29. Jane Addams's "garbage phaeton" is discussed in Allen F. Davis, *Spearheads for Reform*, p.154.
30. On the "garbage lady" see Suellen M. Hoy, " 'Municipal Housekeeping': The Role of Women in Improving Urban Sanitation Practices, 1880–1917," in

Pollution and Reform in American Cities: 1870–1930, edited by Martin Melosi (Austin: University of Texas Press, 1980), p. 174. See also Howard E. Wilson, *Mary McDowell: Neighbor* (Chicago: University of Chicago Press, 1928).

31. The Eliot quote is cited in Peter J. Schmitt, *Back to Nature: The Arcadian Myth in Urban America* (New York: Oxford University Press, 1969), p. 73. See also Dominick Cavallo, *Muscles and Morals: Organized Playgrounds and Urban Reform, 1880–1920* (Philadelphia: University of Pennsylvania Press, 1981); Allen F. Davis, *Spearheads for Reform*, pp. 61–65.
32. On the development of milk stations, see S. Josephine Baker, *Fighting for Life* (New York: Macmillan, 1939), pp. 126–46.
33. The "she brought magnificent weapons to bear" quote is cited in Jane Addams, *My Friend, Julia Lathrop* (New York: Macmillan, 1935), pp. 116–17.
34. See Florence Kelley, "The Sweating System," in *Hull House Maps and Papers*, p. 35. Kelley's research identified 162 separate sweat shops in the nineteenth ward. Kelley's background and early Hull House years are described in her four-part autobiographical sketch in *The Survey*, particularly part two, "My Novitiate" (vol. 58, no. 1, April 1, 1927, pp. 31–35), and part four, "I Go to Work" (vol. 58, no. 5, June 1, 1927, pp. 271–74 et seq.). See also Dorothy Rae Blumberg, *Florence Kelley: The Making of a Social Pioneer* (New York: Augustus M. Kelley, 1966); and Kathryn Kish Sklar's introduction to *Notes of Sixty Years.*
35. The "cost and consequences" quote is from Allen F. Davis, *Spearheads for Reform*, p. 172. See also *Crystal Eastman on Women and Revolution*, edited by Blanche Wiesen Cook (New York: Oxford University Press, 1978) and Samuel P. Hays, *The Reponse to Industrialism, 1885–1914* (Chicago: University of Chicago Press, 1957). The Hull House role in the development of unions is described by Jane Addams in "The Settlement as a Factor in the Labor Movement," in *Hull House Maps and Papers*, p. 187.
36. McDowell wrote that her involvement in the stockyards area "was something I had been looking for all my life, a chance to work with the least skilled workers in our greatest industry; not *for* them as a missionary, but *with* them as a neighbor and seeker after truth." Cited in Allen F. Davis, *Spearheads for Reform*, p. 113. The "neighborhood consciousness" quote is cited in Howard E. Wilson, *Mary McDowell: Neighbor*, p. 66. See also Lea Taylor, "The Social Settlement and Civic Responsibility—The Life Work of Mary McDowell and Graham Taylor," *Social Service Review*, vol. 28, March 1954, pp. 31–40.
37. The Bubbly Creek quotes are from Howard E. Wilson, *Mary McDowell: Neighbor*, p. 159. The "no other neighborhood" quote is from Sophonisba P. Breckenridge and Edith Abbott, "Housing Conditions in Chicago, Ill.: Back of the Yards," *American Journal of Sociology*, vol. 16, no. 4, January 1911, p. 434. There are numerous stories about Bubbly Creek, some undoubtedly apocryphal, including how people fell into the creek and never came out. One

particularly compelling story told of a reporter attempting to row across the creek enveloped by a six-foot bubble. Regardless, Bubbly Creek was clearly "a filthy piece of water," as one resident characterized it. See Robert A. Slayton, *Back of the Yards: The Making of a Local Democracy* (Chicago: University of Chicago Press, 1986), p. 28. For other descriptions of Packingtown and related environmental issues, see Edith Abbott, *The Tenements of Chicago, 1908–1935* (Chicago: University of Chicago Press, 1936), and Louise Carroll Wade, *Chicago's Pride: The Stockyards, Packingtown and Environs in the Nineteenth Century* (Urbana: University of Illinois Press, 1987). Upton Sinclair's *The Jungle* (New York: Doubleday, Page, and Co., 1906) presents a powerful description of the area as well.

38. The "certain minimum requirements of well-being" is the phrase used by the Conference of Charities and Correction, to which Jane Addams was appointed president, and is discussed in Allen F. Davis, *Spearheads for Reform*, pp. 194–96. The background to the conflicts concerning passage of the Pure Food and Drug Act are described in James Whorton, *Before Silent Spring: Pesticides and Public Health in Pre-DDT America* (Princeton, N.J.: Princeton University Press, 1974). Sinclair's perspective on the writing of *The Jungle* is described in Rachel Scott, *Muscle and Blood* (New York: Monthly Review Press, 1972).
39. For a discussion of the Progressive Party and the various reform and conservationist tendencies, see Robert H. Wiebe, *The Search for Order, 1877–1920* (New York: Hill and Wang, 1967), and Samuel P. Hays, "The Politics of Reform in Municipal Government in the Progressive Era," *Pacific Northwest Quarterly*, vol. 55, no. 4, October 1964, pp. 157–69. See also Allen F. Davis, *Spearheads for Reform*, and Grant McConnell, *Private Power and American Democracy* (New York: Knopf, 1966).
40. The "socializing democracy" concept is spelled out in Jane Addams, *Twenty Years at Hull House*, p. 310. See also Staughton Lynd, "Jane Addams and the Radical Impulse," *Commentary*, vol. 32, no. 1, July 1961, pp. 54–59, and *The Social Thought of Jane Addams*, edited by Christopher Lasch (Indianapolis: Bobbs-Merrill, 1965). The "search for order" and professionalizing impulses in the Progressive Era are described in Robert Wiebe's *The Search for Order*. Wiebe defines a split within the progressive camp by 1905 between one group that used "the language of the budget, boosterism, and social control," while the other spoke of "economic justice, human opportunities, and rehabilitated democracy." Wiebe contrasts these approaches as "efficiency-as-economy" versus "efficiency-as-social service" (p. 176).
41. Francis Hackett, "Hull House—A Souvenir," *The Survey Mid-Monthly*, vol. 54, no. 5, June 1925, p. 277.
42. The "disease of the working classes" quote is from an Alice Hamilton speech, "Occupational Conditions of Tuberculosis," reproduced in Graham Taylor, "Industrial Viewpoint," *Charities and the Commons*, vol. 16, no. 5, May 5, 1906,

pp. 205–206. See also David Rosner and Gerald Markowitz, *Deadly Dust: Silicosis and the Politics of Occupational Disease in Twentieth-Century America* (Princeton, N.J.: Princeton University Press, 1991). On the movement to create union hospitals, see Alan Derickson, " 'To Be His Own Benefactor': The Founding of the Coeur d'Alene Miners' Union Hospital, 1891," in *Dying for Work: Workers' Safety and Health in Twentieth-Century America*, edited by David Rosner and Gerald Markowitz (Bloomington: Indiana University Press, 1989), pp. 1–15.

43. The "hot, humid, noisy" quote is from Edward H. Beardsley, *A History of Neglect: Health Care for Blacks and Mill Workers in the Twentieth-Century South* (Knoxville: University of Tennessee Press, 1987), p. 62. The environmental conditions of the sweats are described in Florence Kelley, "The Sweating System," in *Hull House Maps and Papers*, p. 35–36. See also Carl Gersuny, *Work Hazards and Industrial Conflict* (Hanover, N.H.: University Press of New England, 1981), pp. 20–21.
44. For background on these two organizations, see Nancy Schrom Dye, *As Equals and as Sisters: Feminism, the Labor Movement, and the Women's Trade Union League of New York* (Columbia: University of Missouri Press, 1980); Mary Elizabeth Pidgeon, *Toward Better Working Conditions for Women: Methods and Policies of the National Women's Trade Union League of America*, Bulletin No. 252 (Washington, D.C.: U.S. Department of Labor, Women's Bureau, 1953); Allis Rosenberg Wolfe, "Women, Consumerism, and the National Consumers League in the Progressive Era, 1900–1923," *Labor History*, vol. 16, 1975, pp. 378–92; Erma Angevine, *A History of the National Consumers League, 1899–1979* (Washington, D.C.: National Consumers League, 1979); Josephine Goldmark, *Impatient Crusader: Florence Kelley's Life Story* (Urbana: University of Illinois Press, 1953). Goldmark, also a key figure in the National Consumers League, described the formation of the league as "the new organization designed to bring the power of consumers to bear upon the improvement of working conditions" (p. 51).
45. The investigation of munitions plants is described in Gertrude Seymour, "Industrial Poison in Munitions Plants," *The Survey*, vol. 38, June 30, 1917, pp. 283–85. The radium dial painting and tetraethyl lead fights are described in Josephine Goldmark, *Impatient Crusader*.
46. The "research adjunct" comment was made by WHB organizing secretary Charlotte Todes and is cited in David Rosner and Gerald Markowitz, "Safety and Health as a Class Issue: The Workers' Health Bureau of America During the 1920s," in *Dying for Work*, edited by David Rosner & Gerald Markowitz, p. 54. See also Harriet Silverman, "The Workers' Health Bureau," *The Survey*, vol. 50, no. 10, August 15, 1923, pp. 539–40, and Angela Nugent, "Organizing Trade Unions to Combat Disease: The Workers' Health Bureau, 1921–1928," *Labor History*, vol. 26, Summer 1985, pp. 423–46.
47. The "organized labor" quote is from Harriet Silverman, "The Workers' Health Bureau," p. 540.

48. The "fumes and dusts" quote is from Harriet Silverman, "The Workers' Health Bureau," p. 540.
49. See Roy Lubove, "The Roots of Urban Planning," in *The Urban Community: Housing and Planning in the Progressive Era*, edited by Roy Lubove (Englewood Cliffs, N.J.: Prentice-Hall, 1967), p. 14.
50. For background on the RPAA, see Carl Sussman, ed., *Planning the Fourth Migration: The Neglected Vision of the Regional Planning Association of America* (Cambridge, Mass.: MIT Press, 1976); Donald L. Miller, *Lewis Mumford: A Life* (New York: Weidenfeld and Nicolson, 1989); Roy Lubove, *Community Planning in the 1920s: The Contribution of the Regional Planning Association of America* (Pittsburgh: University of Pittsburgh Press, 1963); Lewis Mumford, "The Theory and Practice of Regionalism," *The Sociological Review*, vol. 20, no. 1, January 1928, pp. 18–33; John Friedmann and Clyde Weaver, *Territory and Function: The Evolution of Regional Planning* (Berkeley: University of California Press, 1979), pp. 29–35.
51. Lewis Mumford, "Regions—To Live In," *The Survey*, vol. 54, no. 3, May 1, 1925, p. 151.
52. Ebenezer Howard, *Garden Cities of Tomorrow* (London: Faber and Faber, 1946), p. 48.
53. See C. B. Purdom, "New Towns for Old: 1. Garden Cities—What They Are and How They Work," in *The Survey*, vol. 54, no. 3, May 1, 1925, pp. 169–73.
54. The "cycle of ecological balance" quote is from Donald L. Miller, *Lewis Mumford: A Life*, p. 17. The "cosmopolitan city of scale" concept is discussed in Benton MacKaye, *The New Exploration: A Philosophy of Regional Planning* (Urbana: University of Illinois Press, 1962), p. 66.
55. The Appalachian trail idea was first spelled out in Benton MacKaye's article "An Appalachian Trail: A Project in Regional Planning," in the *Journal of the American Institute of Architects*, October 1921, pp. 325–30. It is further elaborated in MacKaye's article "Regional Planning" in *The Sociological Review*, vol. 20, no. 4, October 1928, pp. 293–99, and in his book *The New Exploration*, also published in 1928. The Stein quote is from Donald L. Miller, *Lewis Mumford: A Life*, p. 205. The "to make the earth more habitable" quote is cited in Mary Susan Cole, "Catherine Bauer and the Public Housing Movement, 1926–1937" (Ph.D. diss., George Washington University, 1975), p. 47.
56. Edith Elmer Wood's writings on public housing include *The Housing of the Unskilled Wage Earner: America's Next Problem* (New York: Macmillan, 1919) and *Recent Trends in American Housing* (New York: Macmillan, 1931). On Catherine Bauer's background, see Mary Susan Cole, "Catherine Bauer and the Public Housing Movement," and Eileen A. Reilly, "Catherine Bauer and the Genesis of the United States Public Housing Movement: A Woman Who Transformed the Environment and the Environment Which Transformed Her" (senior thesis, Princeton University, April 15, 1981).
57. Letter from Catherine Bauer to Lewis Mumford, July 22, 1937, cited in

Donald L. Miller, *Lewis Mumford: A Life*, p. 334. See also Catherine Bauer, "Housing: Paper Plans or a Workers' Movement?" in *America Can't Have Housing*, edited by Carol Aronovici (New York: Museum of Modern Art, 1934), pp. 20–23.

58. David M. Potter, *People of Plenty: Economic Abundance and the American Character* (Chicago: University of Chicago Press, 1954), p. 177. The ultimate purpose of the American economy, the chairman of President Dwight Eisenhower's Council of Economic Advisors remarked, is "to produce more consumer goods. This is the goal. This is the object of everything we are working at; to produce things for consumers." Cited in Arthur Schlesinger, *The Politics of Hope* (Boston: Houghton Mifflin, 1963), p. 83.
59. See Peter Spitz, *Petrochemicals: The Rise of an Industry* (New York: Wiley, 1988), and Jeffrey L. Meikle, "Plastic: Material of a Thousand Uses," in *Imagining Tomorrow: History, Technology, and the American Future*, edited by Joseph J. Corn (Cambridge, Mass.: MIT Press, 1986).
60. Harold Stassen, a special assistant to President Dwight Eisenhower on disarmament issues, wrote in 1955 in the *Ladies Home Journal* that nuclear energy had the potential to create a world "in which there is no disease . . . where hunger is unknown . . . a world where no one stokes a furnace or curses the smog, where air is everywhere as fresh as on a mountain top and the breeze from a factory as sweet as from a rose. . . . Imagine the world of the future . . . the world that nuclear energy can create for us." *Ladies Home Journal*, August 1955, vol. 72, p. 48, cited in Stephen L. Del Sesto, "Isn't the Future of Nuclear Energy Wonderful?" in *Imagining Tomorrow*, edited by Joseph J. Corn, pp. 58–76. Fourteen years later, in 1969, a Westinghouse booklet on nuclear energy suggested that nuclear fission offered "power seemingly without end. Power to do everything man is destined to do." The booklet is cited in William R. Burch, Jr., *Daydreams and Nightmares: A Sociological Essay on the American Environment* (New York: Harper & Row, 1971), p. 4.
61. Robert Moses' role is discussed in Robert A. Caro, *The Power Broker: Robert Moses and the Fall of New York* (New York: Knopf, 1974).
62. The "smell of poison," "harsh, gritty town," and "On its outskirts" quotes are from a *New Yorker* profile of Donora, "The Fog," written by Berton Roueché, vol. 26, no. 31, September 30, 1950, pp. 33–51; On the background to the Donora episode, see Clarence A. Mills, "The Donora Episode," *Science*, vol. 111, no. 2873, January 20, 1950, p. 68; H. H. Schrenk et al., *Air Pollution in Donora, Pa.: Epidemiology of the Unusual Smog Episode of October 1948*, Public Health Bulletin No. 306 (Washington, D.C.: Public Health Service, 1949); James G. Townsend, "Investigation of the Smog Incident in Donora, Pa., and Vicinity," *American Journal of Public Health*, vol. 40, no. 2, February 1950, pp. 183–89; Leslie Silverman and Philip Drinker, "The Donora Episode: A Reply to Clarence A. Mills," *Science*, vol. 112, no. 2899, July 21, 1950, pp. 92–93.

63. The Arthur Kill blob is discussed in Frank Graham, Jr., *Disaster by Default: Politics and Water Pollution* (New York: M. Evans & Co., 1966), pp. 90–96.
64. The problems of landfills in this period are detailed in Environmental Protection Agency, *An Environmental Assessment of Gas and Leachate Problems at Land Disposal Sites* (530/SW-110) (Washington, D.C.: Environmental Protection Agency, 1973), and Environmental Protection Agency, *Closing Open Dumps* (530/SW-61ts) (Washington, D.C.: Environmental Protection Agency, 1971). See also William E. Small, *Third Pollution: The National Problem of Solid Waste Disposal* (New York: Praeger, 1971), and Louis Blumberg and Robert Gottlieb, *War on Waste* (Washington, D.C.: War on Waste, 1989), pp. 15–21.
65. See Clancy Sigal, *Going Away: A Report, A Memoir* (Boston: Houghton Mifflin, 1962).

Chapter 3

1. The "For the first time" and "so thoroughly distributed" quotes are from Rachel Carson, *Silent Spring* (New York: Fawcett, 1964), p. 24.
2. Carson's National Book Award acceptance speech is reprinted in Paul Brooks, *The House of Life: Rachel Carson at Work* (Boston: Houghton Mifflin, 1972), pp. 127–29.
3. Arthur Kallet and F. J. Schlink, *100,000,000 Guinea Pigs: Dangers in Everyday Foods, Drugs, and Cosmetics* (New York: Vanguard Press, 1932). For background on pre–World War II insecticide use, see James Whorton, *Before Silent Spring: Pesticides and Public Health in pre-DDT America* (Princeton, N.J.: Princeton University Press, 1974).
4. Huckins also sent Carson a copy of her letter to the *Boston Herald* about the incident in which she described how her birds had "died horribly, and in the same way" from the DDT spraying. "Their bills were gaping open, and their splayed claws were drawn up to their breasts in agony," Huckins wrote. The letter is reproduced in Paul Brooks, *The House of Life*, p. 232. The question of research and purchased expertise is discussed in Frank Graham, Jr., *Since Silent Spring* (Boston: Houghton Mifflin, 1970), pp. 55–68.
5. Cited in Loren Eisley, "Using a Plague to Fight a Plague," *Saturday Review*, September 29, 1962, vol. 45, no. 39, p. 18.
6. The two Carson quotes are from *Silent Spring*, p. 262. For discussions of Carson as a writer, see Carol B. Gartner, *Rachel Carson* (New York: Frederick Ungar, 1983), and Paul Brooks, *Speaking for Nature: How Literary Naturalists from Henry Thoreau to Rachel Carson Have Shaped America* (Boston: Houghton Mifflin, 1980), as well as Brooks's *The House of Life*, which also includes excerpts of Carson's writings.
7. Cited in Frank Graham, Jr., *Since Silent Spring*, p. 49.
8. The "hysterically overemphatic" and "mystical attachment" quotes are from

"Pesticides: The Price for Progress," *Time*, September 28, 1962, vol. 80, no. 13, pp. 45–48. The "ignorance or bias" quote is from William J. Darby, "Silence, Miss Carson," *Chemical and Engineering News*, vol. 40, no. 40, October 1, 1962, p. 62. For a discussion of Carson's critics, see Paul Brooks, *The House of Life*, pp. 293–307, and Frank Graham, Jr., *Since Silent Spring*, pp. 48–68. See also "Industry Maps Defense to Pesticide Criticisms," *Chemical and Engineering News*, vol. 40, no. 33, August 13, 1962, pp. 23–25. The gender-based attacks against Carson are discussed in H. Patricia Hynes, *The Recurring Silent Spring* (New York: Pergamon Press, 1989). Even articles that appeared to laud Carson's contributions, such as a 1963 commentary in *Science* regarding the report on pesticides by the President's Science Advisory Committee, still complained that Carson "can be legitimately charged with having exceeded the bounds of scientific knowledge for the purpose of achieving shock" (D. S. Greenberg, "News and Comments," *Science*, vol. 140, no. 3569, May 24, 1963, p. 878).

9. The "thanks to a woman" quote is from Edwin Diamond, "The Myth of the Pesticide Menace," *Saturday Evening Post*, September 28, 1963, pp. 16–18. The "priestess of nature" quote is from Clarence Cottam, "A Noisy Reaction to Silent Spring," *Sierra Club Bulletin*, vol. 48, no. 1, January 1963, p. 4.
10. See letters in the *Sierra Club Bulletin*, vol. 48, no. 3, March 1963, p. 18, and the April-May 1963 issue as well. Despite the acknowledgment of controversy, the feedback generated by *Silent Spring* was far less than the heated debate that had gone on for several months concerning "which climbing classification system [the club] should use—NCCS (National Climbing Classification System) or decimals." This was reflected by more than two pages of letters in the June 1963 issue of the *Bulletin* (vol. 48, no. 6, pp. 8–9, 12).
11. David Brower, *For Earth's Sake: The Life and Times of David Brower* (Salt Lake City: Peregrine Smith Books, 1990), p. 215. Another example of conservationist ambivalence toward *Silent Spring* can be found in an article in *National Wildlife*, the publication of the National Wildlife Federation, suggesting that Carson's book "might have gone too far" in its effort to "shock the public" about pesticide dangers. The reviewers argued that by frightening the public "into believing there is not a legitimate use of chemicals," Carson could cause "unneeded restrictions which would hamstring future research and progress. It might mean that we would never have another DDT, the chemical miracle that rescued millions of lives from hunger and disease." See R. G. Lynch and Cliff Ganschow, "Pesticides: Man's Blessing or Curse?" *National Wildlife*, vol. 1, February-March 1963, pp. 10–17. On the Sierra Club approach to pesticides, see also William Siri, "Reflections on the Sierra Club, the Environment, and Mountaineering, 1950s–1970s" (Berkeley: University of California Regional Oral History Office, 1979).
12. Carson's statement is cited in Rachel Carson, "Silent Spring—III," *The New Yorker*, vol. 38, no. 19, June 30, 1962, p. 67.

13. Lewis Herber, *Our Synthetic Environment* (New York: Knopf, 1962), p. XIV.
14. Lewis Herber, "The Problems of Chemicals in Food," *Contemporary Issues*, vol. 3, no. 12, June-August 1952, pp. 206–41; Lewis Herber, *Our Synthetic Environment*, p. 26.
15. The "bane" quote is from Lewis Herber, *The Crisis in the Cities* (Englewood Cliffs, N.J.: Prentice-Hall, 1965), p. 15. The "ecological burden" quote is from Murray Bookchin, *The Limits of the City* (New York: Harper & Row, 1974), p. 92. Many of the ideas for Bookchin's 1974 book on the city first appeared in *Contemporary Issues*, vol. 10, no. 39, August-September 1960, pp. 191–216.
16. Murray Bookchin, *The Limits of the City*, p. 3.
17. Lewis Herber, *The Crisis in the Cities*, p. 194. See also Murray Bookchin, "Toward a Liberatory Technology," in *The Case for Participatory Democracy*, edited by C. George Benello and Dimitrios Roussopoulis (New York: Viking, 1971), pp. 95–139; Murray Bookchin, "Ecology and Revolutionary Thought," *Anarchos*, vol. 1, no. 1, February 1968; Murray Bookchin, *Remaking Society: Pathways to a Green Future* (Boston: South End Press, 1990), p. 187.
18. The "Angry Middle-Aged Man" characterization is from Paul Goodman, *Growing Up Absurd* (New York: Vintage, 1960), p. 56.
19. Paul and Percival Goodman, *Communitas: Means of Livelihood and Ways of Life*, rev. ed. (New York: Vintage, 1960).
20. The "ways of work and leisure" and "organic relation of work, living, and play" quotes are from Paul and Percival Goodman, *Communitas*, pp. 5, 125.
21. The "a parallel system" quote is from Paul and Percival Goodman, *Communitas*, p. 6. The "entire environment" quote is from Paul Goodman, *People or Personnel: Decentralizing and the Mixed System* (New York: Random House, 1965), p. 56.
22. The quotes concerning the limits of the garden city can be found in Paul and Percival Goodman, *Communitas*, p. 8; see also *Utopian Essays and Other Practical Proposals* (New York: Random House, 1972), pp. 33–34. The "rush into production with neat solutions" quote is from Paul Goodman, *The New Reformation: Notes of a Neolithic Conservative*, which is reproduced in part in *Technology and Man's Future*, edited by Albert H. Teich (New York: St. Martin's Press, 1981), p. 341. See also Paul Goodman, "Can Technology Be Humane?" in *The Ecological Conscience: Values for Survival*, edited by Robert Disch (Englewood Cliffs, N.J.: Prentice-Hall, 1970), pp. 103–117.
23. The "mere possibility of an alternative" quote is from Paul Goodman, *People or Personnel*, p. 178. Goodman's utopian ideas are laid out in *Utopian Essays and Other Practical Proposals*. The proposal to ban cars in Manhattan was first laid out in *Communitas* and further developed in Goodman's other writings, including the novel *Making Do* (New York: Macmillan, 1963), which has a chapter entitled "Banning the Cars from New York." This idea later became a rallying cry for the Urban Underground, a budding New Left organization of urban planners in which the author participated. One of its members eventually

became deputy commissioner of transportation in New York City and unsuccessfully sought to develop a modified version of Goodman's plan for midtown Manhattan forty years after it had first been proposed.

24. Marcuse had already become embroiled in campus politics at Brandeis in 1965 in a free speech issue involving one of the faculty. A few years later, he became even more controversial when he supported his former student and Communist Party member Angela Davis. This action, as well as the media's increased targeting of the New Left as violence-prone, with Marcuse a supposed advocate of such violence, brought California Governor Ronald Reagan and the California Legislature into the fray. Ultimately, the chancellor at the University of California at San Diego decided not to renew Marcuse's contract in 1970, although the author of *One-Dimensional Man* continued to write and lecture and otherwise remain an influential critic of the postwar order. See Alain Martineau, *Herbert Marcuse's Utopia*, translated by Jane Brierly (Montreal: Harvest House, 1986), pp. 16–20. The "guru of the student rebels" quote is from "One-dimensional Philosopher," *Time*, March 22, 1968, p. 38. See also Irving Kristol, "Improbable Guru of Surrealistic Politics," *Fortune*, July 1969, p. 191, and Herbert Gold, "California Left, Mao, Marx and Marcuse!" *Saturday Evening Post*, October 27, 1968, p. 57.
25. The "manipulation of needs" quote is from Herbert Marcuse, *One-Dimensional Man: Studies in the Ideology of Advanced Industrial Society* (Boston: Beacon Press, 1966), p. 3.
26. The "peaceful production of the means of destruction," "ever-more-effective domination of man and nature," and "tends to be fatal to this universe" quotes can be found, respectively, in Herbert Marcuse, *One-Dimensional Man*, pp. ix, 17, and 166.
27. The "one-dimensional thought and behavior" quote is from Herbert Marcuse, *One-Dimensional Man*, p. 12. The "second nature of man" quote is from Herbert Marcuse, *An Essay on Liberation* (Boston: Beacon Press, 1969) p. 11.
28. The "break this containment" and Walter Benjamin quotes are from Herbert Marcuse, *One-Dimensional Man*, pp. xv and 257. By the early 1970s, Marcuse, despite the eclipse of the New Left, began to develop a more optimistic assessment of the development of new discourse, based in part on feminist and environmentalist critiques. This position is laid out in his 1972 book *Counterrevolution and Revolt* (Boston: Beacon Press) and is also explored in an interview with Marcuse conducted by the author and Clare Spark for radio station KPFK in Los Angeles that aired on March 21, 1973. See also Robert Pippin, Andrew Feenberg, and Charles P. Webel, eds., *Marcuse: Critical Theory and the Promise of Utopia* (South Hadley, Mass.: Bergin and Garvey, 1988).
29. Students for a Democratic Society, *The Port Huron Statement* (New York: Students for a Democratic Society, August 1962), pp. 3–4. For background on the SDS, see Todd Gitlin, *The Sixties: Years of Hope, Days of Rage* (New York:

Bantam, 1987); Kirkpatrick Sale, *SDS* (New York: Random House, 1973); James Miller, *"Democracy is in the Streets": From Port Huron to the Seige of Chicago* (New York: Simon & Schuster, 1987).

30. See Paul Boyer, "From Activism to Apathy: The American People and Nuclear Weapons, 1963–1980," *Journal of American History*, vol. 70., no. 4, March 1984, pp. 821–44. See also "Fallout: The Silent Killer," *Saturday Evening Post* (two-part series), August 29, 1959, pp. 26, 89, and September 5, 1959, p. 86. As a high school student in New York City in the late 1950s and early 1960s, the author participated in the growing high school movement against nuclear testing and "civil defense" routines, established to protect students from a nuclear attack. The large number of high school participants willing to undertake civil disobedience, and the new understanding that the protest movements were about values as much as politics, caught the liberal and Old Left leadership of some of the protest groups, such as the Committee for a Sane Nuclear Policy (SANE) and Student Peace Union (SPU), by surprise. See also Barbara Epstein, *Political Protest and Cultural Revolution: Non-Violent Direct Action in the 1970s and 1980s* (Berkeley: University of California Press, 1991), and Robert Divine, *Blowing on the Wind: The Nuclear Test Ban Debate, 1954–1960* (New York: Oxford University Press, 1978).

31. The "vision and program" quote is from Students for a Democratic Society, *The Port Huron Statement*, p. 63. The statement goes on to say: "If we appear to seek the unattainable, as it has been said, then let it be known that we do so to avoid the unimaginable."

32. The focus on radiation from above-ground nuclear tests rather than the production of nuclear weaponry as such (including its environmental impacts) left the antinuclear movement vulnerable when the 1963 Test Ban Treaty appeared to have eliminated that environmental danger. See Frances B. McCrea and Gerald E. Markle, *Minutes to Midnight: Nuclear Weapons Protest in America* (Newbury Park, Ca.: Sage Publishers, 1989).

33. Clark Kerr's analysis of the "multiversity" was developed in a series of lectures he delivered at Harvard University in 1963 that was reprinted in *The Uses of the University*, 3rd ed. (Cambridge, Mass.: Harvard University Press, 1982). The "factory that turns out a certain product" quote is from Mario Savio's speech at a student sit-in in December 1964, "An End to History," which is reproduced in *The New Left: A Documentary History*, edited by Massimo Teodori (Indianapolis: Bobbs-Merrill, 1969), pp. 158–61. Savio's well-known dramatic and evocative statement about students being cogs in the machine and their readiness to throw their "bodies upon the gears and the wheels . . . to make it stop" is cited in Todd Gitlin, *The Sixties*, p. 291f. Savio also recognized the importance of the civil rights movement experience for himself and other leaders of the Free Speech Movement. In a speech during a sit-in demonstration at Sproul Hall, he defined the group's action and its relationship to the civil

rights movement as "another phase of the same struggle" and suggested they both challenged an organized status quo and a "depersonalized, unresponsive bureaucracy." The speech is reproduced in Mario Savio, Eugene Walker, and Raya Dunayevskaya, "The Free Speech Movement and the Negro Revolution" (Detroit: News and Letters, July 1965).

34. At the beginning of *The Graduate*, the protaganist, a recent college graduate played by Dustin Hoffman, is advised to go into plastics by the father of the young student he eventually falls in love with. In the exchange, an ironic point-counterpoint is established between the two generational figures. The term *plastic* itself had, in fact, already begun to enter common usage as referring to something false or artificial, less than a decade after the technology had begun to be celebrated as symbolizing the advances of the new postwar urban and industrial order. See "Plastic: Material of a Thousand Uses," p. 77.

35. The "technological progress" quote is from Edward J. Nell, "Automation and the Abolition of the Market," "Praxis and the New Left" supplement to *New Left Notes*, August 7, 1967. The "Daily Smog Smear" and "the middle class controls" quotes are from Bob Gottlieb and Marge Piercy, "Movement for a Democratic Society: Beginning to Begin to Begin," in *The New Left: A Documentary History*, edited by Massimo Teodori, pp. 403–11.

36. See "On the Use of Herbicides in Vietnam," *Scientist and Citizen*, June-July 1968, pp. 118–22. See also Sam Love, "Vietnam: The True Cost of War," *Environmental Action*, vol. 2, no. 21, March 20, 1971, pp. 10–12, and Barry Weisberg, *Ecocide in Vietnam: The Ecology of War* (San Francisco: Canfield Press, 1970).

37. "The direct source of pollution," the authors of this pamphlet argued, "is not the individual consumer, but industrial corporations (smog and water pollution and consumption) and agribusiness (pesticides and fertilizers). It's not 'drivers' who cause air pollution and it's not 'cars'; it's General Motors, and Ford, and Chrysler." See Los Angeles Power & Light, "Where There's Pollution, There's Profit" (Santa Monica, Ca.: Los Angeles Power & Light, 1970).

38. The "deterioration of the natural environment" quote is from Barry Weisberg, "The Politics of Ecology," *Liberation*, January 1970, reproduced in *The Ecological Conscience*, edited by Robert Disch, p. 154.

39. See, for example, Katherine Barkley and Steve Weissman, "The Eco-Establishment," *Ramparts*, vol. 8, no. 11, November 1970, and Steven Weissman, "Why the Population Bomb is a Rockefeller Baby," *Ramparts*, vol. 8, no. 5, May 1970, pp. 43–47 (distributed by the ecology-oriented New Left publishing collective, The Glad Day Press, of Ithaca, New York). See also the manifesto of Ecology Action East, "The Power to Destroy, the Power to Create," *Rat*, January 1970, reprinted in *The Ecological Conscience*, edited by Robert Disch, p. 163. On the wilderness/elitism issue, see Jon Weiner and Dick Cluster, "The Politics of Conservation," *The Old Mole* (a New Left

alternative paper from the Boston area), May 23–June 5, 1969, and Peter Marcuse, "Is the National Parks Movement Anti-Urban?" *Parks and Recreation*, vol. 6, no. 7, July 1971, pp. 16–21+.

40. Todd Gitlin, in his book *The Sixties*, cites an internal memo, written by the Student Non-Violent Coordinating Committee (SNCC) organizer Jane Stembridge, contrasting some of the politically oriented revolutionaries of SNCC with the rural black poor and their "closeness with the earth" and "closeness with each other in the sense of community developed out of dependence" (pp. 164–65). The "alienation" quote is from Laurence Veysey, *The Communal Experience: Anarchist and Mystical Counter-Cultures in America* (New York: Harper & Row, 1973), p. 197.
41. The "our movement must encompass" quote is taken from a paper written by Richard Flacks and delivered at the June 1965 SDS Convention, as reprinted in Paul Jacobs and Saul Landau, *The New Radicals: A Report with Documents* (New York: Vintage Books, 1966), p. 163.
42. The first issue (January 1970) of *Mother Earth News* stated that it would be a publication "edited by, and expressly for, today's influential 'hip' young adults. . . . Heavy emphasis is placed on alternative life styles, ecology, working with nature, and doing more with less." See also Robert Rodale, "The New 'Back to the Land' Movement," *Organic Gardening and Farming*, September 1969, pp. 21–24, and Jerry Minnich, "The Campus Whole Earth Coop," *Organic Gardening and Farming*, September 1970, pp. 73–77. On the food cooperative movement, see Daniel Zwerdling, ". . . But Will It Sell?" *Environmental Action*, vol. 7, no. 21, March 13, 1976, pp. 3–9, and William Ronco, *Food Co-ops—An Alternative to Shopping in Supermarkets* (Boston: Beacon Press, 1974).
43. For background on the underground press, see Lawrence Leamer, *The Paper Revolutionaries: The Rise of the Underground Press* (New York: Simon & Schuster, 1972); Abe Peck, *Uncovering the Sixties: The Life and Times of the Underground Press* (New York: Pantheon, 1985); David Armstrong, *A Trumpet to Arms: Alternative Media in America* (Los Angeles: J. P. Tarcher, 1981).
44. The *Rat* and Keith Lampe quotes are cited in Warren J. Belasco, *Appetite for Change: How the Counterculture Took on the Food Industry, 1966–1988* (New York: Pantheon, 1989), p. 21.
45. Roszak especially feared the "adolescentization of dissent" among the young, though he also praised their "healthy instincts" in their challenge to the "technocratic society" and its "culture of expertise." Roszak concluded that "the strange youngsters who don cowbells and primitive talismans and who take to the public parks or wilderness to improvise outlandish communal ceremonies are in reality seeking to ground democracy safely beyond the culture of expertise." See Theodore Roszak, *The Making of a Counter Culture: Reflections on the Technocratic Society and Its Youthful Opposition* (Garden City,

N.Y.: Doubleday, 1969), pp. 41, 265. See also Charles A. Reich, *The Greening of America* (New York: Random House, 1970).

46. See, for example, "Trouble in Hippieland," *Newsweek*, October 30, 1967, pp. 84–90; Horace Sutton, "Summer Days in Psychedelphia," *Saturday Review*, August 19, 1967, pp. 36–41; "Love on Haight," *Time*, March 17, 1967, p. 27; "Where Have All the Flowers Gone?" *Time*, October 13, 1967, pp. 30–31; Keith Melville, *Communes in the Counter Culture: Origins, Theories, Styles of Life* (New York: William Morrow, 1972), pp. 66–67.
47. The Motherfucker group was founded by SDS organizers to create an affinity group on the Lower East Side consisting of cultural rebels drawn from guerrilla theater and anarchist circles. It also included Herbert Marcuse's stepson, Tom Neuman, who was influential in helping Marcuse establish links with both New Left and counterculture groups, including the Motherfuckers. The garbage action is depicted in the film *Garbage*, produced by the alternative documentary film group Newsreel and released in 1968. It remains a cult classic of the era. See also Todd Gitlin, *The Sixties*, pp. 239–41.
48. Gary Snyder, *Earth House Hold: Technical Notes & Queries to Fellow Dharma Revolutionaries* (New York: New Directions, 1969), p. 113.
49. The "questions about the quality of our lives" and "non-negotiable demands of the Earth" quotes are cited in Warren J. Belasco, *Appetite for Change*, p. 21. The *Berkeley Tribe*, an underground paper, described People's Park as "the beginning of the Revolutionary Ecology Movement. It is the model of the struggle we are going to have to wage in the future if life is going to survive at all on this planet. In the Park we blended the new culture and the new politics that was developing in Berkeley for almost a decade. The revolutionary culture gives us new communal, eco-viable ways of organizing our lives, while peoples' politics give us the means to resist the System." Cited in *Berkeley Tribe*, "Blueprint for a Communal Environment," in *Sources: An Anthology of Contemporary Materials Useful for Preserving Personal Sanity While Braving the Great Technological Wilderness*, edited by Theodore Roszak (New York: Harper & Row, 1972), p. 393.
50. The *New York Times* article, "Communes Spread as the Young Reject Old Values," by Bill Kovach, appeared December 17, 1970, and is cited, along with the other estimates, in Judson Jerome, *Families of New Eden: Communes and the New Anarchism* (New York: Seabury Press, 1974), pp. 16–17. See also Rosabeth Moss Kanter, "Communes in Cities," in *Coops, Communes and Collectives: Experiments in Social Change in the 1960s and 1970s*, edited by John Case and Rosemary C. R. Taylor (New York: Pantheon, 1979), pp. 112–35.
51. All quotes are from Paul Goodman, "The Diggers in 1984," *Ramparts*, September 1967, pp. 28–30.
52. See "The Young Eco-activists," *Time*, August 22, 1969, p. 43, and Elizabeth Rogers, "Protest!" *Sierra Club Bulletin*, vol. 54, no. 11, December 1969, pp. 10–11+.

53. Nelson made these remarks before a convention of the Industrial Union Department of the AFL-CIO shortly after announcing plans for the teach-in. He made the announcement because he felt "the political establishment was just not keyed to the issue" (author's interview with Gaylord Nelson, 1991). See also Edward E. C. Clebsch, "The Campus Teach-in on the Environmental Crisis: 1970," *The Living Wilderness*, vol. 34, no. 109, Spring 1970, p. 10.
54. Author's interview with Gaylord Nelson, 1991.
55. Hayes went on to say, "Our goal is not to clean the air while leaving slums and ghettos, nor is it to provide a healthy world for racial oppression and war. We wish to make the probability of life greater, and the quality of life higher. Those who share these goals cannot be 'coopted'; they are our allies —not our competitors." The text of this press statement is reprinted in *The Living Wilderness*, vol. 34, no. 109, Spring 1970, pp. 12–13.
56. Author's 1991 interview with Denis Hayes.
57. Barry Weisberg, "The Politics of Ecology," in *The Ecological Conscience*, edited by Robert Disch, p. 159. Weisberg also quotes an Earth Day organizer as saying "The country is tired of SDS and ready to see someone like us come to the forefront."
58. Edgar Wayburn, "Survival Is Not Enough," editorial in the *Sierra Club Bulletin*, vol. 55, no. 3, March 1970, p. 2.
59. See the editorial by Michael McCloskey in the *Sierra Club Bulletin*, vol. 55, no. 6, June 1970, p. 2. See also Connie Flateboe, "Environmental Teach-in," *Sierra Club Bulletin*, vol. 55, no. 3, March 1970, pp. 14–15.
60. In the Conservation Foundation's 1970 *Annual Report*, Howe called the Earth Day participants "instinctive environmentalists—those activists who direct their zeal toward actions necessary to man's peace with Earth." Background information from author's 1992 interview with Sydney Howe and 1991 interviews with Gaylord Nelson and Denis Hayes.
61. On Nixon's 1970 State of the Union speech, see "Summons to a New Cause," *Time*, February 2, 1970, pp. 7–8.
62. At this meeting, Nixon suggested that the "current ecology fad" wouldn't last, especially among students, but that he shared the "same basic goals" of the conservationists. See "On the White House Meeting," editorial by the Sierra Club's president, Phillip S. Berry, in the *Sierra Club Bulletin*, vol. 55, no. 4, April 1970, p. 2.
63. A review of eighteen different trade and industry publications that analyzes industry positions in relation to Earth Day was undertaken for this book by Leslea Meyerhoff.
64. On the Monsanto approach, see "From Pollution to Profit," *Time*, August 8, 1969, p. 36.
65. Dow provided $5000 to the Michigan group, and Ford assumed all its printing costs. The national office of Environmental Action, however, turned down offers of $15,000 from Mobil and $5000 from Ford. "We're not allowing the

major polluters to buy in," proclaimed Stephen Cotton, a friend of Denis Hayes from Harvard Law School, who had been recruited to help plan the event. See "Utilities Will Talk to Youth on Earth Day," *Electrical World*, vol. 173, no. 22, April 6, 1970, pp. 36–39, and "Earth Day Broom Sweeps in on Business," *Business Week*, April 18, 1970, pp. 22–24.

66. The "as a platform" quote is from *Electrical World*, April 6, 1970, pp. 38–39. See also *Business Week*, April 18, 1970, p. 23, and "Earth Day: Not All That Bad for Utilities," *Electrical World*, May 4, 1970, pp. 25–26.
67. The "carries with it" and "a lot of young idealists" quotes are from Clifford B. Reeves, "Ecology Adds a New P.R. Dimension," *Public Relations Journal*, vol. 26, no. 6, June 1970, pp. 6–9. See also "Earth Day Draws Industry Involvement," *Engineering News-Record*, April 23, 1970, p. 13. The Conservation Foundation incident was described by Sydney Howe in a 1992 interview with the author.
68. "And the Day After," *Science News*, May 2, 1970, p. 432; "A Memento Mori to the Earth," *Time*, May 4, 1970, p. 4; "A Giant Step—Or a Springtime Skip?" *Newsweek*, May 4, 1970, pp. 26–28.
69. See "The Oregon Experiment After Twenty Years," *RAIN*, vol. 14, no. 1, Winter-Spring 1991, pp. 32–41, and Christopher Alexander et al., *The Oregon Experiment* (New York: Oxford University Press, 1975).
70. The Earth Day speeches are reproduced in *Earth Day—The Beginning, A Guide for Survival*, compiled and edited by the national staff of Environmental Action (New York: Bantam Books, 1970).
71. See "Summons to a New Cause," *Time*, February 2, 1970, p. 7, and, in the same issue, "Fighting to Save the Earth From Man," pp. 56–62.

Chapter 4

1. This confidential memo of January 6, 1981, was from Robert L. Allen, vice-president of the Henry P. Kendall Foundation, to funders and environmental organization leaders. It included a section entitled "Premise" that laid out the points to be discussed at the first meeting of the group of environmental leaders who would come to be known as the Group of Ten. General background on the Group of Ten comes from the author's 1991 interviews with Russell Peterson, Robert Allen, Louise Dunlap, Jay Hair, Janet Brown, and William Turnage.
2. The "indisputable achievements" quote is from William Reilly's essay "A View Towards the Nineties" in *Crossroads: Environmental Priorities for the Future*, edited by Peter Borelli (Washington, D.C.: Island Press, 1988), p. 97. The "establish diplomatic relations" quote is cited by Neil Pierce in "Will the 1980s Turn Out to Be 2nd Environmental Decade?" *Los Angeles Times*, April 22, 1980. On the Reagan transition period, see Peter Wiley and Robert Gottlieb, *Empires in the Sun: The Rise of the New American West* (Tucson: University of Arizona Press, 1985), p. 286.

3. Author's 1992 interview with Robert Allen.
4. The description of Allen's role is based on the author's 1991 interviews with Robert Allen, Louise Dunlap, and Russell Peterson. While Allen and the Kendall Foundation played the instrumental role with the Group of Ten, some larger foundations, especially the Ford Foundation and the Rockefeller Family Fund, had crucial roles with other environmental groups and projects. The competition for foundation money was fierce, becoming at times the basis for tension between groups. For example, Thomas Kimball, the executive director of the National Wildlife Federation between 1961 and 1981, at one point asked Laurance Rockefeller to serve on the group's board. Rockefeller declined because of other commitments. "It didn't stop him from serving on other boards," Kimball bitterly commented years later, referring to Rockefeller's relationship with the Conservation Foundation. "Big money wants control; it's as simple as that," Kimball declared (author's 1991 interview with Thomas Kimball). For a discussion of the role of foundations in the mainstream environmental movement, see Edward A. Ames, "Philanthropy and the Environmental Movement in the United States," *The Environment*, vol. 1, no. 1, Spring 1981, pp. 11–19.
5. Author's 1991 interview with Louise Dunlap. William Turnage, then head of The Wilderness Society, recalled how he and Audubon's Peterson, both of whom had no previous experience with the mainstream environmental groups prior to their appointment as CEOs, were shocked at the intensity of the competition and rivalries. "The Sierra Club and Wilderness Society really disliked each other in that period," Turnage said by way of example, characterizing their relationship as that of an "armed truce" (author's 1991 interview with William Turnage).
6. The rise during the 1970s of the business CEO groups, particularly with respect to environmental and resource issues, is described in Bob Gottlieb and Peter Wiley, "The New Power Brokers Who Are Carving Up the West," *Straight Creek Journal*, vol. 9, no. 12, March 20, 1980, pp. 1–3.
7. The words "rationalize the field" were used by Edward Ames, a key environmental funder (author's 1992 interview with Edward Ames); also 1991 interviews with Robert Allen, Louise Dunlap, and Russell Peterson.
8. Robert Allen's January 6, 1981, memo on the Iron Grill meeting; also the author's 1991 interviews with Allen and Peterson.
9. Memo from Russell Peterson to the Group of Ten, December 28, 1981, and memo from Russell Peterson to Rupert Culter, December 2, 1981, in Russell Peterson papers provided the author.
10. The regional conference idea was first proposed at the second meeting of the Ten. Former Wilderness Society head Stewart Brandborg was hired to coordinate the effort, while several of the groups from the Ten assumed responsibility for hosting the events. Though designed to promote grassroots participation in the movement, the conference agendas, with their emphasis

on expanding lobbying capabilities, were framed by the Ten. See "A Proposal for Eight Regional Conferences to Further Environmental Quality," December 1981; memos from Russell Peterson to the Group of Ten, January 27, 1981, and December 2, 1981; memo from Stewart Brandborg, regional conference coordinator, to Group of Ten, January 12, 1982. In Russell Peterson papers.

11. See John H. Adams et al., *An Environmental Agenda for the Future* (Washington, D.C.: Island Press, 1985). See also memos from Russell Peterson to the Group of Ten, March 5, 1983, and July 20, 1983. The Group of Ten also helped fund an environmental critique of the early Reagan years that was eventually published as *A Season of Spoils: The Reagan Administration's Attack on the Environment*, by Jonathan Lash et al. (New York: Random House, 1984). Also the author's 1991 interviews with Robert Allen and Russell Peterson.
12. John H. Adams et al., *An Environmental Agenda for the Future*, p. 4.
13. According to Russell Peterson, prior to its 1981 mailing attacking Reagan, Audubon had never raised more than $84,000 on a direct-mail piece. The Reagan attack piece not only netted $985,000 but generated more than 30,000 responses, unprecedented in Audubon's history (author's 1991 interview with Peterson).
14. On James Watt's role, see Barry W. Walsh, "After Watt," *Journal of Forestry*, April 1985, pp. 212–17; Eleanor Randolph, "Watt Courts Sportsmen to Back Policies," *Los Angeles Times*, June 10, 1981, p. 1; Paul J. Culhane, "Sagebrush Rebels in Office: Jim Watt's Land and Water Politics," in *Environmental Policy in the 1980s: Reagan's New Agenda*, edited by Norman J. Vig and Michael E. Kraft (Washington, D.C.: CQ Press, 1984), pp. 293–317.
15. Concerns about the NWF's relationship with the Reagan administration were a constant theme with the Ten throughout much of its first three years. A January 18, 1982, memo from B Team staff member William A. Butler, for example, refers to the ambivalence of the NWF board toward the Group of Ten and their unwillingness to participate in the spring offensive against the Reagan administration. A May 27, 1983, memo from B Team alternate member Leslie Dach refers to a B Team meeting called by the NWF's Pat Parentau to discuss "NWF's emerging relationship with the White House," initiated by the hunting and wildlife organization as a "conscious decision to distance themselves from the rest of the environmental community." Dach said of the NWF approach: "They felt that their institutional needs and identity required that they project a different tone from the rest of the environmental community," and that the tone used by some of the Ten indicated a kind of "ecological illiteracy." The NWF/Reagan administration relationship was the subject of a sharp note from Russell Peterson to NWF head Jay Hair on June 29, 1984, in which Peterson urged that Hair "stop being sucked in by the Reagan administration. You are helping to give the impression that there is something meritorious about what the Reagan administration is doing in the environmental

area." Earlier, in a May 10, 1983, memo from Hair to the Group of Ten following inquiries about NWF actions, the NWF leader had sought to allay Group of Ten fears, suggesting that it was the "diversity" of orientation and views, as symbolized by the different approach taken by the NWF, that was "one of the great strengths of the environmental movement." See also memo from Jack Lorenz to the Group of Ten, July 6, 1984, and letter from William Turnage to Jack Lorenz, July 11, 1984, in Russell Peterson papers.

16. On the chemical company/Group of Ten meetings, see February 9, 1983, memo from Russell Peterson to the Group of Ten environmental leaders on a proposed March 9, 1983, meeting with the chemical company executives and an August 24, 1984, joint memo from Peterson and Louis Fernandez of Monsanto on the activities of the working group of chemical executives and the Group of Ten CEOs, in Russell Peterson papers.
17. The "political anomaly" quote is from Lynton K. Caldwell, *Science and the National Environmental Policy Act: Redirecting Policy Through Procedural Reform* (University: University of Alabama Press, 1982), p. 52. See also Richard A. Liroff, *A National Policy for the Environment: NEPA and Its Aftermath* (Bloomington: Indiana University Press, 1976).
18. In a special twenty-fifth anniversary issue of the *Natural Resources Journal* devoted to NEPA, Senator Henry Jackson wrote that the changes in the period prior to NEPA's passage signified that "traditional economic indices are no longer viewed as the sole measures of progress. We are entering an era in which the qualitative values and aesthetic factors are considered as important as material well being." See Henry Jackson, "Environmental Policy and the Congress," in *Enclosing the Environment: NEPA's Transformation of Conservation into Environmentalism*, edited by Channing Kuty (Albuquerque: *Natural Resources Journal*, 1985), p. 36.
19. Some bond and investment houses, such as Prudential-Bache, have established annual bond-buyer conference sessions with representatives of the environmental industry, including pollution control technology firms, waste collection and landfill operators, and integrated waste management companies. Author's notes, proceedings from the Environmental Control Industry Conference, Tucson, Arizona, February 27–28, 1991. See also "The Cost of Cleaning Up: Financing the Environment Forum," *Institutional Investor* (reprint by Waste Management Inc., 1990) and John Crudele, "Trash Issues Likely to Be at Top of Heap for Investors," *Los Angeles Times*, March 18, 1990, p. D-1. For the estimate of revenues for the waste industry, see Eugene J. Wingerter, "Where the Waste Industry Is Going," *Waste Age*, April 1990, p. 281.
20. For background on the debate concerning air pollution issues, see Charles O. Jones, *Clean Air: The Policies and Politics of Pollution Control* (Pittsburgh: University of Pittsburgh Press, 1975), and John C. Esposito (project director), *Vanishing Air: Ralph Nader's Study Group Report on Air Pollution* (New York: Grossman Publishers, 1970).

21. The "activist period" quote and background discussion of GASP can be found in Charles O. Jones, *Clean Air*, pp. 149–52. See also Lynton K. Caldwell et al., *Citizens and the Environment: Case Studies in Popular Action* (Bloomington: Indiana University Press, 1976), pp. 130–33, and Deborah Baldwin, "GASP-ing for Air," *Environmental Action*, vol. 7, no. 11, pp. 12–13.
22. See Barbara Reid, "Building an Eco-Alliance," *ZPG National Reporter*, vol. 3, no. 4, April 1971, p. 1. Also the author's 1991 interviews with Joseph Browder, Louise Dunlap and Denis Hayes and 1992 interview with George Coling.
23. In his introduction to *The Politics of Pollution*, by J. Clarence Davies III (Indianapolis: Bobbs-Merrill, 1970), Muskie wrote that it was "easy to blame pollution only on the large economic interests, but pollution is a by-product of our consumption-oriented society. Each of us must bear his share of the blame." Nixon, similarly, was concerned that his environmentalist rhetoric could be misinterpreted as indicating an anti-industry posture and made it clear in several speeches that the environmental issue must not be used "to destroy the industrial system which made this great country what it is." Cited in Charles Hardin, "Observations on Environmental Politics," in *Environmental Politics*, edited by Stuart S. Nagel (New York: Praeger, 1974), p. 185.
24. The six criteria pollutants included nitrogen oxides, ozone (which is formed from a combination of nitrogen oxide and hydrocarbons), particulate matter, carbon monoxide, lead, and sulfur oxides. Hazardous air pollutants (also known as noncriteria pollutants) include metals such as cadmium and chromium, organics such as polychlorinated biphenyls (PCBs), various classes of dioxins, and acid gases such as hydrogen chloride. See Edward J. Calabrese and Elaina M. Kenyon, *Air Toxics and Risk Assessment* (Chelsea, Mich.: Lewis Publishers, 1991).
25. The "advocate an environmental position" quote is from John C. Whitaker, *Striking a Balance: Environment and Natural Resources Policy in the Nixon-Ford Years* (Washington, D.C.: American Enterprise Institute for Public Policy Research, 1976), p. 50. The "air pollution, water pollution" quote is cited in Jack Lewis, "The Birth of EPA," *EPA Journal*, November 1985, p. 9. The *EPA Journal* article pointed out that EPA programs in air quality, water quality, solid wastes, pesticides, and radiation issues displaced existing programs in other agencies involving more than 5600 government employees. See also Marc K. Landy, Marc J. Roberts, and Stephen R. Thomas, *The Environmental Protection Agency: Asking the Wrong Questions* (New York: Oxford University Press, 1990), and Daniel A. Dreyfus and Helen M. Ingram, "The National Environmental Policy Act: A View of Intent and Practice," in *Enclosing the Environment*, edited by Channing Kuty, p. 55.
26. For background on the EPA, see Alfred A. Marcus, *Promise and Performance: Choosing and Implementing Environmental Policy* (Westport, Conn.: Greenwood Press, 1980).

27. The Browder quote is from an interview with him in 1980 and is cited in Peter Wiley and Robert Gottlieb, *Empires in the Sun*, pp. 294–95. The revolving-door relationships are discussed in Deborah Baldwin, "Environmentalists Open the Revolving Door," *Environmental Action*, vol. 9, no. 6, July 16, 1977, pp. 13–15. See also Deborah Baldwin, "Environmentalism Goes Underground: An Irreverent Look at 1978," *Environmental Action*, vol. 10, no. 15, December 2, 1978, pp. 4–5, and Carl Pope, "On Having Friends in High Places," *Environmental Action*, vol. 9, no. 16, December 17, 1977, pp. 16–17.
28. When Joseph Browder was asked in a 1977 interview whether he accomplished more inside rather than outside the Carter administration, he replied, "It's not accurate to say you're inside or outside the government. You're part of it by the simple act of trying to make it work." See Michael C. Lipske, "Washington Outlook—Administration Brings Cheer to Environmentalists," June 1977, p. 213, cited in James A. Tober, *Wildlife and the Public Interest: Nonprofit Organizations and Federal Wildlife Policy* (New York: Praeger, 1989), p. 164.
29. Author's 1991 interview with Gaylord Nelson.
30. Cited in John C. Esposito (project director), *Vanishing Air*, p. 264.
31. For background on the use of expertise in the Clean Air Act debates, see Thomas C. Jorling, "Environmentalism at the Center of Public Life," *Amicus Journal*, vol. 1, no. 2, Fall 1979, pp. 35–37.
32. The "ecology of the cities" quote is from the author's 1992 interview with Sam Love. Why the Ford Foundation rejected the proposal from Environmental Action remains a subject of dispute among some of the key parties involved. Denis Hayes argued that Ford head McGeorge Bundy had been uncomfortable with the group's "radical ideas" and that the foundation program officer handling the grant, Edward Ames, told Environmental Action staff member Bob Lilly that both Bundy and Ford director Robert McNamara vetoed the grant because of Hayes's role in the group. Ames recalls the episode differently, suggesting that Bundy actually had a favorable impression of Hayes, but that the grant was denied because it had too much of a "citizen action" focus. In terms of other funding decisions discussed in this chapter, it is clear that Ford was most interested in the development of more professional-type environmental organizations as distinct from the kind of 1960s-type interest in direct action and mobilization that Environmental Action still clearly expressed. Yet Environmental Action continued to remain ambivalent about its own organizational definition. In a revealing organizational promotion shortly after the Earth Day events, a fundraising ad for the group suggested that "more action" rather than another Earth Day was now needed. This "action," however, was defined as donations to support research work, lobbying, and litigation. See "Earth Day Failed," *Environmental Action*, vol. 2, no. 1, May 14, 1970, p. 11; also author's interviews with Denis Hayes (1991), Edward Ames (1992), and letter from

Denis Hayes to Chuck Savitt of August 16, 1992. On the eco-tage issue, see Sam Love and David Obst, *Ecotage!* (New York: Basic Books, 1972).

33. The Hayes quote is from the author's 1992 interview with Denis Hayes. For background on the Dirty Dozen campaigns, see "Introducing the . . . Clean Up Congress Act," *Environmental Action*, vol. 7, no. 16, December 20, 1975, pp. 12–14, and Tom Redburn, "The Only Good Dirty Dozen Is a Defeated Dirty Dozen," *Environmental Action*, vol. 7, no. 22, March 27, 1976, pp. 3–10.
34. For background on the EDF's history, see Marion Lane Rogers, *Acorn Days: The Environmental Defense Fund and How It Grew* (New York: Environmental Defense Fund, 1990). This book, which includes Rogers' text as well as contributing essays by early staff members and leaders of the EDF, covers the period from the BTNRC days through the mid-1970s.
35. The "partnership" quote is from Roderick Cameron (the EDF's first executive director), "View from the Front Office," in Marion Lane Rogers, *Acorn Days*, p. 35. The Yannacone lawsuit episode is described in Charles F. Wurster, "The Power of an Idea," in *Acorn Days*, pp. 49–50.
36. The Audubon Society convention episode is described in Charles Wurster, "The Power of an Idea," in *Acorn Days*, pp. 52–53. See also Frank Graham, Jr., and Carl W. Buchheister, *The Audubon Ark*, p. 232. Though the EDF was effectively segregated from Audubon, the passage of the anti-DDT position was sufficient to cause one long-time Audubon Society member, DuPont executive Crawford Greenewalt, to resign from the organization.
37. The "crucible of the adversarial process" quote is from Roderick Cameron, "View from the Front Office," in *Acorn Days*, p. 37.
38. *Fortune* magazine characterized the EDF as "legally daring but scientifically cautious," cited in *Acorn Days*, p. 41. See also Roderick Cameron's statement in "Five Who Care," one of *Look* magazine's articles for its Earth Day issue, April 21, 1970, p. 38. Cameron argued that the new movement needed to combine the use of the courts "as an impartial forum" with the need to "demonstrate to the hilt" in order to pressure regulators and legislators.
39. Information on the five "Gurus" and background on the Ford Foundation's role is from the author's 1992 interview with Edward Ames. See also Jeffrey M. Berry, *Lobbying for the People: The Political Behavior of Public Interest Groups* (Princeton, N.J: Princeton University Press, 1977), pp. 51–52. The Litigation Review Committee is discussed in the EDF's June 26, 1970, minutes, whose text is cited in *Acorn Days*, p. 72. The "legally, scientifically and practically" quote is from *Acorn Days*, p. 73. The "EDF's very early policy" quote is from the commentary "Drafted into the Movement," by Amyas Ames, in *Acorn Days*, p. 63.
40. Background on the EDF's role is from the author's interviews with Janet Brown (1991) and Tom Graff (1987).
41. The demand-based model for the utility industry is described in David Roe, *Dynamos and Virgins* (New York: Randon House, 1984).

42. For background on the Storm King controversy, see Allan R. Talbot, *Power Along the Hudson: The Storm King Case and the Birth of Environmentalism* (New York: Dutton, 1972). Duggan's remarks are from Natural Resources Defense Council, *Twenty Years Defending the Environment: NRDC 1970–1990* (New York: Natural Resources Defense Council, 1990), pp. 9–10.
43. The "bringing professionalism" quote is from the author's 1991 interview with Frances Beinecke. The dispute with the IRS is discussed in Jeffrey M. Berry, *Lobbying for the People*, pp. 51–52.
44. Ayres, one of the original Yale group, has remained the NRDC's resident air expert. Author's 1991 interview with John Adams.
45. The Adams quote is from the author's 1991 interview with him. When Southern California Edison narrowed its choice for CEO, its two finalists were Bryson and Edison's vice-president, Michael Peevey. Peevey, the director of the industry-oriented California Council on Economic and Environmental Balance lobbying group before becoming an Edison executive, represented a parallel, though contrasting, corporate leadership route associated with the construction of the environmental policy system. See Larry B. Stammer and Michael Parrish, "Ex-Environment Activist Will Take Helm at Edison," *Los Angeles Times*, October 1, 1990. Author's 1991 interview with John Bryson.
46. Author's 1991 interview with Joseph Browder; also 1991 interviews with Louise Dunlap and Marion Eddy.
47. One key EPI staff member, Peter Carlson, who had played a central role in shaping the EPI's water policy activities, was let go by the organization after funders shifted their priorities away from water policy. "We are dependent on foundation support for certain activities," EPI head Michael Clark remarked at the time, "and if that funding is no longer available, we don't have the resources to sustain that work" (author's 1987 interviews with Michael Clark and Peter Carlson).
48. Background on the League of Conservation Voters is from the author's 1991 interviews with Marion Edey and Joseph Browder.
49. On the Fallon election, see "Political Conservation Group Is Campaigning for 13 Congressional Candidates," *New York Times*, August 16, 1970, p. 49, and Ben A. Franklin, "Rep. Fallon Loses in Maryland Race," *New York Times*, September 17, 1970. Background on the Aspinall defeat is from the author's 1979 interview with Wayne Aspinall for Peter Wiley and Robert Gottlieb, *Empires in the Sun*, pp. 128–34; also the author's 1991 interview with Marion Edey.
50. By the 1974 elections, the LCV had achieved broad recognition among the mainstream groups for its efforts. In that election, thirteen of seventeen LCV-endorsed candidates won (and Environmental Action was successful in eight of their twelve Dirty Dozen campaigns). Of those seventeen LCV candidates, six were Republicans. The only unsuccessful Republican LCV-endorsed challenger was Newt Gingrich, who would subsequently win a congressional race,

emerge as a leading figure in his party and within Congress, and ultimately become a major antagonist of the mainstream groups. See Marion Edey, "Mandate for the Environment," *Sierra Club Bulletin*, vol. 60, no. 3, March 1975, pp. 5–6+. On the LCV voting charts, see, for example, Steven Pearlman, *The Presidential Candidates for 1984: Their Records and Positions on Energy and the Environment* (Washington, D.C.: League of Conservation Voters, 1984), and *How Congress Voted on Energy and the Environment*, another annual voting chart put out by the organization.

51. Author's 1991 interview with Marion Edey.
52. In terms of its international activities, only Greenpeace rivaled FOE's international organization. As the mainstream groups developed a more pronounced global orientation in the 1980s and 1990s, they still operated as domestic players with a global perspective and were frequently criticized for an environmental imperialism in their approach to such issues as global warming. FOE largely escaped those characterizations due to the relative autonomy of its international groups. See Aldemaro Romero, "International Conservation Leadership and the Challenges of the '90s," in *Voices from the Environmental Movement: Perspectives for a New Era*, edited by Donald Snow (Washington, D.C.: Island Press, 1992), pp. 168–84; also author's 1991 interview with Janet Brown.
53. On the FOE/David Brower split, see "FOE Election: Board Majority Retained, New Directors Elected," *Not Man Apart*, vol. 16, no. 2, March-April 1986, p. 2, and vol. 16, no. 3, May-June 1986, p. 4; *High Country News*, July 23, 1984, and Ed Marston, "Campaigning Begins in the FOE Contest," April 14, 1986; Larry B. Stammer, "Environmentalists' Friends of the Earth Rocked as Founder Feuds with Directors," *Los Angeles Times*, March 2, 1986; *Earth Island Journal*, vol. 4, no. 1, March 1986. The "willingness to take risks" and "EPI's expertise" quotes are from an interview with Michael S. Clark in *Not Man Apart*, November 1988–January 1989, pp. 4–5.
54. See Susan R. Schrepfer, "Sierra Club Executive Director: The Evolving Club and the Environmental Movement, 1961–1981," Michael McCloskey Oral History, Sierra Club History Series (Berkeley: University of California Regional Oral History Office, 1983), p. 79. The Philip Berry comment to the club board is in the minutes of the Sierra Club board of directors, February 14–15, 1970, and is cited in Michael P. Cohen, *The History of the Sierra Club*, p. 442.
55. The "real battle" quote is from John D. Hoffman, "The Club, The Cause and the Courts: Environmental Law in 1976," *Sierra Club Bulletin*, vol. 62, no. 2, February 1977, p. 47. See also Julie Cannon, "New Standing for the Environment," *Sierra Club Bulletin*, vol. 58, no. 9, October 1973, pp. 14–16+.
56. The "acceptable accommodations" quote is from a February 1977 article for the *Sierra Club Bulletin* (vol. 62, no. 2, p. 21), "Are Compromises Bad?" by Michael McCloskey. In the article, McCloskey sought to define the concept of

compromise as key to the Sierra Club's effectiveness and "the means by which legitimate interests in a democracy come to understand that they are being given fair consideration. Accommodations among contending interests build confidence that our institutions are listening. If we want these considerations for ourselves, we must accord them to others." On the Sierra Club nuclear position, see Laurence I. Moss, "The True Costs of Energy," *Sierra Club Bulletin*, vol. 58, no. 10, November-December 1973, p. 20, and "Environmental Groups Will Not Oppose," *Nuclear News*, vol. 14, no. 6, June 1971, pp. 41–42. On the National Coal Policy Project, see "Summary of the Mining Task Force Report," in National Policy Project, *Where We Agree: Report of the National Coal Policy Project, Summary and Synthesis*, July 1979, pp. 17–27, and "Strange Bedfellows," *Environmental Action*, vol. 8, no. 17, January 15, 1977, p. 13.

57. A May 1975 article by Robert A. Irwin in the *Sierra Club Bulletin* (vol. 60, no. 5, p. 17), "News of the Members and Their Club," described how the club's organizational structures, including regional groups, chapters, special sections, task forces, various committees, offices, services, executive committees, departments, publications, as well as the board of directors and the club council, played a role in determining club policy. The article also pointed out that "the splintering of chapters and proliferation of groups paradoxically has hastened an opposite process: consolidation." The next year the organization created the Sierra Club Committee on Political Education as "a central place for the development and implementation of general guidelines" with respect to club involvement in political action. These tendencies, however, most strengthened the role of the professional staff in the central organization rather than the organization's chief executive officer. See "Q & A on SCOPE," *Sierra Club Bulletin*, vol. 61 no. 8, September 1976, p. 14. See also Susan R. Schrepfer, "Sierra Club Executive Director," and the author's interviews with Michael Fischer (1992), William Turnage (1991), and Russell Peterson (1991).
58. See Donald L. Rheem, "Environmental Action: A Movement Comes of Age," *Christian Science Monitor*, January 13, 1987, pp. 17–19.
59. Fischer himself resigned for personal reasons in 1992 and was replaced by long-time staff member Carl Pope, signifying the momentary triumph of staff forces in the continuing battles among the various sources of power in the club.
60. For background on the Audubon transition, see Carl Buchheister and Frank Graham, Jr., "From the Swamps and Back: A Concise and Candid History of the Audubon Movement," *Audubon*, vol. 75, no. 1, January 1973, pp. 7–45. On the debate about the focus of the publication, see the "Reader's Turn Column" in *Audubon*, vol. 76, nos. 2 and 4, May 1974 and July 1974. One response in the July issue came from an irascible Edward Abbey. The novelist, then completing his classic work *The Monkey Wrench Gang*, wrote (with magazine deletions): "Unless us (ineligible) bird-lovers make more effort to save the environment, there won't be any (expletive deleted) birds left. Give us more,

not less, ecology articles. If I see any more (expletive deleted) bird pictures in *Audubon*, I am going to hang the (expletive deleted) magazine on the barn door and use it for target practice."

61. Peterson's background is described in the speech ("The Evolution of a Generalist") given by him at Williams College on April 23, 1986.
62. The "who considered the Society's mission" quote is from Frank Graham, Jr., with Carl W. Buchheister, *The Audubon Ark*, p. 257.
63. Author's 1991 interview with Louise Dunlap; also author's 1991 interview with William Turnage. On the clash with Watt, see Russell Peterson, "A Declaration of War," *Audubon*, vol. 83, no. 3, May 1981, p. 113.
64. At the outset of the Berle era, Audubon had more than 400,000 members (second only to the National Wildlife Federation among the mainstream groups), 265 staff, and a $26.5 million budget. See Donald L. Rheem, "Environmental Action: A Movement Comes of Age," pp. 16–17.
65. The "who wouldn't fit in socially" quote is from Frank Graham, Jr., with Carl W. Buchheister, *The Audubon Ark*, p. 305.
66. Berle's "diverse and far flung" quote is from *Audubon Activist*, September 1986, p. 2. Audubon's changing identity was often reflected by the changing location of the central offices of the organization. In the Berle era, the group relocated from a Fifth Avenue home to a more corporate-oriented midtown Manhattan high-rise. See Frank Graham, Jr., "Audubon is Back on Broadway," *Audubon*, vol. 93, no. 2, March 1991, pp. 138–42.
67. For background on the Brandborg period, see Stephen Fox, "We Want No Straddlers," *Wilderness*, Winter 1984, pp. 5–19.
68. The "each battle as if it was the last battle" quote is from the author's 1991 interview with Frances Beinecke, a former board chair of The Wilderness Society.
69. The "knock heads" quote is from the author's 1991 interview with William Turnage.
70. Author's 1991 interviews with William Turnage and Frances Beinecke. In 1980, under Turnage, The Wilderness Society received a three-year, $620,000 grant from the Richard King Mellon Foundation to create its National Resource Policy Analysis unit, which later evolved into the group's Resources Planning and Economics Department, which had the task of reviewing all 124 national forest plans prepared by the U.S. Forest Service; see William Turnage, "Mellon Grant Breaks New Ground," *The Living Wilderness*, vol. 44, no. 15, September 1980, p. 39, and author's 1992 interview with Louis Blumberg.
71. As Turnage was being replaced, the organization's search committee decided it needed an "analytic, rational manager more than a firebrand activist type," as the head of the search committee put it. Frampton seemed the most attractive candidate because of his management skills and his interest in the expertise functions of the organization (author's 1992 interview with Edward Ames). On

Frampton's approach, see George T. Frampton, Jr., "A Greeting to the Society's Members," *Wilderness*, vol. 49, no. 172, Spring 1986, pp. 8–9, and George T. Frampton, Jr., "Conservation Up Front," *Wilderness*, vol. 53, no. 189, Spring 1990, p. 2.

72. "The commercial exploiters of America who have wrought this sinful destruction are nationally organized," Dilg wrote in 1922 about the need for a new organization. "Therefore thoughtful and principled sportsmen know that they themselves must come into national organization. There is no other way if we are to save our God-given heritage from the land of the despoilers." For background on the IWL's early years, see the 60th anniversary issue of *Outdoor America*, the publication of the Izaak Walton League, vol. 47, no. 2, March-April 1982, especially Carol Dana, "An Invitation to Unite," p. 6–7+, and "A 60-Year Battle to Preserve the Outdoors," pp. 8–12. The "of destruction and not of preservation" quote is from the first issue of *Outdoor America*, August 1922, and is reprinted in the special 60th anniversary issue, p. 4.
73. Cited in Thomas B. Allen, *Guardian of the Wild: The Story of the National Wildlife Federation* (Bloomington: Indiana University Press, 1987), p. 31. See also David L. Lendt, *Ding: The Life of Jay Norwood Darling* (Ames, Iowa: Iowa State University Press, 1979), p. 85.
74. In the Grand Canyon fight, both the Utah and Arizona chapters of the NWF supported the Bureau of Reclamation's plans, primarily on the grounds that a new dam site, like the one constructed at Lake Powell, would offer significant recreational opportunities. Though the national leadership prevailed, this happened only after the issue was debated at a contentious national meeting. Support for the Wilderness Act was arranged after Kimball's predecessor, Ernest Swift, a friend of The Wilderness Society's Howard Zahniser, convinced Zahniser, who was personally critical of hunting impacts on wilderness, to add provisions to the legislation allowing hunting in wilderness areas. On a number of other occasions, however, hunting-oriented NWF positions, such as its support for clearcutting in national forests in order to promote certain game species, conflicted directly with the positions of other conservationist groups. See Thomas B. Allen, *Guardian of the Wild*, p. 147; also author's 1991 interview with Thomas.
75. The "extremists and kooks" comments are by Thomas Kimball and are cited in an article (Jeff Stansbury, "How *Not* to Hold an Energy Conference," *Environmental Action*, vol. 3, no. 15, December 11, 1971, p. 3) describing a joint energy-environment conference held by the NWF and the electric power industry in the fall of 1971. The NWF's slogan change occurred with the April-May 1970 Earth Day issue of *National Wildlife*.
76. When Arnett was president of the NWF in 1976, he had already publicly indicated his displeasure with the organization's emphasis on nonhunting issues and legal activities. See "Taking the Controversy Out of Conservation,"

Environmental Action, vol. 8, no. 5, July 3, 1976, p. 10. The "revert to its good-old-boy" quote is from the author's 1991 interview with Jay Hair. In a *Time* profile of Arnett (Frederic Golden, "A Sharpshooter at Interior," April 16, 1984, p. 43), Hair derisively spoke of Arnett's "*bwana*, great white hunter" image.

77. Hair later characterized the Ten's *An Environmental Agenda for the Future* as a "lowest common denominator piece of mush" relegated to a staff-level function. Its 1985 publication, he concluded, was a "clear signal that the Group of Ten process was intellectually bankrupt." Author's 1991 interview with Jay Hair.
78. The "fairly aggressive outreach" and "enlisting the entrepreneurial zeal" quotes are from Thomas B. Allen, *Guardian of the Wild*, pp. 8 and 109, respectively. By 1991, the Clean Sites board of directors continued to include major corporate representation (eight of twenty-one directors, including several chemical company officials) as well as three former EPA heads and two Group of Ten CEOs, Jay Hair of the NWF and Peter Berle of Audubon. See Clean Sites Inc., *Annual Report*, 1991; also Geraldine Cox, "Back to Basics," Chemical Manufacturers Association, *Journal of the Air Pollution Control Association*, vol. 37, no. 9, September 1987, pp. 1027+. Also author's 1991 interview with Jay Hair.
79. Hair's $250,000 salary was significantly higher than the salaries of most other mainstream group leaders. See Keith Schneider, "Pushed and Pulled, Environment Inc. Is on the Defensive," *New York Times*, March 29, 1992, section 4, p. 1.
80. The 1990 Clean Air Act set up a marketable permits system in which individual emitters of sulfur dioxide would be granted a designated number of annual "allowances," or permits, based on their historical emissions output, with the ability to trade such permits if the plants either exceeded or had emissions figures lower than the designated figure. The "33/50" program was designed by the EPA to reduce, initially by 33 percent and eventually by 50 percent, a set of seventeen toxic chemicals and was undertaken by industry on a voluntary basis. Critics of "33/50" complained that industry sought to include hazardous waste incineration as a form of reduction, directly contradicting the intent of the concept of toxics reduction. See "Title V—Permits," *Clean Air Act Amendments of 1990*, Public Law 101–546 104 Stat. pp. 2635–48. On the criticism of "33/50," see "White House Stalls Pollution Prevention Law," *Working Notes on Community Right-to-Know*, March-April 1992, p. 1.
81. The career-training concept is broadly parallel to the analysis of the evolution of the labor movement's leadership-training function from an adversarial to a labor-management relations approach. See Donald Snow, ed., *Inside the Environmental Movement: Meeting the Leadership Challenge* (Washington, D.C.: Island Press, 1992).

Chapter 5

1. The description of Penny Newman's activities in October 1989 is based on the author's week-long travels with Newman as described in "Penny Newman: Toxics Activist," a profile by the author that appeared in the *Los Angeles Weekly*, February 23–March 1, 1990, pp. 21–27.
2. For background on the McToxics Campaign, see Karen Stults, "Tarnish on the Golden Arches," *Business and Society Review*, Fall 1987, no. 63, pp. 18–21; Brian Lipsett and Karen Stults, "High School Students Take on Ronald McDonald," *Everyone's Backyard*, vol. 7, no. 2, Summer 1989, p. 2; James Ridgeway and Dan Bischoff, "Ronald McToxic: The Children's Crusade to Ban the Box," *The Village Voice*, vol. 35, no. 24, June 12, 1990; Amy Brison, "McCycling: Public Service or Private Interest?" *Waste Age*, January 1990, pp. 91–92; "McDonald's Pullout: CFC Issue Hits Home," *Modern Plastics*, October 1987, pp. 15–16; author's 1988 interviews with Karen Stults and Will Collette and 1991 interviews with Brian Lipsett and Lois Gibbs.
3. The "basically air" quote is from Karen Stults, "Tarnish on the Golden Arches." On the McDonald's/EDF agreement, see Richard Denison, "McDonald's Says No to Foam: Why and How the Environment Benefits," (Washington, D.C.: Environmental Defense Fund, December 1990); Alan Kovski, "McDonald's Decision on Polystyrene: Cooperation Instead of Confrontation," *Oil and Environment*, December 4, 1990; John Holusha, "Packaging and Public Image: McDonald's Fills a Big Order," *New York Times*, November 2, 1990, p. 1.
4. The McDonald's/EDF agreement produced substantial anger and resentment from several of the antitoxics activists most involved in the McToxics Campaign. Relations between the EDF and the Citizen's Clearinghouse for Hazardous Wastes (CCHW) had already been strained prior to the McDonald's situation, due to conflicting positions on key waste issues, such as incinerator ash, as well as differences in organizational style, temperament, and objectives. The EDF saw the agreement as another example of its "win-win" approach, while the CCHW saw the EDF as cashing in on the McToxics Campaign and taking credit for the outcome. Some groups, such as Greenpeace, sought to support both efforts: "CCHW loaded the bases, and then EDF drove in the run with a single," was the way Greenpeace Toxics Campaign staffer Scott Brown characterized it in a 1991 interview. See also "McDonalds Surrenders," *Everyone's Backyard*, vol. 8, no. 6, December 1990, pp. 2–3; David Moberg, "Pulling McToxic Weeds from the Hamburger Patch," *In These Times*, December 5–11, 1990, p. 7; author's 1991 interviews with Richard Denison, Jan Beyea, Lois Gibbs, and Brian Lipsett.
5. For background on the Stringfellow events, see G. J. Trezek, *Engineering Case Study of the Stringfellow Superfund Site* (Washington, D.C.: Office of Technology

Assessment, August 1984); and G. J. Trezek, *Cleanup Strategies for the Stringfellow Site* (Washington, D.C.: Office of Technology Assessment, April 1987); Tina May and Jim Carlton, "Stringfellow: A Town's Nightmare," *Santa Ana Register*, August 26, 1984; Dean Baker, Sander Greenland, James Mendlein, Patricia Harmon, "A Health Study of Two Communities Near the Stringfellow Waste Disposal Site," *Archives of Environmental Health*, vol. 43, no. 5, September-October 1988, pp. 325–34; Robert Gottlieb, *A Life of Its Own: The Politics and Power of Water* (San Diego: Harcourt Brace Jovanovich, 1988), pp. 235–40.

6. For background on the Lavelle Superfund scandals, see Jonathan Lash et al., *A Season of Spoils*, pp. 82–164.
7. On the EPA's approach to Superfund cleanup, see Office of Technology Assessment, *Coming Clean: Superfund's Problems Can Be Solved*, OTA-ITE 433 (Washington, D.C.: Government Printing Office, October 1989). See also Stephen U. Lester, "Reilly and Superfund: Cleaning Up or Selling Out?" *Everyone's Backyard*, vol. 8, no. 1, January-February 1990, p. 11. The "another way of saying" quote is from the author's 1989 interview with Penny Newman and is cited in the author's "Penny Newman: Toxics Activist" article for the *Los Angeles Weekly.*
8. The "plug up the toilet" phrase has become a favorite expression among CCHW organizers (author's 1992 interviews with Lois Gibbs and Penny Newman). On the attitudes of residents who have often been characterized as NIMBYites, see Michael Edelstein, *Contaminated Communities: Social and Psychological Impacts of Residential Toxic Exposure* (Boulder: Westview Press, 1988).
9. For background on the formation of CCHW, see Citizen's Clearinghouse for Hazardous Wastes, *History of the Grassroots Movement for Environmental Justice, Five Years of Progress, 1981–1986*, CCHW Fifth Anniversary Convention, Arlington, Virginia, May 31–June 1, 1986; also author's 1990 interviews with Will Collette and Lois Gibbs.
10. On the criticism of Waste Management, see Charlie Cray, *Waste Management Inc.: An Encyclopedia of Environmental Crimes and other Misdeeds* (Washington, D.C.: Greenpeace, December 1991). On the Waste Management/NWF ties, see the February 4, 1988, letter from Jay Hair to Theresa Freeman reprinted in the *CCHW Action Bulletin*, no. 18, May 1988, p. 3; see also "Reilly-Gate?" *CCHW Action Bulletin*, no. 24, November 1989, p. 1.
11. Newman's speech, "We Are the Power," is reproduced in *Everyone's Backyard*, vol. 7, no. 4, Winter 1989, p. 3.
12. For background on Alinsky's organizing strategies, see Saul D. Alinsky, *Reveille for Radicals* (New York: Vintage, 1969); Saul D. Alinsky, *Rules for Radicals: A Practical Primer for Realistic Radicals* (New York: Vintage, 1972); Robert Bailey, Jr., *Radicals in Urban Politics: The Alinsky Approach* (Chicago: University of Chicago Press, 1979).

13. The Nader Task Force Reports referred to in the text include Joseph A. Page and Mary Win O'Brien, *Bitter Wages* (New York: Grossman, 1973); James Phelan and Robert Pozen, *The Company State* (New York: Grossman, 1973); William C. Osborn, *The Paper Plantation: Ralph Nader's Study Group on the Pulp and Paper Industry in Maine* (New York: Grossman, 1974); and John C. Esposito (project director), *Vanishing Air: Ralph Nader's Study Group Report on Air Pollution* (New York: Grossman, 1970). In terms of his organizational network, Nader used the $425,000 from his General Motors settlement to establish the first PIRG based in Washington, D.C. According to PIRG biographer Kelley Griffin, the award was thirty times greater than any previous invasion-of-privacy court settlement. See Kelley Griffin, *More Action for a Change* (New York: Dembner Books, 1987), p. 4. See also Thomas Whiteside, *The Investigation of Ralph Nader: General Motors vs. One Determined Man* (New York: Pocket Books, 1972).
14. The student PIRGs eventually tied into a network of statewide organizations as well as the Washington, D.C.–based U.S. PIRG, which had been established prior to the Oregon speech. The fee system that became a primary source of funding for the state PIRGs was based on a campus-wide student fee that students vote on to assess themselves and that all students pay unless a student requests otherwise. Through the 1980s and early 1990s, as the PIRGs became increasingly effective and often sharply opposed by university officials and board members, efforts were made to eliminate this type of fee system, most notably within the University of California system. See Pamela Klein, "CAL-PIRG Purge," *L.A. Weekly*, October 19–25, 1990, p. 16, and letter from David Pierpont Gardner to Stephanie Pincetl, October 12, 1990. For background on the development of the PIRG model, see Ralph Nader and Donald Ross, *Action for a Change: A Student's Manual for Public Interest Organizing* (New York: Grossman, 1971). See also Kelley Griffin, *More Action for a Change*, and Steve Williams, "Nader Seeks Activists in Public's Interest," *Environmental Action*, vol. 2, no. 20, March 6, 1971, p. 14.
15. *Scientist and Citizen* subsequently changed its name to *Environment* and remained an important science- and policy-oriented journal dealing with environmental issues.
16. The "apparently insuperable barrier" quote is from Barry Commoner, *Science and Survival* (New York: Viking, 1967), p. 108.
17. The *Time* cover story, "Ecologist Barry Commoner: The Emerging Science of Survival," ran on February 2, 1970. The "enveloping cloud of science and technology" quote is from Barry Commoner, *Science and Survival*, p. 108. The "deeper public understanding" quote is from Barry Commoner, *The Closing Circle* (New York: Knopf, 1972), pp. 10–11.
18. The ecological laws discussion is found in Barry Commoner, *The Closing Circle*, pp. 32–47, as is the "rising miasma of pollution" (p. 47). The "new coherency" quote is from a review of *The Closing Circle* by Environmental Action

staff member Sam Love in *Environmental Action*, vol. 3, no. 13, November 13, 1971, p. 14. For a further discussion of the incompatibility of postwar technologies with the natural environment, see Barry Commoner, "The Ecological Facts of Life," in *The Ecological Conscience*, edited by Robert Disch, pp. 2–16, and Barry Commoner, "Soil and Fresh Water: Damaged Global Fabric," *Environment*, vol. 12, no. 4, April 1970, pp. 4–11.

19. Barry Commoner, *The Poverty of Power: Energy and the Economic Crisis* (New York: Knopf, 1976).
20. See the review of *The Poverty of Power* in the *Sierra Club Bulletin* (vol. 62, no. 3, March 1977, pp. 31–33) entitled "Is Socialism the Answer?" written by the chair of the Sierra Club's Economics Committee, Richard Tybout.
21. See Barry Commoner et al., *Development and Pilot Test of an Intensive Municipal Solid Waste Recycling System for the Town of East Hampton* (Center for the Biology of Natural Systems, submitted to the New York State Energy Research and Development Authority, 1988). See also Barry Commoner, *Making Peace with the Planet* (New York: Knopf, 1990), pp. 103–40.
22. Montague's Ph.D. thesis at the University of New Mexico, "Ralph Nader and Barry Commoner and the Public Interest Movement," identified Nader and Commoner as primary influences on the new forms of environmental politics. Montague and his wife, Katherine Montague, subsequently discussed toxics issues in their books *Mercury* (San Francisco: A Sierra Club Battlebook, 1971) and *No World Without End: The New Threats to our Biosphere* (New York: Putnam, 1977) in ways that anticipated some of Commoner's later writings.
23. Montague changed the name of his publication to *Rachel's Hazardous Waste News* in 1986 due to a copyright issue. "Rachel" referred to an on-line database system Montague had established to aid grassroots groups. Author's 1990 interview with Peter Montague.
24. See Louis Blumberg and Robert Gottlieb, "Wising Up to the Experts," *NAPEC Quarterly*, vol. 3, no. 1, March 1992, p. 7; also author's 1990 interviews with Paul and Ellen Connett.
25. For background on the local antinuclear movements of the 1960s, see Steven Ebbin and Raphael Kasper, *Citizen Groups and the Nuclear Power Controversy: Uses of Scientific and Technological Information* (Cambridge, Mass.: MIT Press, 1974), pp. 9–14; Richard S. Lewis, *The Nuclear Power Rebellion: Citizens vs. the Atomic Industrial Establishment* (New York: Viking, 1972); Cathy Wolf, "The Roots of the Anti-Nuclear Movement," in *Accidents Will Happen: The Case Against Nuclear Power*, edited by Lee Stephens and George R. Zachar (New York: Harper & Row, 1979), pp. 289–96; Jerome Price, *The Antinuclear Movement* (Boston: Twayne Publishers, 1982).
26. On the thermal pollution issue, see, for example, the January 20, 1969, feature story in *Sports Illustrated* by Robert Boyle, "The Nukes Are in Hot Water," and the March 1969 rejoinder to that article, "Are the Nukes in Hot Water?" in the

nuclear industry's trade publication, *Nuclear News*, vol. 12, no. 3, March 1969, pp. 12–15. For background on Gofman and Tamplin and the AEC, see "Standards: Who Should Set Them . . . And How High (or Low?)" *Nuclear News*, vol. 13, no. 4, April 1970, pp. 20–21, and John W. Gofman and Arthur Tamplin, *Poisoned Power: The Case Against Nuclear Power Plants* (Emmaus, Pa.: Rodale Press, 1971). For background on the AEC's troubles, see Robert Gillette's four-part series in *Science*, vol. 177, nos. 4051–54, September 1, 8, 15, 22, 1972, pp. 771–76, 867–71, 970–75, 1080–82.

27. See Dorothy Nelkin, *Nuclear Power and Its Critics: The Cayuga Lake Controversy* (Ithaca, N.Y.: Cornell University Press, 1971).
28. The Sierra Club approach was challenged by some within the organization who worried that the position had been pushed through by antinuclear militants without serious debate, and who believed that those key club members who were rational scientists would eventually be able to achieve a reconciliation of differences with supporters of nuclear energy. Though the moratorium approach remained policy and some club chapters became involved in antinuclear activities, the central organization's leadership still sought during the 1970s to temper the organization's opposition by maintaining a dialogue with the nuclear industry and various regulatory agencies. For the Sierra Club position see *Sierra Club Bulletin*, vol. 59, no. 2, February 1974, p. 15; Sid Moglewer, "Nuclear Safety: The Margin of Ultimate Error," *Sierra Club Bulletin*, vol. 59, no. 6, June 1974, pp. 25–26; "Sierra Club to Study Nuclear Problems," *Nuclear News*, vol. 17, no. 12, September 1974, pp. 45–46.
29. The name "Clamshell Alliance" was adopted in reference to the potential impact of thermal pollution on the clam population along the coast. For background on the Clamshell Alliance, see Gail Robinson, "Clam's Consensus," *Environmental Action*, vol. 10, no. 6, July 15, 1978, pp. 9–11.
30. Bell's remarks are cited in Barbara Epstein, *Political Protest and Cultural Revolution: Nonviolent Direct Action in the 1970s and 1980s* (Berkeley: University of California Press, 1991), p. 68.
31. Barbara Epstein, *Political Protest and Cultural Revolution*, p. 74.
32. Murray Bookchin has linked the affinity group concept of a friendship-based direct-action group to the Spanish anarchist movement of the 1930s. Affinity groups became quite popular during the late 1960s among both the New Left and the counterculture. The organization Up Against the Wall Motherfucker, for example, was a classic affinity group of the period. See Murray Bookchin, *Post-Scarcity Anarchism* (Berkeley: Ramparts Press, 1971).
33. The description of the 1979 San Luis Obispo demonstration against the Diablo Canyon plant is based on the author's own notes from his participation at the event. See also Barbara Epstein's *Political Protest and Cultural Revolution* (pp. 92–124) for a description of the development of the Abalone Alliance.
34. The "occasion, rather than the impetus" quote is from Barbara Epstein,

Political Protest and Cultural Revolution, pp. 92–124. The author was also present at the San Luis Obispo demonstration.

35. On the utility rate controversies, see "A Dark Future for Utilities," *Business Week*, no. 2587, May 28, 1979, pp. 108–24, and Wayne H. Sugai, *Nuclear Power and Ratepayer Protest* (Boulder: Westview Press, 1987).
36. For background on the RFF survey, see Robert Mitchell, *Public Opinion on Environmental Issues: Results of a National Public Opinion Survey* (Washington, D.C.: U.S. Council on Environmental Quality, 1980), cited in Michael K. Heiman, "Hazardous Waste Facility Siting: From Not in My Backyard to Not in Anybody's Backyard," paper presented at the 30th Annual Conference of the Association of Collegiate Schools of Planning, October 27–30, 1988, Buffalo, New York.
37. For background on 1970s hazardous waste issues, see Samuel Epstein, Lester O. Brown, and Carl Pope, *Hazardous Waste in America* (San Francisco: Sierra Club Books, 1982).
38. Neil Seldman, in *The United States Recycling Movement, 1968–1986: A Review* (Washington, D.C.: Institute for Local Self-Reliance, October 1986), estimates that as many as 3000 new recycling centers were established in the Earth Day period.
39. For background on the passage of RCRA, see William L. Kovacs and John F. Klucsik, "The New Federal Role in Solid Waste Management: The Resource Conservation and Recovery Act," *Columbia Journal of Environmental Law*, vol. 3, p. 205, March 1977. On the passage of TSCA, see Luther Carter, "Toxic Substances: Five-Year Struggle for Landmark Bill May Soon Be Over," *Science*, vol. 194, no. 4260, October 1, 1976, pp. 40–42, and Epstein, Brown, and Pope, *Hazardous Waste in America*, pp. 179–256.
40. For background on the Love Canal events, see Lois Marie Gibbs, as told to Murray Levine, *Love Canal: My Story* (Albany: State University of New York Press, 1982); Adeline Gordon Levine, *Love Canal: Science, Politics, and People* (Lexington, Mass.: Lexington Books, 1982); author's 1992 interview with Lois Gibbs.
41. For a critique of the Superfund process, see Office of Technology Assessment, *Are We Cleaning Up? 10 Superfund Case Studies* (a special report of OTA's assessment on Superfund implementation), OTA-ITE-362 (Washington, D.C.: Government Printing Office, 1988).
42. The "more responsible" environmentalists concept is spelled out in a report developed for the California Waste Management Board: Cerrell Associates, J. Stephen Powell, senior associate, "Political Difficulties Facing Waste-to-Energy Conversion Plant Siting," Waste-to-Energy Technical Information Series (Sacramento: California Waste Management Board, 1984). This report, widely cited within the antitoxics movement, is best known for its comments about how to successfully site incinerator plants by selecting low-income or

rural communities whose residents had lower educational levels and would presumably offer less resistance. The report, however, also discusses the importance of cultivating the "more responsible" environmental groups, which would be more willing to accept these high-tech plants as part of an overall integrated waste management strategy if recycling were included as one of its elements.

43. Thomas Kean's remarks ("environmental flying squads" and "explain the truth") are from his keynote address to the annual meeting of the Air Pollution Control Association in June 1987, reprinted as Thomas Kean, "Dealing with NIMBY," *Journal of the Air Pollution Control Association*, vol. 37, no. 9, September 1987, pp. 1025–26. By the early 1990s most consultants and "risk communicators" had concluded that "you're never going to get support in local site-specific areas" for incinerator facilities, as Ed Reilly, a top waste-to-energy industry consultant, put it. Thus, the task for consultants was to separate local residents from the rest of the larger community through stage-managed campaigns involving "luncheon circuit, guest columnists, visits with editorial boards, university lectures." "In other words," Reilly asserted, "you control the dialogue, instead of letting the activists or the community do so." Cited in Barry Siegel, "Spin Doctors to the World," *Los Angeles Times Magazine*, November 24, 1991, pp. 23–24. See also Lillie Craig Trimble, "What Do Citizens Want in Siting of Waste Management Facilities?" *Risk Analysis*, vol. 8, no. 3 (1988), pp. 375–78; Clinton C. Kemp, "The Tussle for Public Support," *Solid Waste and Power*, February 1987; Terry A. Trumbell, "Using Citizens to Site Solid Waste Facilities," *Public Works*, August 1988; Carolyn Konheim et al., "Get Citizens Involved in Siting—And Do It Early," *Waste Age*, March 1988, pp. 37–42.

44. Information on the Superdrive for Superfund campaign is from the author's 1991 interviews with Michael Picker and Ted Smith. See also Peter Montague, "What We Must Do—A Grass-roots Offensive Against Toxics in the '90s," *The Workbook*, vol. 14, no. 3, July-September 1989, pp. 90–113.

45. See David Allen, "Preventing Pollution," in *Fighting Toxics: A Manual for Protecting Your Family, Community, and Workplace*, edited by Gary Cohen and John O'Connor (Washington, D.C.: Island Press, 1990).

46. The "crazy stunts" quote is from Robert Hunter, *Warriers of the Rainbow: A Chronicle of the Greenpeace Movement* (New York: Holt, Rinehart, and Winston, 1979), p. 252. "The oldest urge of all" quote is from David McTaggart with Robert Hunter, *Greenpeace III: Journey Into the Bomb* (London: Collins, 1978), p. 79.

47. The "bogged down" quote is from Paul Watson as told to Warren Rogers, *Sea Shepherd: My Fight for Whales and Seals* (New York: W. W. Norton, 1982), p. 156. From the outset, the Sea Shepherd Society maintained an exclusive focus on the "pure act." Although the organization espoused nonviolence, it showed an

affinity for acts of sabotage, or "eco-tage," which turned the organization into a sea-based equivalent of both Earth First!, the most prominent of the eco-tage sympathizers during the 1980s, and the animal rights movement, one of whose financial supporters, writer Cleveland Amory, also helped finance the organization. See Dick Russell, "The Monkeywrenchers," *The Amicus Journal*, vol. 9, no. 4, Fall 1987, pp. 28–42; Rik Scarce, *Eco-Warriors: Understanding the Radical Environmental Movement* (Chicago: Noble Press, 1990), pp. 97–113; Lynn Franklin, "View from the Sea Shepherd," *Oceans*, November 1983, pp. 64–66.

48. For background on the *Rainbow Warrior* episode, see Michael King, *Death of the Rainbow Warrior* (Auckland, N.Z.: Penguin, 1986); Isabelle Gidley and Richard Shears, *The Rainbow Warrior Affair* (Sydney: Unwin, 1985); Michael Brown and John May, *The Greenpeace Story* (London: Dorling Kindersley, 1989).
49. The "essentially soft" quote is from the author's 1991 interview with Richard Grossman. "Daredevils for the Environment" is the title of an October 2, 1988, *New York Times Magazine* profile of Greenpeace by Michael Harwood.
50. For background on deep ecology, see Bill Devall and George Sessions, *Deep Ecology: Living As If Nature Mattered* (Salt Lake City: Peregrine Smith Books, 1985); Arne Naess, *Ecology, Community and Lifestyle: Outline of an Ecosophy* (New York: Cambridge University Press, 1989); Kirkpatrick Sale, "Deep Ecology and its Critics," *The Nation*, May 14, 1988, pp. 670–75.
51. On bioregionalism, see Kirkpatrick Sale, "Bioregionalism—A Sense of Place," *The Nation*, October 12, 1985, pp. 336–39, and Peter Berg, "Coming of Age in Bio-topia," *Not Man Apart*, vol. 16, no. 5, September-October 1986, p. 3.
52. The "harder-nosed approach" quote is from Jamie Malanowski, "Monkey-Wrenching Around," *The Nation*, May 2, 1987, p. 568. For further background on Earth First!, see Dave Foreman, *Confessions of an Eco-Warrior* (New York: Harmony Books, 1991), pp. 13–23; Dick Russell, "The Monkeywrenchers," *The Amicus Journal*, vol. 9, no. 4, Fall 1987, pp. 28–42; Jim Robbins, "Hurling Sand into Society's Gears," *High Country News*, April 25, 1988, pp. 14–16; Kenneth Brower, "Earth First! People Second," *Los Angeles Reader*, December 16, 1988, pp. 1, 8–10. On the RARE II process, see Janet Marinelli, "Missing a RARE Opportunity," *Environmental Action*, vol. 10, nos. 18–19, February 1979, pp. 4–9.
53. The "practice against the destruction of the wild" quote is from Dave Foreman and Bill Haywood, editors, *Ecodefense: A Field Guide to Monkeywrenching* (Tucson: Ned Ludd Book, 1987), p. 3. Foreman's remarks about his organization are from a 1987 speech of his cited in Christopher Manes, *Green Rage: Radical Environmentalism and the Unmaking of Civilization* (Boston: Little, Brown, 1990), p. 72. See also Dave Foreman, "The New Conservation Movement," *Wild Earth*, Summer 1991, pp. 6–9.

54. The "just don't get caught" and "all decisions, even ahead of human welfare" quotes are taken from a description of the 1987 Round River Rendezvous in Don Morris, "Earth First! No Wimps," *Earth First!* vol. 7, no. 7, August 1, 1987, p. 19.
55. For background on the conflict between Bookchin and Foreman, see *Defending the Earth: A Dialogue Between Murray Bookchin and Dave Foreman*, edited by Steve Chase (Boston: South End Press, 1991), which also includes the discussion about AIDS and Third World starvation; The "eco-wars" against Brazil idea is spelled out in Christopher Manes, *Green Rage*, p. 34. The concept that nuclear war could be defined as an anti-industrial cleansing action was put forth at the International Green Movements and the Prospects for a New Environmental/Industrial Politics in the U.S. conference at the University of California at Los Angeles, April 17–19, 1986 (author's notes). For background on the conspiracy charges against Foreman, see Michael A. Lerner, "The FBI vs. The Monkeywrenchers," *Los Angeles Times Magazine*, April 15, 1990, pp. 10–21.
56. The "into a leftist group" quote is from the "Dear Friends" letter from Dave Foreman and Nancy Morton announcing their withdrawal from Earth First! The split in the organization also reflected differences in style and political temperament as demonstrated by one pro-Foreman letter writer, who exclaimed in a letter to *Earth First!* that the organization's majority faction had become a "bunch of whiney, feminist, geeko computer programmers all trying to turn EF! into a dress wearing, letter writing, debating society of wimps [which would drive] EF! into the Greenpeace sellout corner." See the "Dear Shit for Brains" letter in *Earth First!*, September 23, 1991.
57. On the German and European Green movements, see Werner Hulsberg, *The German Greens: A Social and Political Profile* (London: Verso, 1988), and S. E. Clarke and Margit Mayer, "Responding to Grassroots Discontent: Germany and the U.S.," *International Journal of Urban and Regional Research*, vol. 10, no. 3 (1986), pp. 401–17.
58. See Charlene Spretnak and Fritjof Capra, *Green Politics* (New York: Dutton, 1984).
59. See, for example, the description of the first conference of the Greens, held in Amherst, Massachusetts, in June 1987, in Mark Satin, "The Green Conference," *Not Man Apart*, vol. 17, no. 5, September-October 1987, p. 3.
60. Carl Anthony's comments are part of a special section on the Greens in *Greenpeace*, July-August 1991, pp. 19–20.
61. Robert D. McFadden, "Millions Join Battle for a Beloved Planet," *New York Times*, April 23, 1990, p. 1.
62. Information on the incinerator ash fight is from the author's 1991 interviews with Lois Gibbs and Brian Lipsett. On the conflict concerning the approach toward pesticide legislation, see Victoria Elenes, "Farmworker Pesticide

Exposures: Interplay of Science and Politics in the History of Regulations" (Master's thesis, University of Wisconsin, 1991), pp. 53–55, and Christopher J. Bosso, *Pesticides and Politics* (Pittsburgh: University of Pittsburgh Press, 1987).

63. Information on the Indiana Earth Day events is from the author's 1991 interviews with Sue Greer and Craig Grabow.
64. Ruth Caplan, the executive director of Environmental Action, linked the Earth Day 1990 focus on the individual with George Bush's "1000 points of light" theme regarding voluntarism and criticized the omission of certain key issues, most notably nuclear issues, as indicative of the narrow design of the event (author's 1991 interview with Ruth Caplan).
65. For a critique of the 1990 Earth Day events from an alternative environmental perspective, see Gary Cohen, "It's Too Easy Being Green: The Gains and Losses of Earth Day," *New Solutions*, vol. 1, no. 2, Summer 1990, pp. 9–12.

CHAPTER 6

1. For background on the conference, see Robin Lee Zeff, Marsha Love, and Karen Stults, eds., *Empowering Ourselves: Women and Toxics Organizing* (Arlington, Va.: Citizen's Clearinghouse for Hazardous Wastes, 1987); also author's 1992 interviews with Lois Gibbs and Penny Newman.
2. Author's 1992 interview with Cora Tucker. See also William Gibson, "Cora Tucker: Organizing for a Better America," *The Egg: A Journal of Eco-Justice*, vol. 8, no. 2, Summer 1988, p. 13, and Anne Witte Garland, " 'Good Noise': Cora Tucker," in *Women Activists: Challenging the Abuse of Power* (New York: The Feminist Press, 1988), pp. 120–31.
3. The "we know more about our community" quote is from Cora Tucker's keynote address at the Women in Toxics Organizing conference and is reproduced in Zeff, Love, and Stults, eds., *Empowering Ourselves*, p. 5.
4. The issue of how a threat to property becomes the equivalent of a threat to home has also been a central concern of anti–nuclear power protesters. Robert Jay Lifton, in a psychological study of the impact of the Three Mile Island events on local residents, describes how "families everywhere are anchored by their home, which is both a physical structure and a profound symbol of family integrity and viability. Therefore, a major source of trauma at TMI was its threat to family homes in terms of both those meanings." Cited in Phil Brown and Edwin J. Mikkelson, *No Safe Place: Toxic Waste, Leukemia, and Community Action* (Berkeley: University of California Press, 1990), pp. 113–14. Public health analyst Barbara Berney has described how "property rights and the value of property have been central to almost all political fights in America. But when rich people and corporations do it, nobody calls that NIMBY!" (letter from Barbara Berney to the author, July 10, 1992).
5. Author's 1989 interview with Penny Newman. See also Nicholas Freudenberg

and Ellen Zaltzberg, "From Grassroots Activism to Political Power: Women Organizing Against Environmental Hazards," in *Double Exposure: Women's Health Hazards on the Job and at Home*, edited by Wendy Chavkin (New York: Monthly Review Press, 1984), pp. 246–72.

6. See Lois Gibbs, *Women and Burnout* (Arlington, Va.: Citizen's Clearinghouse for Hazardous Wastes, n.d.), and Robbin Lee Zeff, "Not in My Backyard/Not in Anyone's Backyard: A Folkloristic Examination of the American Grassroots Movement for Environmental Justice" (Ph.D. diss., Indiana University, September 1989), pp. 171–206.
7. Background on the Kenosha events is from the author's 1990 interviews with Nancy Sibert, chair of Kenoshans Against Medical Waste Incinerators, and Sally Teets.
8. Annette Kolodny, *The Land Before Her: Fantasy and Experience of the American Frontiers, 1630–1860* (Chapel Hill: University of North Carolina Press, 1984), p. xiii.
9. Of the Sierra Club's 182 charter members, only 5 were women. See Susan R. Schrepfer, *The Fight to Save the Redwoods: A History of Environmental Reform, 1917–1978* (Madison: University of Wisconsin Press, 1983), p. 10.
10. The "hunting big game in the wilderness" quote is from *American Big-Game Hunting: The Book of the Boone and Crockett Club*, edited by Theodore Roosevelt and George Bird Grinnell (New York: Forest and Stream Publishing Company, 1893), p. 14. The constitution of the club, including Article II describing the "Objects of the Club" is reproduced in this same book, pp. 337–39. See also, Theodore Roosevelt, *The Wilderness Hunter: An Account of the Big Game of the United States and Its Chase with Horse, Hound, and Rifle* (New York: Putnam, 1893); Roderick Nash, *Wilderness and the American Mind*, rev. ed. (New Haven: Yale University Press, 1974), pp. 149–51; Paul Russell Cutright, *Theodore Roosevelt: The Making of a Conservationist* (Urbana: University of Illinois Press, 1985).
11. The "wild rugged life of the outdoors" and "meant to live" quotes are from Joseph Knowles, *Alone in the Wilderness* (London: Longmans, Green, 1914), pp. 4 and 286, respectively. The "resolution, manliness" quote is from *American Big-Game Hunting*, p. 14. Roderick Nash's *Wilderness and the American Mind* analyzes the Joseph Knowles story from the perspective of how the nostalgia about frontier loss came to be transformed into a wilderness cult (pp. 141–43). See also "Tackling Nature in the Raw," *The Literary Digest*, vol. 47, no. 10, September 6, 1913, pp. 394–97; "Two Months a Caveman," *The Literary Digest*, vol. 47, no. 16, October 18, 1913, pp. 722–24; "Alone in the Wilderness: A Review," *The Nation*, vol. 98, no. 2538, February 19, 1914, p. 188.
12. The wilderness colony idea is discussed in Joseph Knowles, *Alone in the Wilderness*, pp. 286–90.
13. The "environmental surrogate" phrase is used by Jeffrey P. Hantover in "The

Boy Scouts and the Validation of Masculinity," in *The American Man*, edited by Elizabeth H. Pleck and Joseph L. Pleck (Englewood Cliffs, N.J.: Prentice-Hall, 1980), pp. 285–301; the Daniel Beard quote is from the same essay, p. 293. The "Scout movement is most potent" quote is from Walter Prichard Eaton, "Boy Scouts: Training the Men of Tomorrow in Courage, Health, and Efficiency," *Colliers National Weekly*, vol. 49, no. 20, August 3, 1912, p. 14. See also Frank B. Arthurs, "Boy Scouts Building for Manhood," *The Outing Magazine*, vol. 57, no. 3, December 1910, pp. 276–84.

14. The "girls of this generation" quote is from Hartley Davis, "The Camp-Fire Girls," *The Outlook*, vol. 101, May 25, 1912, p. 189. The "service in the home" quote is from Mrs. Theodore H. Price, "Girl Scouts," *The Outlook*, vol. 118, March 6, 1918, pp. 366–67. See also Winthrop D. Lane, "The Camp-Fire Girls: A Readjustment of Women," *The Survey*, vol. 28, no. 7, May 18, 1912, pp. 320–22.
15. Zane Grey's comments are from his essay "Vanishing America" in the *Izaak Walton League Monthly*, September 1922, and is cited in Donald N. Baldwin, *The Quiet Revolution: Grass Roots of Today's Wilderness Preservation Movement* (Boulder: Pruett Publishing, 1972), p. 149. The "masculinity crisis" concept is explored in Joe L. Dubbert, "Progressivism and the Masculinity Crisis," in *The American Man*, edited by Pleck and Pleck, pp. 303–20.
16. See Theodore Roosevelt, "The American Woman: A Mother," *Ladies Home Journal*, vol. 22, July 1905, pp. 3–4.
17. The "garbage-pail and overfurnished rooms" and "governed by the best knowledge" quotes are from Ellen H. Richards, *The Cost of Living, as Modified by Sanitary Science* (New York: John Wiley & Sons, 1915), pp. 26 and 28. The "become imbued with the scientific spirit" quote is cited in Caroline L. Hunt, *The Life of Ellen H. Richards* (Washington, D.C.: American Home Economics Association, 1958), p. 160.
18. See Ellen H. Richards and Alpheus G. Woodman, *Air, Water, and Food: From a Sanitary Standpoint* (New York: John Wiley & Sons, 1901); Ellen H. Richards, *Euthenics: The Science of Controllable Environment* (Boston: Whitcomb & Barrows, 1904); Caroline L. Hunt, *The Life of Ellen H. Richards*, pp. 93–106.
19. See Carolyn Merchant, "Earthcare: Women and the Environmental Movement," *Environment*, vol. 23, no. 5, June 1981, p. 7; The "artificial environment" quote is from Ellen H. Richards, *Sanitation in Daily Life* (Boston: Whitcomb & Barrows, 1907), p. v.
20. See Robert Clarke, *Ellen Swallow: The Woman Who Founded Ecology* (Chicago: Follett Publishing Company, 1973).
21. Carolyn Merchant, in an essay on women in the conservation movement, points out that conservation organizations, including the women's clubs, spoke of a "conservation of womanhood, the home, and the child," and that a woman's role in the conservation movement, as Mrs. Carl Vrooman of the

Daughters of the American Revolution put it, was to establish "an atmosphere that makes ideas sprout and grow and ideals expand and develop and take deeper root in the subsoil of the masculine mind." Merchant also quotes a leader of the Women's National Rivers and Harbors Congress who argued that women were the "conservers of the race." See Carolyn Merchant, "Women of the Progressive Conservation Movement: 1900–1916," *Environmental Review*, vol. 8, no. 1, Spring 1984, pp. 73–74.

22. The review of the four organizations was undertaken by Carol Kuester, who found that, up through the 1980s, the groups had few if any women in their leadership. One minor exception was Audubon during the World War II period, when women constituted about 25 percent of the directors and officers of the organization. However, the figures for female participation in Audubon actually declined slightly during the mid- to late 1980s (five women among twenty-four council members in 1986, for example), while increasing slightly for the other three organizations (six women among eighteen directors and three women among sixteen vice-presidents for the Sierra Club in 1989 compared to one woman among twelve directors and no women among five vice-presidents in 1950; five women among twenty-five council members and no women among the five officers of The Wilderness Society in 1986 compared to three women among thirty-three council members and no women among the eighteen officers in the organization's first three decades; and five women on the twenty-six-member board of The National Wildlife Federation in 1991 compared to no women for its twenty-five member board as late as 1970).
23. The 1908 article by Alice Hamilton is "Industrial Diseases: With Special Reference to the Trades in Which Women Are Employed," *Charities and the Commons*, vol. 20, September 5, 1908, pp. 655–59. For background on the employment of women in the early twentieth century, see "What Are the Facts About Women in Industry?" *Charities and the Commons*, vol. 16, no. 1, April 7, 1906, pp. 87–88.
24. On the Triangle Shirtwaist fire and other organizing issues, see Rose Schneiderman and Lucy Goldthwaite, *All for One* (New York: Paul S. Erickson, 1967).
25. The "great many new" quote is from Alice Hamilton, *Women Workers and Industrial Poisons*, *Bulletin of the Women's Bureau*, No. 57 (Washington, D.C.: U.S. Department of Labor, 1926), p. 1. The "we must remember" quote is from Hamilton's 1908 "Industrial Diseases" article, pp. 658–59. See also Patricia Vauter Klein, " 'For the Good of the Race': Reproductive Hazards from Lead and the Persistence of Exclusionary Policies Toward Women," in *Women, Work, and Technology*, edited by Barbara Drygulski Wright (Ann Arbor: University of Michigan Press, 1987), pp. 101–17.
26. See Muller v. Oregon, 208 U.S. 412 (1908). Hamilton also argued that certain industries' exclusionary policies masked their unwillingness to improve working conditions—for example, the printing trades, in which the danger to health

was "avoidable." Instead of excluding women, the logical action would be to "institute such sanitary measures in printing shops as will make them safe for both sexes." Cited in Alice Hamilton, *Women in the Lead Industries*, Bureau of Labor Statistics Bulletin No. 253 (Washington, D.C.: U.S. Department of Labor, February 1919), p. 25.

27. The "practical effect" quote is from Wilma Ruth Slaight, *Alice Hamilton: First Lady of Industrial Medicine*, p. 177. The "restrictions on the conditions of labor" quote is cited in Jeanne Mager Stellman, *Women's Work, Women's Health: Myths and Realities* (New York: Pantheon Books, 1977), p. 36.
28. See Anna M. Baetjer, *Women in Industry: Their Health and Efficiency* (Philadelphia: W. B. Saunders, 1946), p. 255.
29. "A Maternity Policy for Industry," issued by the U.S. Department of Labor in 1944, is reproduced in Anna M. Baetjer, *Women in Industry*, pp. 191–95. See also Children's Bureau and the Women's Bureau of the U.S. Department of Labor, "Standards for Maternity Care and Employment of Mothers in Industry," *Journal of the American Medical Association*, vol. 120, no. 1, September 5, 1942, pp. 55–56.
30. See Vilma Hunt, *Occupational Health Problems of Pregnant Women*, SA-5305–75 (Washington, D.C.: U.S. Department of Health, Education, and Welfare, 1975), and Andrea Hricko, with Melanie Brunt, *Working for Your Life: A Woman's Guide to Job Health Hazards* (Berkeley: Labor Occupational Health Program/ Public Citizens' Health Research Group, 1976). See also Vilma Hunt, *The Health of Women at Work* (Evanston, Ill.: Program on Women, 1977).
31. Bingham's comments are from her preface to *Double Exposure*, edited by Wendy Chavkin, pp. ix-xi.
32. The American Cyanamid episode is discussed in Joan Bertin, "Reproductive Hazards in the Workplace," in *Reproductive Laws for the 1990s*, edited by Sherrill Cohen and Nadine Taub (Clifton, N.J.: Humana Press, 1989), pp. 277–305; Wendy B. Williams, "Firing the Woman to Protect the Fetus: The Reconciliation of Fetal Protection with Employment Opportunity Goals under Title VII," *Georgetown Law Journal*, vol. 69, 1981, pp. 641–704; Joan E. Bertin, "People Protection, Not 'Fetal Protection,' " *New Solutions*, Summer 1991, p. 5; William J. Curran, "Dangers for Pregnant Women in the Work Place," *New England Journal of Medicine*, vol. 312, no. 3, January 17, 1985, pp. 164–65. Eula Bingham's "exclude women" quote is from her preface to *Double Exposure*, edited by Wendy Chavkin, p. x.
33. The Bingham quote is from her preface to *Double Exposure*, p. x. Patricia Hynes has argued that "the exclusion of women in the lead industries—despite evidence that the wives of male lead workers suffered reproductive disorders—may be the key to the uncritical use of lead in industrial societies," since by putting women out of the workplace, "industry appeared to fulfill its obligation to the health of future generations." See H. Patricia Hynes, "Lead Contamination: A Case of 'Protectionism' and the Neglect of Women," in

Reconstructing Babylon: Essays on Women and Technology, edited by H. Patricia Hynes (London: Earthscan Publications, 1990), p. 22.

34. The "hazards, not workers" and "employers should not be permitted" quotes are from Joan E. Bertin, "People Protection, Not 'Fetal Protection,'" p. 5.
35. On the Johnson Controls case, see Linda Greenhouse, "Court Backs Right of Women to Jobs with Health Risks," *New York Times*, March 21, 1991, p. 1, and Suzanne L. Mager, "Courting Disaster: Judges, Workers and the Case of Johnson Controls," *New Solutions*, Summer 1990, pp. 6–8.
36. On the emergence of consciousness raising, see Sara Evans, *Personal Politics: The Roots of Women's Liberation in the Civil Rights Movement and the New Left* (New York: Vintage, 1980), pp. 156–92.
37. See Helen King Burke, "Sierra Club Outreach to Women: Women's Issues in the Environmental Movement," interview by Waverly Lowell, Sierra Club Oral History Project (San Francisco: Sierra Club, 1980), and Gail Robinson, "A Woman's Place Is in the Movement," *Environmental Action*, vol. 9, no. 22, March 25, 1978, pp. 12–13. The EPA publication is EPA Office of Public Awareness, *Women and the Environment . . . Women as Agents of Change* (Washington, D.C.: Government Printing Office, November 1977). See also Douglas Costle, "Women and the Environment," *EPA Journal*, November-December 1978, p. 4.
38. Author's 1991 interviews with Louise Dunlap and Joseph Browder and 1987 interviews with Peter Carlson and Mike Clark.
39. Author's 1991 interview with Janet Brown.
40. The Atomic Industrial Forum had a coordinator of women's activities and a speaker's bureau aimed at various women's organizations. See Gail Robinson, "A Woman's Place Is in the Movement," p. 12. See also Lin Nelson, "Promise Her Everything: The Nuclear Power Industry's Agenda For Women," *Feminist Studies*, vol. 10, Summer 1984, pp. 291–314, and Dorothy Nelkin, "Nuclear Power as a Feminist Issue," *Environment*, vol. 23, no. 1, January-February 1981, pp. 14–20+.
41. The *Newsweek* quote is from the November 13, 1961, article "The Women Protest," p. 21.
42. For background on Women Strike for Peace, see Amy Swerdlow, "Pure Milk, Not Poison: Women Strike for Peace and the Test Ban Treaty of 1963," in *Rocking the Ship of State: Toward a Feminist Peace Politics*, edited by Adrienne Harris and Ynestra King (Boulder: Westview Press, 1989), pp. 225–37.
43. The call to action for the Women's Pentagon March was in the form of a Unity Statement written primarily by long-time peace activist, feminist, and poet Grace Paley with input from other participants in the action. A section of the Unity Statement is reprinted in Ynestra King, "If I Can't Dance in Your Revolution, I'm Not Coming," in *Rocking the Ship of State*, edited by Adrienne Harris and Ynestra King, pp. 287–88.
44. Lois Gibbs's speech is reproduced in edited form in "Action from Tragedy,"

edited by Celeste Wesson, in the "Feminism and Ecology" issue of *Heresies*, no. 13, 1981. See Susan Griffin, *Woman and Nature: The Roaring Inside Her* (New York: Harper & Row, 1978), and Rosemary Radford Ruether, *New Woman, New Earth: Sexist Ideologies and Human Liberation* (New York: Seabury Press, 1975). On the Women and Life on Earth Conference, see Ynestra King, "If I Can't Dance in Your Revolution, I'm Not Coming," pp. 284–90.

45. For background on ecofeminism, see Ynestra King, "Toward an Ecological Feminism and a Feminist Ecology," in *Machina Ex Dea: Feminist Perspectives on Technology*, edited by Joan Rothschild (New York: Pergamon Press, 1983), pp. 118–29; Karen J. Warren, "Feminism and Ecology: Making Connections," *Environmental Ethics*, vol. 9, no. 1, Spring 1987, pp. 3–44; Katherine Davies, "What is Ecofeminism?" *Women and Environments*, vol. 10, no. 3, Spring 1988, pp. 4–8; *Reweaving the World: The Emergence of Ecofeminism*, edited by Irene Diamond and Gloria Feman Orenstein (San Francisco: Sierra Club Books, 1990); "Feminism and Ecology" issue of *Heresies*, no. 13, 1981. See also the special issue "Women and the Environment," edited by Carolyn Merchant, *Environmental Review*, vol. 8, no. 1, Spring 1984.

Chapter 7

1. The "Oh, maybe you are thinking" quote is from Alice Hamilton, *Exploring the Dangerous Trades*, p. 152.
2. The "dirty, hot, and unpleasant" quote is from Sterling D. Spero and Abram L. Harris, *The Black Worker: The Negro and the Labor Movement* (New York: Columbia University Press, 1931), p. 156. See also David Rosner and Gerald Markowitz, *Deadly Dust: Silicosis and the Politics of Occupational Disease in Twentieth-century America* (Princeton, N.J.: Princeton University Press), p. 70.
3. See Martin Cherniack, *The Hawk's Nest Incident: America's Worst Industrial Disaster* (New Haven: Yale University Press, 1986). Cherniack's account of the events at Gauley Bridge is the most comprehensive description of this episode and has been used as a primary reference for the discussion of those events in this chapter.
4. The quotes from the court testimony are recounted in House of Representatives, *Investigation Relating to Health Conditions of Workers Employed in the Construction and Maintenance of Public Utilities, January 16–February 4, 1936*, 74th Cong., 2d sess. (Washington, D.C.: Government Printing Office, 1936), p. 203.
5. Cherniack's calculations and "The death toll" quote are from his book *The Hawk's Nest Incident*, pp. 104–105. The disproportionate number of black deaths also reflected where workers were employed. While nearly three times the number of black workers worked either inside the tunnel or both inside and outside, there were an equal number of black and white workers who worked exclusively outside the tunnel and were thus least exposed to the dust emissions.

6. The first of *The New Masses* articles to appear was a thinly disguised fictional account of the Gauley Bridge events entitled "Man on a Road," written by Albert Maltz (vol. 14, no. 2, January 8, 1935, pp. 19–21). This was followed in the next two issues by a two-part article, "Two Thousand Dying on a Job," written by Phillipa Allen under the pen name Bernard Allen (vol. 14, no. 3, January 15, 1935, pp. 18–19, and vol. 14, no. 4, January 22, 1935, pp. 13–14).
7. The occupational health–related reexploration of the Gauley Bridge events can be found in Joseph A. Page and Mary-Win O'Brien, *Bitter Wages: Ralph Nader's Study Group Report on Disease and Injury on the Job* (New York: Grossman, 1972), pp. 59–63; David Rosner and Gerald Markowitz, *Deadly Dust*, pp. 98–102; Fred A. Wilcox, "The Hawk's Nest Incident: A Review," *The Amicus Journal*, vol. 9, no. 4, Fall 1987, pp. 43–44. For a perspective that links the events more directly to ethnicity, see Pamela Tau Lee, "An Overview: Workers of Color and the Occupational Health Crisis" (Paper presented at the People of Color Environmental Leadership Summit, Washington, D.C., October 24–27, 1991), and Charles Lee, "The Integrity of Justice: The Evidence of Environmental Racism," *Sojourners*, vol. 19, no. 2, February-March 1990, p. 22. Interestingly, some mainstream environmental groups, including the West Virginia chapter of the Sierra Club and Friends of the Earth, sought in 1987 to challenge the renewal of the hydroelectric license of the Hawk's Nest project for causing "significant adverse impacts" on one of West Virginia's "best warmwater rivers." See "A Human and Environmental Tragedy at Hawk's Nest," *Not Man Apart*, vol. 18, no. 1, January-February 1988, p. 7.
8. The name *Chicano*, as it evolved in the middle to late 1960s, has often been used to mean U.S. residents of Mexican descent. Though the distinctions between Chicanos and Mexicanos (those Mexicans who have recently crossed over the border) have often blurred, they became more significant during the period after the UFW had become a recognized bargaining agent for farmworkers and the use of illegal Mexican labor was expanded by growers partly to undermine union efforts at representation. During the late 1960s, however, the power of the farmworker campaign was most directly associated with its unifying racial identity themes. See Laura Pulido, "Latino Environmental Struggles in the Southwest" (Ph.D., diss., University of California at Los Angeles, 1991).
9. On farmworker pesticide-related illnesses, see Irma West, "Occupational Disease of Farm Workers," *Archives of Environmental Health*, vol. 9 (1964), pp. 92–98; Laura Pulido's interview with Eliseo Medina in "Latino Environmental Struggles"; Bureau of Occupational Health and Environmental Epidemiology, California Department of Public Health, "Occupational Disease in California Attributed to Pesticides and Other Agricultural Chemicals," in House of Representatives, *Occupational Safety and Health Act of 1969, Hearings Before the Select Subcommittee of the Committee on Education and Labor*, 91st Congress (Washington, D.C.: Government Printing Office, 1969). See also U.S. Senate, *Migrant and Seasonal Farmworker Powerlessness: Pesticides and the Migrant Farmworker*,

Senate Subcommittee on Migrant Labor, 91st Congress, September 29, 1969. On the "pesticide clouds," see Laura Pulido's interview with Dolores Huerta in "Latino Environmental Struggles."

10. On growers' fears about a pesticide-linked boycott, see Laura Pulido's interview with Ralph Abascal in "Latino Environmental Struggles."
11. Laura Pulido's interview with Jerry Cohen in "Latino Environmental Struggles." See also Ruth Harmer, "Poison, Profits, and Politics," *The Nation*, August 25, 1969, pp. 134–37.
12. The UFW's first DDT suit argued that DDT was a toxic food additive and should be banned under the provisions of the 1959 Delaney Amendment prohibiting the introduction of cancer-causing substances in food products. The UFW/EDF/CRLA "marriage of convenience" is described in an April 4, 1991, letter from Charles Wurster to Laura Pulido. See also Pulido's interview with Ralph Abascal in "Latino Environmental Struggles."
13. The "We will not tolerate the systematic poisoning" quote is from Cesar Chavez's January 14, 1969, letter to Martin J. Zaninovich (Archives of Labor and Urban Affairs, United Farm Workers, Office of the President Collection, Box 46, Folder 6). On the *El Malcriado* coverage, see the articles from October 15 to November 30, 1969, also in the UFW archives. See also "Chavez Charges Pesticide Health Hazards, Calls Statewide Parley," *Fresno Bee*, January 22, 1969.
14. For a representative bargaining agreement with health and safety language, see "Collective Bargaining Agreement Between the United Farm Workers Organizing Committee and Wonder Palms Ranch, March 31, 1970" (Bancroft Library, University of California at Berkeley); see also Laura Pulido, "Latino Environmental Struggles," pp. 223–35.
15. On the response by industry to information about lead impacts, see William Graebner, "Hegemony Through Science: Information Engineering and Lead Toxicology," in *Dying for Work*, edited by David Rosner and Gerald Markowitz, pp. 121–39, and Richard Rabin, "Warnings Unheeded: A History of Child Lead Poisoning," *American Journal of Public Health*, vol. 79, no. 12, December 1989, pp. 1668–74.
16. The "lead world" concept is described in John C. Ruddock, "Lead Poisoning in Children: With Special Reference to PICA," *Journal of the American Medical Association*, vol. 82, no. 21 (1924), pp. 1682–84.
17. The term "predilection for the poor" is used in Jane S. Lin-Fu, "The Evolution of Childhood Lead Poisoning as a Public Health Problem," in *Lead Absorption in Children: Management, Clinical, and Environmental Aspects*, edited by J. Julian Chisolm, Jr., and David M. O'Hara (Baltimore: Urban and Schwarzenberg, 1982), p. 3. The "inextricably related" quote is from a 1967 Department of Health, Education, and Welfare study, *Lead Poisoning in Children*, DHEW Pub. No (HSA) 78–5142, by Jane S. Lin-Fu (Washington, D.C.: Government Printing Office, 1967).

18. The Scientists' Committee for Public Information statement, "A Call for Help," is reprinted on page 49 of the April 1968 issue of *Scientist and Citizen*, an issue devoted to the lead problem. See also Mark W. Oberle, "Lead Poisoning: A Preventable Childhood Disease of the Slums," *Science*, vol. 165, no. 3897, September 5, 1969, pp. 991–92.
19. For background on the CCELP, see Ann Koppelman Simon, "Citizens vs. Lead in Three Communities: Chicago," *Scientist and Citizen*, April 1968, pp. 58–59. See also Barbara Berney, "Round and Round It Goes: The Epidemiology of Lead, 1950–1990" (Boston University, 1991, unpublished manuscript).
20. See the statement by the Philadelphia Citywide Coalition Against Childhood Lead Poisoning in *Lead-Based Paint Poisoning*, Hearing Before the Subcommittee on Health of the Committee on Labor and Public Welfare, U.S. Senate, 91st Congress, 2d sess., November 23, 1970 (Washington, D.C.: Government Printing Office, 1970). See also Jonathan M. Stein, "An Overview of the Lead Abatement Program Response to the Silent Epidemic," in *Low-Level Lead Exposure: The Clinical Implications of Current Research*, edited by H. L. Needleman (New York: Raven Press, 1980), pp. 279–84.
21. See Nicholas Freudenberg and M. Golub, "Health Education, Public Policy and Disease Prevention: A Case History of the New York City Coalition to End Lead Poisoning," *Health Education Quarterly*, vol. 14, no. 4, Winter 1987, pp. 387–401. See also Barbara Berney, "Round and Round It Goes," pp. 8–14.
22. On the relationship of leaded gasoline to blood lead levels, see I. H. Billick, *Predictions of Pediatric Blood Lead Levels from Gasoline Consumption* (Washington, D.C.: United States Department of Housing and Urban Development, 1982), cited in Barbara Berney, "Round and Round It Goes," pp. 20–21. On the Clean Air Act implementation process, see David Schoenbroad, "Why Regulation of Lead Has Failed," in *Low-Level Lead Exposure: The Clinical Implications of Current Research*, edited by H. L. Needleman, pp. 259–65.
23. The *New England Journal of Medicine* editorial, "Exposure to Lead in Childhood—The Importance of Prevention," is by Kathryn R. Mahaffey (vol. 327, no. 18, October 29, 1992, pp. 1308–9).
24. See Karen L. Florini, George D. Krumbharr, Jr., and Ellen Silbergeld, *Legacy of Lead: America's Continuing Epidemic of Childhood Lead Poisoning* (Washington, D.C.: Environmental Defense Fund, March 1990). Among the recent studies, see California Department of Health Services, *Childhood Lead Poisoning in California: Causes and Prevention* (interim report) (Sacramento: California Department of Health Services, April 1989), and Neil Maizlich et al., "Elevated Blood Lead in California Adults, 1987: Results of a Statewide Surveillance Program Based on Laboratory Reports," *American Journal of Public Health*, vol. 80, no. 8, August 1990, pp. 931–34.
25. For background on the mining on Navajo lands, see Harold Tso and Laura Mangum Shields, "Navajo Uranium Operations: Early Hazards and Recent

Interventions," *New Mexico Journal of Science*, vol. 20, no. 1, June 1980, pp. 11–17.

26. On radiation impacts, see the ground-breaking study by Leon S. Gottlieb and Luverne Husen, "Lung Cancer Among Navajo Uranium Miners," *Chest*, vol. 81, April 1982, pp. 449–52. See also Chris Shuey et al., "The 'Costs' of Uranium: Who's Paying with Lives, Lands, and Dollars," *The Workbook*, vol. 10, no. 3, July/September 1985, and Tony Davis, "Uranium Has Decimated Navajo Miners," *High Country News*, June 18, 1990, p. 1.
27. For background on the Rio Puerco spill, see Jana Bommersbach, "The Nation's Worst Radioactive Spill Continues to Plague Southwest," *New Times Weekly*, August 29, 1979, p. 1, and Jana Bommersbach, "Uranium Spill: Too Many Shortcuts and Lax Controls," *New Times Weekly*, October 3–9, 1979, p. 7; Sandra Blakeslee, "Radioactive Waste Spill Called Worst Ever in U.S.," *Los Angeles Times*, August 22, 1979.
28. On the current efforts to mine on Havasupai and Hualapai lands, see Cate Gilles with Lena Bravo and Don Watahomigie, "Uranium Mining at the Grand Canyon: What Costs to Water, Air, and Indigenous People?" *The Workbook*, vol. 16, no. 1, Spring 1991, pp. 2–17. See also Tonantzin Land Institute, "Havasupai, Forest Service at Loggerheads Over Canyon Mine," *Tribal Peoples Survival*, Summer 1988, p. 2. Michael Begay's remarks were made at a presentation at the UCLA Urban Planning Program's Brown Bag Seminar, May 17, 1991 (author's notes).
29. The "many environmentalists now believe" quote is from James Noel Smith, "The Coming of Age of Environmentalism in American Society," in *Environmental Quality and Social Justice in Urban America*, edited by James Noel Smith (Washington, D.C.: Conservation Foundation, 1972), p. 5; See also, in this same book, Peter Marcuse, "Conservation for Whom?" pp. 17–36.
30. The firing of Sydney Howe was described by Howe in the author's 1992 interview with him. The Sierra Club poll is discussed in Don Coombs, "The Club Looks at Itself," *Sierra Club Bulletin*, vol. 57, no. 7, July-August 1972. Paul Swatek's comments are from an editorial on the poll results in the January 1973 issue of the *Sierra Club Bulletin*, vol. 58, no. 1, pp. 16–17. A National Center for Voluntary Action (NCVA) survey of mainstream environmental groups at about the same time pointed to similar results. The largest number of respondents listed their organization's priority concern as wilderness or natural area preservation, with urban-based issues such as sanitation and rat and pest control at the bottom of the list. Interestingly, nearly twenty years later, a similar survey of mainstream environmental leaders undertaken by the Conservation Fund had comparable results. The NCVA survey is discussed in Clem L. Zinger et al., *Environmental Volunteers in America* (Washington, D.C.: National Center for Voluntary Action, 1973), pp. 5–23; the Conservation Fund survey results are included in *Inside the Environmental Movement: Meeting the*

Leadership Challenge, edited by Donald Snow (Washington, D.C.: Island Press, 1992).

31. Whitney Young's remarks are quoted in the *Trenton Evening Times*, February 16, 1970, and cited in Harold Sprout, "The Environmental Crisis in the Context of American Politics," in *The Politics of Eco-suicide*, edited by Leslie L. Roos, Jr. (New York: Holt, Rinehart and Winston, 1971), p. 46. Faramelli's remarks are cited in James Noel Smith, "The Coming of Age of Environmentalism in American Society," p. 2.
32. The action of the San Jose State students was bitterly criticized by the campus Black Student Union, which saw the incident as a slap in the face to poor people who couldn't afford to buy a car. See Paul Ruffins, "Divided We Fall," *New Age Journal*, March-April 1990, p. 46.
33. See William T. Hornaday, *Our Vanishing Wildlife: Its Extermination and Preservation* (New York: New York Zoological Society, 1913), pp. 101–102.
34. Madison Grant's nativist writings include *The Alien in Our Midst or 'Selling Our Birthright for a Mess of Pottage,'* edited by Madison Grant and Charles Stewart Davison (New York: Galton Publishing Co., 1930); *The Conquest of a Continent, or the Expansion of Races in America* (New York: Charles Scribner and Sons, 1933); and *The Passing of the Great Race, or the Racial Basis of European History* (New York: Charles Scribner and Sons, 1916). The Henry Fairfield Osborn address, "Shall We Maintain Washington's Ideal of Americanism?" is included in *The Alien in Our Midst,* pp. 204–209; Osborn also wrote the preface to *The Conquest of a Continent.* The "intellectually, the most important nativist" quote is from John Higham, *Strangers in the Land: Patterns of American Nativism, 1860–1925* (New Brunswick, N.J.: Rutgers University Press, 1925), p. 155. For a discussion of nativist ideas within preservationist circles, see Stephen Fox, *John Muir and His Legacy*, pp. 346–47, and Susan R. Schrepfer, *The Fight to Save the Redwoods*, pp. 43–44.
35. Joseph L. Fisher, *Conservation for More and More People*, Resources for the Future Reprint Series No. 21 (Washington, D.C.: Resources for the Future, August 1960), p. 82.
36. Paul R. Ehrlich, *The Population Bomb*, rev. ed. (New York: Sierra Club/Ballantine Books, 1971), pp. 1, 44.
37. See Garrett Hardin, "The Tragedy of the Commons," *Science*, vol. 162, no. 3859 (1968), pp. 1243–48, and Garrett Hardin, "Living on a Lifeboat," *Bioscience*, vol. 24, no. 10, October 1974, pp. 561–68. See also *The Population Bomb*, pp. 151, 173–74.
38. The Alabama sterilization incident is discussed in Carl Pope, "Sterilization Controversy," *ZPG National Reporter*, vol. 5, no. 5, August 1973, p. 8, and Helen Rodriguez-Trias, "A Model for Advocacy: From Proposals to Policy," in *Reproductive Laws for the 1990s*, edited by Sherrill Cohen and Nadine Taub (Clifton, N.J.: Humana Press, 1989), pp. 353–57. See also Jack Slater,

"Sterilization: Newest Threat to the Poor," *Ebony*, vol. 28, October 1973, pp. 150–52; Claudia Dreifus, "Sterilizing the Poor," *The Progressive*, vol. 39, December 1975, pp. 13–19; Peters Willson, "Another Look at Sterilization," *ZPG National Reporter*, vol. 9, no. 2, March 1977, p. 1. The Population and World Resources Conference walkout is described in Paul Ehrlich and Richard L. Harriman, *How to Be a Survivor* (New York: Ballantine Books, 1971).

39. The organizational emphasis on immigration occurred most noticeably between 1974, when ZPG members rated immigration as their organization's lowest priority (although 50 percent of the members responding to a survey indicated they still wanted immigration "greatly reduced") and 1976, when immigration became a dominant concern of the group. See "Results of Membership Questionnaire," *ZPG National Reporter*, vol. 6, no. 1, January 1974, p. 2, and Melanie Wirken, "Illegal Immigration: The American Dilemma," *ZPG National Reporter*, vol. 8, no. 7, September 1976, p. 1. See also John Tanton, "Immigration: An Illiberal Concern?" *ZPG National Reporter*, vol. 7, no. 3, April 1975, p. 4, and "Zero Population Growth: Recommendations for a New Immigration Policy," *ZPG National Reporter*, vol. 7, no. 5, June 1975, pp. 4–5.
40. Edward Abbey's discussion of Mexicans can be found in *Down the River* (New York: E. P. Dutton, 1982), pp. 149–53; The "could pose a real threat" quote is from a press release for *The Golden Door* issued by Dermot McEvoy, publicist for Ballantine Books. The discussion of the need for immigration restriction can be found in Paul R. Ehrlich, Loy Bilderback, and Anne H. Ehrlich, *The Golden Door: International Migration, Mexico, and the United States* (New York: Ballantine Books, 1979). See also Thomas F. McMahon, " 'Required Reading' on U.S.-Mexico Relations," *ZPG National Reporter*, vol. 11, no. 7, September 1979, p. 7.
41. For background on FAIR, see its "Goals and Principles Statement" (Washington, D.C.: Federation of American Immigration Reform, 1979). See also FAIR executive director Roger Conner's testimony before the Select Commission on Immigration and Refugee Policy (San Antonio, Texas, December 17, 1979) and Roger Conner, "Immigration in the Era of Limits," *Los Angeles Times*, February 17, 1980. On FAIR's census lawsuit (*FAIR vs. Luther H. Hodges, Jr.*, Civil Action No. 79–3269, U.S. District Court, District of Columbia), see the organization's statement to the press of December 5, 1979, announcing the census suit.
42. Defenders of Wildlife et al., *Blueprint for the Environment: Advice to the President-elect from America's Environmental Community* (Washington, D.C.: Blueprint for the Environment, November 1988), pp. 27–29, and John H. Adams et al., *An Environmental Agenda for the Future*, pp. 33–39. See also "The Population Bomb: An Explosive Issue for the Environmental Movement?" *Utne Reader*, May-June 1988, pp. 78–93, and Jeffrey M. Berry, *Lobbying for the People*, pp. 32–33. On the ZPG membership decline in the 1970s, see "Director Notes

Highlighted in a Good Year," *ZPG National Reporter*, vol. 9, no. 10, December 1977–January 1978, p. 1.

43. The letters were sent to nine organizations since two Group of Ten organizations, the Environmental Policy Institute and Friends of the Earth, had merged prior to the sending of the letter.
44. See the letter from Pat Bryant et al. to Jay Hair et al., January 16, 1990, and the letter from Richard Moore et al. to Jay Hair et al., March 16, 1990. These lettes are reprinted in "The Whiteness of the Green Movement," *Not Man Apart*, vol. 20, no. 2, April-May 1990, pp. 14–17, and author's 1992 interviews with Richard Moore and Dana Alston.
45. The Fred Krupp quote is from Philip Shabecoff, "Environmental Groups Told They Are Racists in Hiring," *New York Times*, February 1, 1990.
46. The "I don't think anybody is as aware" quote is cited in Philip Shabecoff, "Environmental Groups Told They Are Racists in Hiring"; also author's 1991 interview with Jay Hair. Information on the NRDC's approach is from the author's 1991 interview with John Adams; also the presentation by NRDC staff attorney Joel Reynolds at the UCLA Urban Planning Program, Los Angeles, November 3, 1991 (author's notes).
47. The NWF staff figures are discussed by Jay Hair in the author's 1991 interview with him.
48. The "everybody's a professional" remark by Jay Hair is from the author's 1991 interview with him. On NWF board composition, see the National Wildlife Federation's 1990 *Annual Report.*
49. The UEC also received some modest support from private foundations after establishing the Urban Environment Foundation, headed by Sydney Howe. The differences between the UEC, with its movement style, and the foundation group, which sought to maintain a more professional appearance, eventually caused the foundation and the UEC to formally sever relations in the early 1980s, with the foundation changing its name to the Human Environment Conference. Author's 1992 interviews with Sydney Howe and George Coling.
50. For background on the UEC, see "Labor, Environmentalists, Community Action Groups Work Together—Phil Hart's Inspiration" (*Congressional Record*, July 2, 1976, pp. S11521–26) and Michele Tingling, "The Urban Environment," *Environment*, vol. 22, no. 3, April 1980, p. 5+. The New Orleans conference proceedings are described in Urban Environment Conference, Inc., *Taking Back Our Health: An Institute on Surviving the Toxics Threat to Minority Communities* (Washington, D.C.: Urban Environment Conference, Inc., 1985). See also Stephanie Pollack and JoAnn Grozuczak, *Reagan, Toxics and Minorities: A Policy Report* (Washington, D.C.: Urban Environment Conference, September 1984), and author's 1992 interview with George Coling.
51. For background on the North Carolina events, see "Warren County, North Carolina," in Urban Environment Conference, Inc., *Taking Back Our Health*,

pp. 38–39, and "PCB Protest Unites County," *In These Times*, September 29–October 5, 1982, pp. 5–6.

52. For background on the LANCER fight, see Cynthia Hamilton, "Women, Home and Community: The Struggle in an Urban Environment," *Race, Poverty, and the Environment*, vol. 1, no. 1, April 1990, pp. 3–6, and Louis Blumberg and Robert Gottlieb, *War on Waste: Can America Win Its Battle with Garbage?* (Washington, D.C.: Island Press, 1989), pp. 155–88.
53. The "negative land use" and community identity issues for East Los Angeles are explored in Cynthia Pansing, Hali Rederer, and David Yale, "A Community at Risk: The Environmental Quality of Life in East Los Angeles" (Graduate School of Architecture and Urban Planning, University of California at Los Angeles, Client Project Report, August 20, 1989). On the Mothers of East Los Angeles organization, see Mary Pardo, "Mexican American Women Grassroots Community Activists: 'Mothers of East Los Angeles,' " *Frontiers*, vol. 11, no. 1, 1990, pp. 1–7; José Medivil, "Unwanted Projects Dumped on East L.A. Residents," *La Gente de Aztlan*, March-April 1989; presentation by Juana Gutierrez at the UCLA Urban Planning Program, Los Angeles, January 21, 1992 (author's notes).
54. See Charles Lee, *Toxic Wastes and Race in the United States*, A National Report on the Racial and Socio-Economic Characteristics of Communities with Hazardous Waste Sites (New York: Commission for Racial Justice, United Church of Christ, 1987); Robert Bullard, *Dumping in Dixie* (Boulder: Westview Press, 1990); Robert Bullard and Beverly Hendrix Wright, "Environmentalism and the Politics of Equity: Emergent Trends in the Black Community," *Mid-American Review of Sociology*, vol. 12, no. 2, Winter 1987, pp. 21–37. See also Robert Bush, *Hazardous Waste Facilities: Race as a Siting Factor* (Arlington, Va.: Citizen's Clearinghouse for Hazardous Wastes, 1988) and Charles Lee, "The Integrity of Justice," *Sojourners*, February-March 1990, pp. 22–25.
55. Even the EPA, under pressure from environmental justice advocates, explored the issue of environmental equity, arguing in a draft report released in February 1992 that, with a few exceptions, such as lead poisoning and Native American concerns, there has been a "general lack of data on environmental health effects by race and income." See Environmental Protection Agency, *Environmental Equity: Reducing Risk for All Communities*, Draft Report 230-DR-92-002, to the Administrator from the EPA Environmental Equity Workgroup (Washington, D.C.: Government Printing Office, February 1992). The "different links" quote is from Jesus Sanchez, "The Environment: Whose Movement?" *California Tomorrow*, Fall 1988, p. 12.
56. Letter from Richard Moore et al. to John O'Connor et al., May 20, 1990; author's 1992 interviews with Richard Moore, Ted Smith, and Cora Tucker.
57. See Richard Moore, "Toxics, Race, and Class: The Poisoning of Communities" (Paper presented at the Annual Briefing of the Interfaith Impact for

Justice and Peace, Washington, D.C., March 18, 1991); Southwest Organizing Project, "Southwest Network Gathering: Where 'Minorities' Were Majority," *Voces Unidas*, vol. 1, no. 4, p. 1; Arnoldo Garcia, "Environmental Inequities: An Interview with Richard Moore," *Responsive Philanthropy*, Winter 1991, pp. 1–5.

58. On the "cancer alley" marches, see Zack Nauth, "How Toxic Pollution Can Break Down Racial Barriers," *In These Times*, December 14–20, 1988, p. 2; also author's 1992 interview with Richard Miller and Pat Bryant, "A Lily-White Achilles Heel," *Environmental Action*, January-February 1990, pp. 28–29.
59. Dana Alston's remarks are from her October 26, 1991, talk at the "Our Vision of the Future" session of the People of Color Environmental Leadership Summit. The summit's "Principles of Environmental Justice" were revised October 27, 1991, and issued as part of the overall summit packet.

Chapter 8

1. The background to this section is based on a series of interviews with Larry Davis in 1991 and 1992 as well as materials from Davis's own files, which were made available to the author. For background on the Midco situation, see U.S. Environmental Protection Agency, *Midco I and Midco II Gary Indiana: Superfund Remedial Program Fact Sheet*, Region V (Chicago: U.S. EPA, November 1989); "History of Midco, Gary, Indiana," a memorandum from David D. Lamm to Bruce Palin, April 9, 1981, State Board of Health, Indianapolis; Jim Proctor, "Midco Cleanup: A Tough Task," *Post-Tribune*, April 23, 1989.
2. The Burns Harbor plant has been one of the largest users of TCA in the country, with more than 815,000 pounds emitted into the air in 1988, according to the EPA records Davis reviewed. TCA figures by plant and by company are available through the Toxics Release Inventory (TRI) data collected under section 313 of the Emergency Planning and Community Right-to-Know Act of 1986.
3. Background on the USW's involvement in environmental issues comes from the author's 1992 interview with Jack Sheehan.
4. This and the following quotes are from the author's 1991 and 1992 interviews with Larry Davis.
5. Davis's quote is from *Our Children's World: Steelworkers and the Environment*, Report of the USWA (United Steelworkers of America) Task Force on the Environment (Pittsburgh: USWA, August 1990), p. 17. Steelworkers' president Lynn Williams, who helped preside over this most recent shift in emphasis, was a prominent environment-related participant in the Economic Summit conference held on December 14 and 15, 1992, and organized by the incoming Clinton-Gore administration.
6. During the period she lived at Love Canal, Lois Gibbs's husband worked at a

Goodyear Chemical plant, but the community protests never extended to his or other plants in the area. Years later, Gibbs would cite that situation and ask, in the course of presentations to either worker or community groups, who in the audience had spouses working at plants producing or using hazardous materials (if she was speaking to a community group) or had spouses involved in community protests (if she was addressing a group of workers). A majority of the audience—in both settings—almost always raised their hands. Gibbs would then point out that despite such personal connections, workplace/community ties remained minimal (author's 1992 interview with Lois Gibbs). That disjuncture between community and workplace at Love Canal was also the subject of a research paper (drawing on personal information) on the attitudes of chemical workers who lived in the Love Canal area by doctoral student Kathy Kubarski for a fall 1991 class on the history of environmentalism at the UCLA Urban Planning Program taught by the author.

7. See David Rosner and Gerald Markowitz, "Worker's Health and Safety—Some Historical Notes," in *Dying for Work: Workers' Safety and Health in Twentieth-Century America*, edited by David Rosner and Gerald Markowitz (Bloomington: Indiana University Press, 1989), pp. xii–xiii; Graham Taylor, "Industrial Basis for Social Interpretation," *The Survey*, vol. 22, April 3, 1909, pp. 9–11; Mary Brown Sumner, "A Strike for Clean Bread," *The Survey*, vol. 24, June 18, 1910, pp. 483–88.
8. The "woo the public" and "social responsibilities" quotes are from page 54 of a *Time* cover story on Johns-Manville and its CEO, Lewis Brown, that discussed this overall public relations dimension. See "Corporate Soul," *Time*, vol. 33, no. 14, April 3, 1939, pp. 52–58. In the article, Johns-Manville is described on page 52 as having "acquired a position in the public eye as a model Big Business" at the very moment the company sought to suppress information about asbestos impacts.
9. The *British Medical Journal* article is W. E. Cooke, "Fibration of the Lungs Due to the Inhalation of Asbestos Dust," vol. 2 (1924), p. 147. The 1935 report was authored by A. J. Lanza, V. J. McConnell, and J. W. Fehnel and is entitled "Effects of Inhalation of Asbestos Dust on the Lungs of Asbestos Workers," *U.S. Public Health Service Reports*, vol. 50, no. 1, pp. 1–12. Both studies are cited and discussed in David Kotelchuk, "Asbestos: 'The Funeral Dress of Kings' —and Others," in David Rosner and Gerald Markowitz, *Dying for Work*, pp. 192–207.
10. The deposition by Roehmer was taken April 25, 1984, from the court case *Johns-Manville Corp, et al. v. United States of America*, U.S. Claims Court Civ. No. 465–83C, and is cited in Barry Castleman, *Asbestos: Medical and Legal Aspects* (New York: Harcourt Brace Jovanovich, 1984), p. 401. The Kenneth Smith memo, "Industrial Hygiene—Survey of Men in Dusty Areas" is cited in David Kotelchuk, "Asbestos: The 'Funeral Dress of Kings'—and Others,"

p. 203. See also, David E. Lilienfeld, "The Silence: The Asbestos Industry and Early Occupational Cancer Research—A Case Study," *American Journal of Public Health*, vol. 81, no. 6, June 1991, pp. 791–800; James C. Robinson, *Toil and Toxics: Workplace Struggles and Political Strategies for Occupational Health* (Berkeley: University of California Press, 1991), pp. 110–111; Paul Brodeur, *Outrageous Misconduct: The Asbestos Industry on Trial* (New York: Pantheon, 1985).

11. The "powerful symbols" quote is from Barbara Ellen Smith, *Digging Our Own Graves: Coal Miners and the Struggle Over Black Lung Disease* (Philadelphia: Temple University Press, 1987), p. 5. Descriptions of the environmental conditions of mining towns can be found in William Graebner, *Coal-mining Safety in the Progressive Period: The Political Economy of Reform* (Lexington: University of Kentucky Press, 1976). The "insidious undermining of health" quote is from the July 2, 1910, issue of the *Mining and Scientific Press* and is cited in Alan Derickson, *Workers' Health, Workers' Democracy: The Western Miners' Struggle: 1891–1925* (Ithaca, N.Y.: Cornell University Press, 1988), p. 56. See also Alan Derickson, "The United Mine Workers of America and the Recognition of Occupational Respiratory Diseases, 1902–1968," *American Journal of Public Health*, vol. 81, no. 6, June 1991, pp. 782–90.
12. Kerr's speech is cited in Joseph A. Page and Mary-Win O'Brien, *Bitter Wages: Ralph Nader's Study Group Report on Disease and Injury on the Job* (New York: Grossman, 1972), p. 126. See also Alan Derickson, "Down Solid: The Origins and Development of the Black Lung Insurgency," *Journal of Public Health Policy*, vol. 4, March 1983, pp. 25–44.
13. Ralph Nader's *New Republic* article, "They're Still Breathing," was published February 3, 1968 (vol. 158, no. 5), p. 15. The impact of the Farmington disaster on black lung and occupational health–related legislation is discussed in Barbara Ellen Smith, *Digging Our Own Graves*, and Bennett M. Judkins, *We Offer Ourselves as Evidence: Toward Workers' Control of Occupational Health* (Westport, Conn.: Greenwood Press, 1986).
14. The Nader telegram is quoted by the Huntington, West Virginia, *Herald-Dispatch* of January 22, 1969, and is cited in Bennett M. Judkins, *We Offer Ourselves as Evidence*, p. 68.
15. The "join with us in the effort" quote is cited in Joseph A. Page and Mary-Win O'Brien, *Bitter Wages*, p. 138. See also Charles Noble, *Liberalism at Work: The Rise and Fall of OSHA* (Philadelphia: Temple University Press, 1986), pp. 81–82.
16. For background on the OSH Act, see Robert Asher, "Organized Labor and the Origins of the Occupational Safety and Health Act," *Labor's Heritage*, vol. 3, no. 1, January 1991, pp. 54–76; also "New Coalition Results from Work Hazards," *Environmental Action*, vol. 2, no. 7, August 15, 1970, p. 13; Charles Noble, *Liberalism at Work*, pp. 68–98. Despite the promise of OSHA, its early years

were marred by industry and Nixon administration maneuvering. Immediately after the OSH Act was passed, a high-powered industry countermobilization led to what the Steelworkers' Jack Sheehan called "an unbelievable turnaround in Congress," with "bill after bill, speech after speech attacking the legislation and seeking to gut its provisions" (author's 1992 interview with Jack Sheehan). OSHA also became a prime candidate for the 1972 Nixon reelection campaign's "responsiveness program," designed to attract industry support through favors and soft regulations. In this period, agency activities were best characterized by a minimalist approach toward standard setting based on the economic feasibility of such standards, limited numbers of inspections, and the almost complete absence of serious violations cited in those inspections. The responsiveness program was explored at the Senate Watergate Hearings and is discussed in Willard S. Randall and Stephen D. Solomon, *Building Six: The Tragedy at Bridesburg* (Boston: Little, Brown), p. 160.

17. Frank Wallick, editor of the *UAW Washington Report*, a key union publication that came to focus on workplace environment questions, recalled Nader's influence in getting Wallick more deeply involved concerning such issues. After Wallick had written an article about occupational health legislation in 1968, he received a call late one afternoon from Nader. Berating Wallick and the union for their inaction, Nader, in "one of those urgent phone conversations of his [that would occur] when I was ready to head for home," as Wallick recalled the episode, rattled off a series of questions and comments about "consensus standards" and "safety boards." Wallick didn't have the slightest idea what Nader was talking about, but Nader's urgency and passion prompted him to find out. The anecdote is recounted in Franklin Wallick, *The American Worker: An Endangered Species* (New York: Ballantine, 1972), p. 20; also author's 1991 interview with Andrea Hricko.
18. The July 1971 meeting is described in Willard S. Randall and Stephen D. Solomon, *Building Six: The Tragedy at Bridesburg*, pp. 132–33. See also Andrea Hricko and Daniel Pertschuk, "Cancer in the Workplace: A Report on Corporate Secrecy at the Rohm and Haas Company, Philadelphia, Pennsylvania" (Washington, D.C.: Health Research Group, October 2, 1974).
19. W. G. Figueroa, Robert Raszkowski, and William Weiss, "Lung Cancer in Chloromethyl Methyl Ether Workers," *New England Journal of Medicine*, vol. 288, no. 21, May 24, 1973, pp. 1096–97. See also "Letters to the Editor," *New England Journal of Medicine*, vol. 290, no. 17, April 25, 1974, pp. 971–72.
20. See Sheldon Hochheiser, *Rohm and Haas: History of a Chemical Company* (Philadelphia: University of Pennsylvania Press, 1986), pp. 168–77; see also Health Research Group, "Cancer in the Workplace;" also author's 1991 interview with Andrea Hricko.
21. The Wallick quotes are from his book, *The American Worker: An Endangered Species*, p. 32. See also Franklin Wallick, "Factory Pollution: It Doesn't All

Go Up the Chimney," *Environmental Action*, vol. 3, no. 16, January 8, 1972, pp. 3–5.

22. The UAW Safety Department approach and Lloyd Utter quote are from Joseph A. Page and Mary-Win O'Brien, *Bitter Wages*, pp. 128, 223. In later years, the UAW took a more active role in seeking to establish workplace health and safety representatives at the shop level.
23. Even at the height of the anticommunist campaign against the Mine, Mill union, the focus on silicosis, "as deadly a killer as any other form," as one union official put it, remained substantial. See the *Proceedings of the 46th Convention of the International Union of Mine, Mill, and Smelter Workers*, Denver, September 11–15, 1950, p. 47. The affiliation agreement with the Steelworkers Union is reproduced as Appendix II in the *Proceedings of the Special Convention of the International Union of Mine, Mill, and Smelter Workers*, Tucson, January, 16, 1967. For background on the union's efforts regarding silicosis, see David Rosner and Gerald Markowitz, *Deadly Dust: Silicosis and the Politics of Occupational Disease in Twentieth-century America* (Princeton, N.J.: Princeton University Press, 1991).
24. *Peril on the Job: A Study of Hazards in the Chemical Industries* (Washington, D.C.: Public Affairs Press, 1970) served as a forerunner for a number of texts, most notably Dorothy Nelkin and Michael Brown's *Workers at Risk: Voices from the Workplace* (Chicago: University of Chicago Press, 1984) and Studs Terkel's best-selling *Working* (New York: Avon Press, 1975), which presented workers' own commentaries about their jobs and the occupational hazards they experienced. See also Peter Montague and Katherine Montague, *No World Without End: The New Threats to Our Biosphere* (New York: Putnam 1977), pp. 118–20.
25. The "plant situation team" concept is described in Joseph A. Page and Mary-Win O'Brien, *Bitter Wages*, p. 225. See also Ray Davidson, *Challenging the Giants: A History of the Oil, Chemical, and Atomic Workers International Union* (Washington, D.C.: OCAW, 1988), pp. 321–40. Wodka was the OCAW representative contacted by Karen Silkwood in her effort to expose the Kerr-McGee company's unsafe and hazardous plant practices.
26. The New Directions Grant Program was initiated by Eula Bingham, the head of OSHA during the Carter administration. It was designed to support efforts by nongovernment organizations to "increase employer and employee awareness of occupational safety and health." Though the program was small—only about $15 million in grants was distributed among a large number of organizations during its four years of operation—it provided significant support for nearly half the COSH groups, many of which were just getting off the ground. See *The President's Report on Occupational Safety and Health* (Washington, D.C.: U.S. Department of Labor, 1980), cited in Charles Noble, *Liberalism at Work* p. 134. For background on the COSH groups, see Daniel Berman, "Grassroots Coalitions in Health and Safety: The COSH Groups," *Labor Studies Journal*,

Spring 1981, pp. 104–13, and New York Committee on Occupational Safety and Health, "1990 COSH Survey" (New York: New York Committee on Occupational Safety and Health, November 1990).

27. A review of the publications of the major mainstream environmental groups—including the Sierra Club, Audubon Society, Wilderness Society, and National Wildlife Federation—during 1969 and 1970, when the OSH Act was being debated, revealed no articles or news briefs about that debate or about any occupational hazard issue, such as the black lung struggles. The only mainstream groups whose publications included articles on these subjects were Environmental Action and Friends of the Earth.
28. The "Club's extension of its activities" quote is from a *Sierra Club Bulletin* editorial by club president Raymond Sherwin on Sierra Club support for the Shell workers, "A Broader Look at the Environment," vol. 58., no. 4, April 1973, p. 18. Long-time activist George Coling, currently with the Sierra Club, recalled a meeting at which former Sierra Club executive director Mike McCloskey suggested that Sierra Club support for the Shell strike had been only "nominal." Coling, however, remembered how McCloskey's appearance at an OCAW convention, where he seconded the motion in support of the Shell strike, had been an "important moment for those of us trying to generate environmental support for the Shell workers" (author's 1992 interview with George Coling).
29. The "were a unique period" and "labor market bonanza" quotes are from James C. Robinson, *Toil and Toxics: Workplace Struggles for Occupational Health* (Berkeley: University of California Press, 1991), p. 25. The "No work, no food" slogan is cited in Deborah Baldwin, "GASPing for Air," *Environmental Action*, vol. 7, no. 11, October 11, 1975, pp. 12–13.
30. On the pollution control/job creation issue, see Patrick Heffernan, "Jobs and the Environment," *Sierra Club Bulletin*, vol. 60, no. 4, April 1975, pp. 25–26, and Alan S. Miller, "Towards an Environmental/Labor Coalition," *Environment*, vol. 22, no. 5, June 1980, pp. 32–39.
31. The prediction of a loss of 750,000 jobs comes from Erwin Baker, "City OKs Lease of Property for LNG Terminal," *Los Angeles Times*, July 21, 1977, p. 1. The proposed port facility is discussed in several other *Los Angeles Times* stories, including Tom Redburn and Lydia Chavez, "Forecasts Flop; Southland Is Awash in Natural Gas," January 12, 1981, p. 1, and Karen Tumulty, "Point Conception LNG Plant Is Put on Hold," October 5, 1982, p. 3.
32. The founding statement of EFFE called for a "general reevaluation of our human and natural resource policies" and for an end to the "exploitation that environmentalists and labor unions have, heretofore, fought independently." The statement is cited in "What We Think" in the organization's first newsletter, November 1975, p. 1. See also Richard Grossman, "Environmentalists and the Labor Movement," *Socialist Review*, vol. 15, nos. 4 and 5, pp. 63–87.

33. The "eloquently" quote is by Richard Grossman from the author's 1991 interview with him. On the Black Lake conference, see Richard Grossman, "Environmentalists and the Labor Movement," pp. 8–9. See also Deborah Baldwin, "They Said It Couldn't Be Done," *Environmental Action*, vol. 8, no. 2, May 27, 1976, pp. 7–11; Gladwin Hill, "Labor Seeks Ties to Ecology Bloc," *New York Times*, May 9, 1976; Gladwin Hill, "Woodcock Calls for an Alliance of Labor and Environmentalists," *New York Times*, May 4, 1976; also author's 1991 interview with Richard Grossman.
34. See Natural Resources Defense Council et al., *America's Economic Future: Environmentalists Broaden the Industrial Policy Debate*. A Report of the Project on Industrial Policy and the Environment (Washington, D.C.: Natural Resources Defense Council, June 1984), pp. 38, 65.
35. The "1970s were an exciting period" quote is from the author's 1991 interview with Andrea Hricko.
36. On the Reagan administration's approach to OSHA, see Philip J. Simon, *Reagan in the Workplace: Unravelling the Health and Safety Net* (Washington, D.C.: Center for Study of Responsive Law, 1983); Charles Noble, *Liberalism at Work*, pp. 193–96; Charles Noble, "OSHA at 20," *New Solutions*, Spring 1990, pp. 30–42.
37. The industry counterattack on OSHA standards, such as the one for cotton dust, was further aided by a U.S. Supreme Court decision in 1980 regarding an OSHA benzene standard. The Supreme Court ruled in this case that OSHA standards had to be "reasonably necessary" by demonstrating that the risk to occupationally exposed workers would be significantly reduced if a new standard were adopted. As a consequence of the benzene decision, OSHA began to institute quantitative risk assessments as a way to measure the degree of risk in relation to the exposure. This approach paralleled the use of quantitative risk assessment by the EPA and other environmental agencies to establish whether communities faced increased risks from such sources as industry emissions and hazardous waste sites. The OSHA threshold for significant risk (1 excess cancer death per 1000 workers), however, has been far weaker than the prevailing approach at the EPA (1 excess cancer death per 1,000,000 population). Author's 1992 interview with John Froines.
38. On the cotton dust standard, see Bennett M. Judkins, *We Offer Ourselves as Evidence*, p. 160; author's 1991 interview with John Froines.
39. For background on the BASF campaign, see Richard Leonard and Zack Nauth, "Beating BASF: OCAW Busts Union-Buster," *Labor Research Review*, no. 16, pp. 35–49; Zack Nauth, "BASF: Bhopal on the Bayou?" *In These Times*, January 24–30, 1990, pp. 12–13+; author's 1992 interview with Richard Miller.
40. Author's 1992 interview with Richard Miller.
41. See Lois Gibbs, "Labor and Grassroots Activists Seek Common Ground for the '90s," *Everyone's Backyard*, vol. 8, no. 1, January-February 1990, pp. 9–10;

Citizens' Environmental Coalition, "New York State Labor and Environment Network: 1992 Plan of Action" (Albany, N.Y.: Citizens' Environmental Coalition, 1991); "Unions, environmentalists not foes, group says," *Daily Freeman*, November 25, 1991, p. 4; Eric Mann, *L.A.'s Lethal Air: New Strategies for Policy, Organizing and Action* (Van Nuys, Calif.: Labor-Community Strategy Center, 1991).

42. On the SVTC approach, see Ted Smith, "Pollution in Paradise: The Legacy of High-Tech Development," *Silicon Valley Toxics News*, vol. 10, no. 1, Winter 1992, p. 2; Meta Mendel-Reyes, "Coalition Says: Ban Toxics, Not Workers," *Silicon Valley Toxics News*, vol. 5, no. 1 (1987), p. 1; author's 1992 interview with Ted Smith.
43. For background on the Superfund for Workers concept, see Lucinda Wykle, Ward Morehouse, and David Dembo, *Worker Empowerment in a Changing Economy: Jobs, Military Production and the Environment* (New York: Apex Press, 1991). See also Michael Merrill, "No Pollution Prevention Without Income Protection: A Challenge to Environmentalists," *New Solutions*, vol. 1, no. 3, Winter 1991, pp. 9–11.
44. On the Byrd amendment, see *Los Angeles Times*, October 23, 1990, p. 12.
45. The "explicitly directs workers into poverty" quote is from a statement by the OCAW's president, Robert Wages, before the U.S. Senate Committee on Agriculture, Nutrition, and Forestry concerning the employment impact on pesticide manufacturing workers from the Circle of Poison Prevention Act of 1991 (S. 898). On Title XI of the Clean Air Act Amendments, see *Clean Air Act Amendments of 1990: Detailed Summary of Titles* (Washington, D.C.: Environmental Protection Agency, November 1990). See also House of Representatives, *Clean Air Act Amendments of 1990: Conference Report to Accompany S. 1630*, 101st Cong., 2d sess. H. Rept. 101–952, sec. 1101f, pp. 325–28.
46. On the labor party concept, see Anthony Mazzochi, "Working People Must Create Political Alternatives," *OCAW Reporter*, January-February 1990, p. 6, as cited in Lucinda Wykle et al., *Worker Empowerment in a Changing Economy* p. 80.
47. Barry Commoner's *New Yorker* article "The Environment," which appeared on July 15, 1987, was influential in the development of the toxics use reduction idea. The antitoxics groups were also influenced by the U.S. Congress's Office of Technology Assessment report entitled *Serious Reduction of Hazardous Waste: For Pollution Prevention and Industrial Efficiency*, OTA-ITE-317 (Washington, D.C.: Government Printing Office, September 1986). The OTA report provided a detailed critique of the EPA's waste minimization approach, which failed to adequately distinguish among reduction, recycling, and treatment strategies. For further background on the conceptual distinctions among strategies, see University of California, Office of the President, *University of California Report on Toxics Reduction* (Berkeley: University of California Report to the

State Legislature, December 1991). For background on the development of the toxics use reduction and pollution prevention concepts, see Ken Geiser, "Toxics Use Reduction and Pollution Prevention," *New Solutions*, Spring 1990, pp. 1–8; Toxics Coordinating Project, *Focus on Toxics Use Reduction* (Sacramento: Toxics Coordinating Project, 1987); *Making Pollution Prevention Pay: Ecology with Economy as Policy*, edited by Donald Huisingh and Vicki Bailey (New York: Pergamon Press, 1982); also author's 1990 interview with Michael Picker.

48. For background on toxics use reduction policies at the state level, see William Ryan and Richard Schrader, *An Ounce of Toxic Pollution Prevention: Rating States' Toxics Use Reduction Laws* (Boston and Washington, D.C.: National Environmental Law Center and the Center for Policy Alternatives, January 1991), and Waste Reduction Institute for Training and Applications Research, *State Legislation Relating to Pollution Prevention: Survey and Summaries* (Minneapolis: Waste Reduction Institute for Training and Applications Research, April 1991).

Conclusion

1. See Students for a Democratic Society, *The Port Huron Statement* (New York: Students for a Democratic Society, August 1962), pp. 23, 24.
2. The national security dimension of resource policy is discussed in President's Materials Policy Commission, *Resources for Freedom, A Report to the President* (Washington, D.C.: Government Printing Office, 1952), vol. 1, p. 2. The same theme is discussed by Samuel Hays in his essay "The Myth of Conservation" in *Perspectives on Conservation*, edited by Henry Jarrett (Baltimore: Johns Hopkins Press, 1958), pp. 40–45. The "availability of a wide variety of resources" quote is from Joseph L. Fisher, "Long-Range Research in Times Like These," in Resources for the Future, *Annual Report* (for the year ending September 30, 1961) (Washington D.C.: Resources for the Future, December 1961), p. 4. Fairfield Osborn's remarks are cited in Peter Collier and David Horowitz, *The Rockefellers: An American Dynasty* (New York: Holt, Rinehart and Winston, 1976), p. 305.
3. The "chemical plague" quote is from Lenny Siegal and Gary Cohen, "America's Worst Environmental Nightmare: The U.S. Military's Toxic Legacy," *New Solutions*, Winter 1991, p. 37.
4. See "Join SEAC!" *Greenpeace*, November-December 1988, vol. 13, no. 6, p. 12.
5. The author was the faculty supervisor for the UCLA study, by Tamra Brink et al., *In Our Backyard: Environmental Issues at UCLA, Proposals for Change, and the Institution's Potential as a Model* (Los Angeles: UCLA Urban Planning Program Comprehensive Project, June 1989). See also April A. Smith and Robert Gottlieb, "Campus Environmental Audits: The UCLA Experience," in *The*

Campus and Environmental Responsibility, edited by David J. Eagan and David W. Orr (San Francisco: Jossey-Bass, New Directions for Higher Education Series, no. 77, Spring 1992), and UCLA Environmental Research Group, *Campus Environmental Audit* (Palo Alto, Ca.: Earth Day, 1990, February 1990). April Smith, one of the coauthors of *In Our Backyard* and the *Campus Environmental Audit*, became the liaison with SEAC and its university accountability campaign. See also April E. Smith and the Student Environmental Action Coalition, *Campus Ecology: A Guide to Assessing Environmental Quality and Creating Strategies for Change* (Los Angeles: Living Planet Press, 1993).

6. The *Forbes* article, by Martin Kihn, is entitled "SDS Jr." (June 10, 1991, pp. 51–52). For background on the SEAC, see Barbara Ruben, "Catalyzing Student Action," *Environmental Action*, January-February 1991, p. 7, and Jim McNeill, "Jonathan Goldman: Student SEACer," *In These Times*, September 18–24, 1991, pp. 4–5.
7. Author's 1992 interview with Heather Abel.
8. The phrase "long march through the institutions" was first used in 1967 by German New Left activist (and German Green cofounder) Rudi Dutschke in describing the New Left's link to what would subsequently be defined by European and U.S. analysts as "the new social movements." (Author's notes from his attendance at the annual convention of the German SDS in Munich, June 1967.)
9. See George Will, "The Green Thrill Has Replaced the Red Scare," *Los Angeles Times*, April 19, 1990. See also Charles Krauthammer, "An Insidious Rejuvenation of the Old Left," *Los Angeles Times*, December 24, 1990, and, along the same lines, Martin W. Lewis, *Green Delusions: An Environmentalist Critique of Radical Environmentalism* (Durham, N.C.: Duke University Press, 1992).
10. Author's 1991 interview with Thomas Kimball.
11. On the "three-way" process, see Judith Redmond and Robert Gottlieb, "The Select Few Who Still Control State's Water Policy Discussions," *Sacramento Bee*, June 1, 1992.
12. Author's 1992 interview with Lois Gibbs.

Index

About the Author

Robert Gottlieb is the author of several books on environmental and resource policy. These include *A Life of Its Own: The Politics and Power of Water; War on Waste: Can America Win Its Battle with Garbage?* (coauthored with Louis Blumberg); *Empires in the Sun: The Rise of the New American West* (coauthored with Peter Wiley); and *Thirst for Growth: Water Agencies as Hidden Government in California* (coauthored with Margaret FitzSimmons). Gottlieb is a long-time environmental and social justice activist whose writings have appeared in numerous New Left and environmental journals as well as the *Wall Street Journal, Los Angeles Times*, and *Unomásuno*. He currently teaches environmental policy and analysis in the UCLA Urban Planning Program and lives with his wife and two children in Santa Monica, California.

Island Press Board of Directors